# REMOTE SENSING AND CLIMATE MODELING: SYNERGIES AND LIMITATIONS

# ADVANCES IN GLOBAL CHANGE RESEARCH

VOLUME 7



*The titles in this series are listed at the end of this volume.*

# REMOTE SENSING AND CLIMATE MODELING: SYNERGIES AND LIMITATIONS

Edited by

Martin Beniston
*Department of Geography,*
*University of Fribourg,*
*Switzerland*

and

Michel M. Verstraete
*Space Applications Institute,*
*Joint Research Centre,*
*Ispra (VA), Italy*

KLUWER ACADEMIC PUBLISHERS
DORDRECHT / BOSTON / LONDON

A C.I.P. Catalogue record for this book is available from the Library of Congress.

ISBN 978-90-481-5648-1 ISBN 978-0-306-48149-9 (eBook)
DOI 10.1007/978-0-306-48149-9

---

Published by Kluwer Academic Publishers,
P.O. Box 17, 3300 AA Dordrecht, The Netherlands.

Sold and distributed in North, Central and South America
by Kluwer Academic Publishers,
101 Philip Drive, Norwell, MA 02061, U.S.A.

In all other countries, sold and distributed
by Kluwer Academic Publishers,
P.O. Box 322, 3300 AH Dordrecht, The Netherlands.

*Printed on acid-free paper*



Softcover reprint of the hardcover 1st edition 2001

# Table of contents

# List of contributors

A. BASIST, NOAA-NCDC, Asheville, North Carolina, USA.

M. BENISTON, Department of Geography, University of Fribourg, Switzerland.

P. A: BRIVIO, Telerilevamento - IRRS, CNR, Milan, Italy.

G. BUCINI, Space Applications Institute, Joint Research Center, Ispra (Varese), Italy.

D. R. CAHOON, NASA/Langley Research Center, Atmospheric Sciences Division, Hampton, Virginia, USA.

M. CHIN, NASA-Goddard, GIT, Greenbelt, Maryland, USA.

C. COLLET, Department of Geography, University of Fribourg, Switzerland.

R. COLOMBO, Telerilevamento - IRRS, CNR, Milan, Italy.

D. DINER, Jet Propulsion Laboratory, Pasadena, California, USA.

V. DJEPA, Wageningen University and Research Centre, Wageningen, The Netherlands.

O. DUBOVIK, NASA-Goddard, SSAI, Greenbelt, Maryland, USA.

T. ECK, NASA-Goddard, Raytheon Corporation, Greenbelt, Maryland, USA.

R. EASTER, Battelle, Pacific Northwest Laboratories, Richland, Washington, USA.

A. FREI, National Snow and Ice Data Center, Boulder, Colorado, USA.

M. D. FLANNIGAN, Canadian Forest Service, Edmonton, Alberta, Canada.

M. A. FOSBERG, IGBP-BAHC Core Project Office, Potsdam, Germany.

P. GEOGDZHAYEV, NASA-GISS, Greenbelt, Maryland, USA.

S. GHAN, Battelle, Pacific Northwest Laboratories, Richland, Washington, USA.

P. GINOUX, NASA-Goddard, Greenbelt, Maryland, USA.

N. GOBRON, Space Applications Institute, Joint Research Center, Ispra (Varese), Italy.

J. G. GOLDAMMER, University of Freiburg, Max-Planck-Institute for Chemistry, Freiburg, Germany.

Y. GOVAERTS, EUMETSAT, Darmstadt, Germany.

S. GOYETTE, Department of Geography, University of Fribourg, Switzerland.

I. F. GRANT, CSIRO Atmospheric Research, Aspendale, Victoria, Australia.

A. HIRAGUSHI, National Institute for Environmental Science, Tsukuba, Japan.

B. HOLBEN, NASA-Goddard, Greenbelt, Maryland, USA.

L. JIA, Lanzhou Institute for Plateau Atmospheric Physics, Chinese Academy of Sciences, Langzhou, China.

R. A. KAHN, Jet Propulsion Laboratory, Pasadena, California, USA.

S. KINNE, NASA-Goddard, UMBC – JCET, Greenbelt, Maryland, USA.

W. KNORR, Max Planck Institute for Biogeochemistry, Jena, Germany.

D. KOCH, NASA-GISS, Greenbelt, Maryland, USA.

Z. L. LI, LSIIT, University Louis Pasteur, Illkirch, France.

U. LOHMANN, Dalhousie University, Halifax, Nova Scotia, Canada.

J. V. MARTONCHIK, Jet Propulsion Laboratory, Pasadena, California, USA.

F. MÉLIN, Space Applications Institute, Joint Research Center, Ispra (Varese), Italy.

M. MENENTI, LSIIT, University Louis Pasteur, Illkirch, France.

M. MERONI, Telerilevamento - IRRS, CNR, Milan, Italy.

M. MISCHENKO, NASA-GISS, Greenbelt, Maryland, USA.

S. NIEMEYER, Swiss Fereal Institute for Snow and Avalanche Research, Davos, Switzerland.

A. W. NOLIN, National Snow and Ice Data Center, Boulder, Colorado, USA.

A. OHMURA, Department of Geography, Swiss Institute of Techonolgy (ETH), Zurich, Switzerland.

B. PINTY, Space Applications Institute, Joint Research Center, Ispra (Varese), Italy.

A. ROESCH, Department of Geography, Swiss Institute of Techonolgy (ETH), Zurich, Switzerland.

F. ROVEDA, EUMETSAT, Darmstadt, Germany.

J. P. SCHULZ, Danish Meteorological Institute, Copenhagen, Denmark.

I. SLUTSKER, NASA-Goddard, SSAI, Greenbelt, Maryland, USA.

A. SMIRNOV, NASA-Goddard, SSAI, Greenbelt, Maryland, USA.

B. J. STOCKS, Canadian Forest Service, Sault-Ste. Marie, Ontario, Canada.

L. STOWE, National Snow and Ice Data Center, Boulder, Colorado, USA.

Z. SU, Wageningen University and Research Centre, NL-6700 Wageningen, The Netherlands.

T. TAKEMURA, University of Tokyo, Japan.

D. TANRE, University of Lille, Department of Physics, Lille, France.

I. TEGEN, Max Planck Institute for Biogeochemistry, Jena, Germany.

O. TORRES, NASA-Goddard, Greenbelt, Maryland, USA.

E. VERMOTE, NASA-Goddard, Greenbelt, Maryland, USA.

M. M. VERSTRAETE, Space Applications Institute, Joint Research Center, Ispra (Varese), Italy.

J. VOGT, Space Applications Institute, Joint Research Center, Ispra (Varese), Italy.

J. WANG, Lanzhou Institute for Plateau Atmospheric Physics, Chinese Academy of Sciences, Langzhou, China.

M. WILD, Department of Geography, Swiss Institute of Techonolgy (ETH), Zurich, Switzerland.

J. L. WIDLOWSKI, Space Applications Institute, Joint Research Center, Ispra (Varese), Italy.

C. WILLIAMS, NOAA-NCDC, Asheville, North Carolina, USA.

B. M. WOTTON, Canadian Forest Service, Sault-Ste. Marie, Ontario, Canada.

G. ZIBOZDI, Joint Research Center, Ispra (Varese), Italy.

# Preface

Michel M. VERSTRAETE[1] and Martin BENISTON[2]
[1]*Space Applications Institute, EC Joint Research Centre, Ispra, Italy*
[2]*Department of Geography, University of Fribourg, Switzerland*

This volume contains the proceedings of the workshop entitled "Satellite Remote Sensing and Climate Simulations: Synergies and Limitations" that took place in Les Diablerets, Switzerland, September 20–24, 1999. This international scientific conference aimed at addressing the current and potential role of satellite remote sensing in climate modeling, with a particular focus on land surface processes and atmospheric aerosol characterization.

Global and regional circulation models incorporate our knowledge of the dynamics of the Earth's atmosphere. They are used to predict the evolution of the weather and climate. Mathematically, this system is represented by a set of partial differential equations whose solution requires initial and boundary conditions. Limitations in the accuracy and geographical distribution of these constraints, and intrinsic mathematical sensitivity to these conditions do not allow the identification of a unique solution (prediction). Additional observations on the climate system are thus used to constrain the forecasts of the mathematical model to remain close to the observed state of the system. Ultimately, these models are useful mainly to predict the future values of environmental variables or to estimate these variables wherever and whenever they are not observed directly. Current validation of global and regional climate models is based on comparison between model outputs of standard meteorological fields and meteorological observations. The main problem with traditional meteorological observations when used to validate models is their poor representation of the grid-point average simulated by a model. Now that comprehensive radiative measurements are available from space platforms, models should produce comparable fields as standard outputs to be confronted with these new observations. Remote sensing from space platforms thus provides a unique opportunity to yield reliable and accurate information in support of global and regional weather or climate models.



*M. Beniston and M.M. Verstraete (eds.),*
*Remote Sensing and Climate Modeling: Synergies and Limitations*, 1–3.

Indeed, space-based platforms permit the systematic and repetitive observation of the planetary surface and the atmosphere, at spatial resolutions generally much higher than those used in modeling. Remote sensing data can be exploited either to provide the initial and boundary conditions required to run climate models, to force these models to remain close to the real atmospheric situation, or to evaluate the accuracy of the forecasts. A number of other scientific and technological issues arise at the interface between climate modeling and remote sensing observations. This conference provided a unique opportunity to review the state of the art in the integration of the information derived from satellite remote sensing technologies in global and regional climate models. Specifically, papers were solicited along the following lines:

- The analysis of satellite remote sensing data to derive environmental variables of direct relevance to specify the initial and boundary conditions of GCMs, including surface albedo, emissivity, temperature and roughness, as well as atmospheric composition, aerosols and cloudiness, among others.
- The direct assimilation of radiative measurements made in space into GCMs, to improve the accuracy of forecasts.
- The evaluation of the effectiveness, reliability and accuracy of the models by comparing their results with independent remote sensing observations.
- The need to produce remote sensing "observations" as standard outputs from models to compare with real remote sensing observations in order to evaluate model performances.
- Scaling issues, in particular the methodological problems posed by combining field observations acquired at the local scale, remote sensing observations relative to small but spatially averaged conditions, and model simulations valid for relatively large areas. Contributions on the interpretation and proper exploitation of related but different concepts, such as skin and bulk temperatures, and interpolation in space and in time to match the needs of models with the data offered by satellite systems were also welcome.
- The development of soil-vegetation-atmosphere transfer schemes (SVATs), and the improvement of these models to take advantage of observations from space.
- The design and implementation of observational strategies optimized to provide the information required by the global and regional climate models at the appropriate resolution and with the necessary accuracy.
- The technical and institutional challenges which hinder or prevent a more exhaustive exploitation of satellite remote sensing data in regional and

global climate models, including combining large data streams of remote sensing data in computationally demanding models, locating and accessing appropriate data, and designing models so that they can effectively take advantage of such observations. Also of interest and direct relevance is a discussion of how to improve the methods of remote sensing data analysis so that their products are compatible with the requirements of models.

A new generation of satellite platforms is in the process of being launched (e.g., Spot-4, Landsat-7, Terra, ENVISAT, ADEOS-II). These platforms do or will embark high performance instruments with improved spatial resolutions, enhanced radiometric accuracy, additional spectral bands and observation directions, and many other new features. At the same time, significant improvements have been made in climate modeling techniques at different scales. Last but not least, computer processing speed and communications capabilities continue to increase dramatically. This convergence creates new opportunities to document the state and evolution of the climate system, at a time when concern about climatic change and impacts has reached new heights. This conference thus provided a timely forum to discuss some of the most critical issues arising at the interface between simulation and the observation of our Earth system.

The chapters that follow contain some of the most interesting papers that were presented at this conference. Clearly, these issues will continue to be relevant for the foreseeable future. The Editors of this volume hope that these manuscripts will contribute to the debates and lead to improvements in model performance and satellite data interpretation.

We would like to acknowledge the financial support of ENAMORS (European Network for the development of Advanced models to interpret Optical Remote Sensing data) and the Swiss National Science Foundation. This funding enabled the conference organizers to support a number of outstanding speakers.

Thanks also to Sylvie Bovel-Yerly who, as always, put this volume into shape in her efficient manner.

# A global vegetation index for SeaWiFS : Design and applications

N. GOBRON, F. MÉLIN, B. PINTY, M. M. VERSTRAETE,
J.-L. WIDLOWSKI and G. BUCINI
*Space Applications Institute, EC Joint Research Centre, Ispra, Italy*

Abstract: Optimized vegetation indices provide a convenient approach to estimate crucial plant properties on the basis of satellite data. This paper describes the steps followed to implement an index optimized to estimate the Fraction of Absorbed Photosynthetically Active Radiation (FAPAR) on the basis of data generated by the SeaWiFS instrument, and the preliminary results obtained. Index values are computed on the basis of top of atmosphere bidirectional reflectance factor values in the blue, red and near-infrared domains, as well as information on the geometry of illumination and observation. Results obtained with SeaWiFS data are used to evaluate the performance of the index. This case study documents the ability of the index to discriminate between various surface types, and its insensitivity to changes in the geometrical conditions of observation and to atmospheric effects. The operational environment set up at SAI to process SeaWiFS data is outlined and selected standard retrievals resulting from a monthly composite analysis are shown as examples of the products generated.

## 1. INTRODUCTION

Vegetation indices are often used as an alternative to more complex algorithms to retrieve surface properties from space. Most of the older indices suffer from various well-known defects such as undesirable dependencies to geophysical variables or processes not of interest, or to the conditions of observation. These drawbacks can be avoided by designing, for instance, optimized spectral indices (Verstraete and Pinty, 1996). The application of these principles to the MERIS, GLI and VEGETATION sensors has been discussed in Govaerts et al. (1999), Gobron et al. (1999), and Gobron et al.



*M. Beniston and M.M. Verstraete (eds.),*
***Remote Sensing and Climate Modeling: Synergies and Limitations*, 5–21.**

(2000a), respectively. The details of the actual implementation of these new indices are given in Gobron et al. (1998) and Verstraete et al. (1998).

New vegetation indices are optimized with the help of a training data set generated with radiation transfer models of the coupled surface-atmosphere system which simulate sensor-like observations over various representative land surface types and for a wide range of atmospheric conditions. These simulations produce a large number of radiance fields at the blue, red and near-infrared wavelengths of the given sensor, which can then be sampled in the angular domain in a way similar to what is done with actual instruments. The models used to generate these radiation fields are also suitable to estimate the Fraction of Absorbed Photosynthetically Active Radiation (FAPAR) for the various land surface types under investigation.

The design of optimal spectral indices is based on a two step procedure. First, the spectral radiances measured in the red and near-infrared bands are rectified to decontaminate them from atmospheric and angular effects. Then, the rectified red and near-infrared bands are combined, via a generic polynomial expression, to yield an index formula that optimally estimates the environmental parameter of interest. The rectification process is based on the use of simultaneous measurements in the blue band to address atmospheric effects and on a parametric bi-directional reflectance model to account for angular (anisotropy) effects.

We developed a new spectral index specifically designed to estimate FAPAR for global applications on the basis of Sea-viewing Wide Field-of-view Sensor (SeaWiFS) data. Although this sensor was originally designed for the observation of ocean color, it permits the monitoring of terrestrial land surfaces thanks to its spectral bands centered at 443 (blue), 670 (red) and 865 nm (near-infrared) and a detector and amplifier design which does not saturate over land.

The implementation and optimization of the vegetation index for SeaWiFS is described in the next section. Its robustness with respect to angular variations in viewing geometry and its performance to characterize land surface patterns are discussed later. The last section presents some of the SeaWiFS products available at SAI for global analyses.

## 2. DESIGN OF THE SEAWIFS VEGETATION INDEX

Our optimized SeaWiFS vegetation index requires, as input, the three Bidirectional Reflectance Factor (BRF) values measured by this sensor in the blue, red and near-infrared spectral regions, in addition to the solar and viewing zenith angles and the relative azimuth angle between the sun and the satellite. The computation of the index requires three polynomial expressions as well as the anisotropy reflectance function delivered by the Rahman, Pinty and Verstraete (RPV) BRF parametric model (Rahman et al., 1993). The values of the RPV model parameters are optimally derived once and for all, using a training BRF data set generated for a large range of simulated geophysical scenarios.

Various geophysical quantities are estimated in the process of implementing the optimized SeaWiFS Vegetation Index (SeaWiFS-VI). First, the TOA channel values are "normalized" by the anisotropic function:

$$\tilde{\rho}(\lambda_i) = \frac{\rho^{toa}(\Omega_0, \Omega_v, \lambda_i)}{F(\Omega_0, \Omega_v, k_{\lambda i}, \Omega_{\lambda i}^{HG}, \rho_{\lambda ic})}$$

where $\lambda_i$ stands for the wavelength (blue, red or near-infrared) of spectral band $i$, and $\rho^{toa}(\Omega_0,\Omega_v,\lambda_i)$ denotes the BRF values measured by the sensor in the spectral band $\lambda_i$, as a function of the actual geometry of illumination ($\Omega_0$) and observation ($\Omega_v$). These angular coordinates are fully defined by the zenith ($\theta$) and relative azimuth ($\phi$) angles for the incoming and exiting radiation, respectively, with respect to a plane-parallel system. The spectral anisotropic reflectance function, $F\,(\Omega_0,\Omega_v,k_{\lambda i},\Omega^{HG}_{\lambda i},\rho_{\lambda ic})$, represents the shape of the radiance field, where the triplet $(k_{\lambda i},\Omega^{HG}_{\lambda i},\rho_{\lambda ic})$ are the RPV's parameters optimized *a priori* for each spectral band $\lambda_i$.

The rectification process of the red and near-infrared bands is performed as follows:

$$\rho_{Rred} = g_1[\tilde{\rho}(\lambda_{blu}), \tilde{\rho}(\lambda_{red})]$$

and

$$\rho_{Rnir} = g_2[\tilde{\rho}(\lambda_{blu}), \tilde{\rho}(\lambda_{nir})]$$

where

$$g_n[\tilde{\rho}(\lambda_i), \tilde{\rho}(\lambda_j)] = P(\lambda_i, \lambda_j) / Q(\lambda_i, \lambda_j)$$

$$P(\lambda_i, \lambda_j) = l_{n1}(\tilde{\rho}(\lambda_i) + l_{n2})^2 + l_{n3}(\tilde{\rho}(\lambda_j) + l_{n4})^2 + l_{n5}\,\tilde{\rho}(\lambda_i)\,\tilde{\rho}(\lambda_j)$$

$$Q(\lambda_i, \lambda_j) = l_{n6}(\tilde{\rho}(\lambda_i) + l_{n7})^2 + l_{n8}(\tilde{\rho}(\lambda_j) + l_{n9})^2 + l_{n10}\,\tilde{\rho}(\lambda_i)\,\tilde{\rho}(\lambda_j) + l_{n11}$$

The polynomial coefficients $l_{nm}$ have been optimized in such a way that the values generated by each spectral polynomial $g_n[\rho\tilde{}(\lambda_{blu}), \rho\tilde{}(\lambda_j)]$ correspond to the bi-directional reflectance factors that would be measured at the top of the canopy, normalized by the spectrally appropriate anisotropic reflectance function. In other words, the rectification process yields estimated values of spectral reflectances emerging at the top of the canopy, optimally decontaminated from atmospheric and angular radiative effects in the sense described in the various publications mentioned earlier.

The SeaWiFS-VI itself is then computed on the basis of these rectified channel values, and its formula is

$$VI = g_0(\rho_{Rred}, \rho_{Rnir}) = \frac{l_{01}\rho_{Rnir} - l_{02}\rho_{Rred} - l_{03}}{(l_{04} - \rho_{Rred})^2 + (l_{05} - \rho_{Rnir})^2 + l_{06}}$$

where the coefficients $l_{0m}$ of polynomial $g_0$ are optimized *a priori* to force SeaWiFS-VI to take on values as close as possible to the FAPAR associated with the plant canopy scenarios used in the training data set. The numerical values of the various coefficients resulting from these successive optimizations are summarized in Tables 1 to 4.

*Table 1.* Optimal values of the parameters for the anisotropic function F

| Channel | Parameter values | | |
|---|---|---|---|
| $\lambda_i$ (nm) | $\rho_{\lambda ic}$ | $k_{\lambda i}$ | $\Theta^{HG}_{\lambda i}$ |
| 443 | 0.23265 | 0.56184 | −0.04125 |
| 670 | −0.44444 | 0.70535 | 0.03576 |
| 865 | 0.63149 | 0.86644 | −0.00102 |

*Table 2.* **Optimal values of the coefficients for the polynomial $g_1$**

| | | |
|---|---|---|
| $l_{11}$=−9.8725 | $l_{12}$=−0.027458 | $l_{13}$=2.9144 |
| $l_{14}$= 0.059376 | $l_{15}$=10.904 | $l_{16}$= 0.0 |
| $l_{17}$= 0.0 | $l_{18}$=0.0 | $l_{19}$=0.0 |
| $l_{110}$= 0.0 | $l_{111}$=1.0 | |

Figure 1 illustrates the results obtained after performing the two step procedure described above. The right panel shows the isolines of the SeaWiFS-VI in the spectral space of the rectified channels centered at 670 and 865 nm. The left panel of the same Figure shows that the SeaWiFS-VI is a reliable estimator of the FAPAR with a root mean square deviation equal to 0.05. It can be seen that the SeaWiFS-VI varies between 0 and 1 over partially to fully vegetated surfaces. Most of the remaining variability between FAPAR and SeaWiFS-VI is induced by the large number and diversity of geophysical scenarios considered. In fact this variability results from conflicting requirements on the simultaneous insensitivity of the SeaWiFS-VI to soil, atmospheric and geometrical effects in the SeaWiFS spectral bands. In the present case, it was found that the signal to noise ratio of the SeaWiFS-VI is equal to 21.26. By comparison, the widely used Normalized Difference Vegetation Index (NDVI), computed on the basis of data from the original channels centered at 670 and 865 nm, exhibits a non-linear relationship with respect to FAPAR and a signal to noise ratio of only 7.04 (Figure 2).

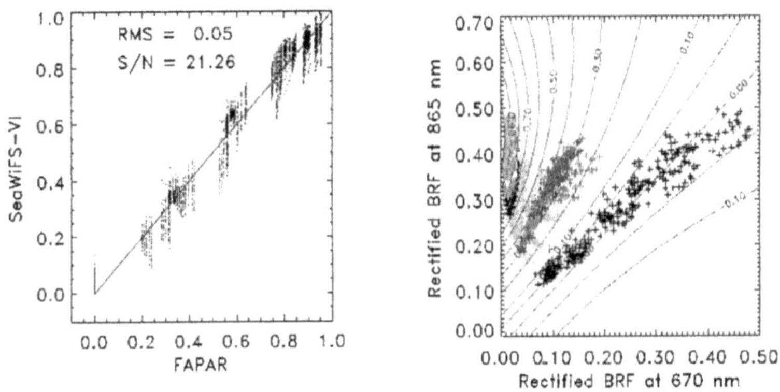


*Figure 1.* The right panel shows the isolines of SeaWiFS-VI in the rectified (670 nm, 865 nm) spectral space together with the spectral rectified bidirectional reflectance factors computed with simulated radiances emerging at the top of the atmosphere. The left panel shows the relationship between the SeaWiFS-VI and the FAPAR values

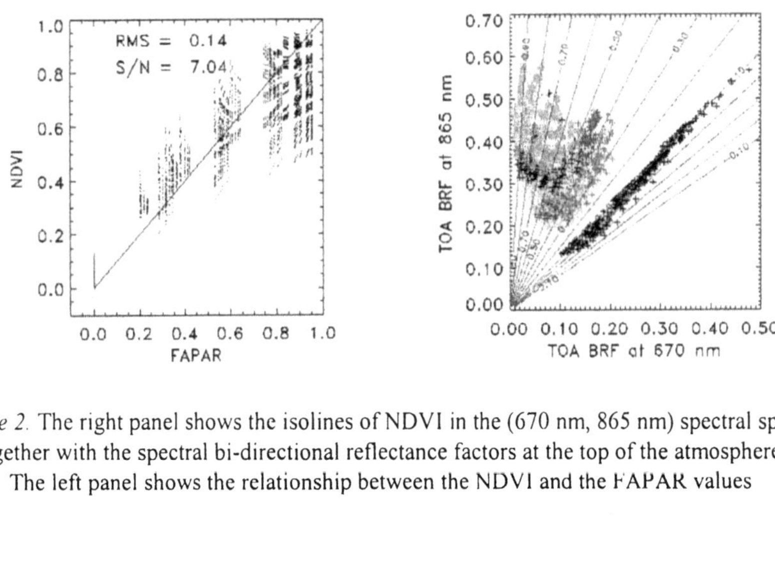


*Figure 2.* The right panel shows the isolines of NDVI in the (670 nm, 865 nm) spectral space, together with the spectral bi-directional reflectance factors at the top of the atmosphere. The left panel shows the relationship between the NDVI and the FAPAR values

*Table 3.* Optimal values of the coefficients for the polynomial $g_2$

| | | |
|---|---|---|
| $l_{21}$=–0.66956 | $l_{22}$=–0.16930 | $l_{23}$=–0.071256 |
| $l_{24}$=–0.090485 | $l_{25}$=–0.81353 | $l_{26}$=–0.035440 |
| $l_{27}$=–1.3438 | $l_{28}$=–0.41673 | $l_{29}$=–0.45123 |
| $l_{210}$=–0.99648 | $l_{211}$=0.0 | |

*Table 4.* Optimal values of the coefficients for the polynomial $g_0$

| | | |
|---|---|---|
| $l_{01}$=0.25130709 | $l_{02}$=0.30589629 | $l_{03}$=–0.0048298 |
| $l_{04}$=–0.3213674 | $l_{05}$=0.31415914 | $l_{06}$=–0.01074418 |

These results demonstrate the significant advances allowed by this approach in the analysis of SeaWiFS-VI. Furthermore, the optimization of the index formula so that it takes values statistically equivalent to the FAPAR permits us

1. to evaluate and monitor the state of land surfaces consistently over the globe in a quantitative physically sound manner,
2. to deliver, to the remote sensing user community, geophysical products relatively independent of atmospheric conditions and of the geometry of illumination and observation, and
3. to process vast amounts of remote sensing data at relatively minor computational costs, without any need for further pre- or post-processing.

For instance, many indices must be computed on the basis of data already partially corrected for atmospheric effects (e.g., Rayleigh scattering, such as in Kaufman and Tanré, 1992 and Huete et al., 1997), or yield values that are not of direct interest to the users.

The applicability of such an optimized index over heterogeneous surfaces, where three-dimensional effects might play a dominant role in controlling the radiation transfer regime, and for various aerosol types, is discussed further in Gobron et al. (2000a). It will be sufficient to state that the application of the same technique to different multispectral single view instruments will allow the development and implementation of high performance compositing methods based directly on FAPAR products, since they are all comparable and independent from the original source of the space data.

## 3. PERFORMANCE OF THE SEAWIFS-VI

SeaWiFS was launched on the SeaStar spacecraft on August 1, 1997. Since mid-September, 1997, it delivers multispectral BRF values collected over all regions of the globe. The wide geographical and long temporal availability of SeaWiFS data enables the verification of 1) the robustness of the

SeaWiFS-VI with respect to large variations in the observation zenith angles and/or rapid changes in atmospheric conditions and 2) the capacity of the index to relate to well-identified ecological patterns.

The polar orbit of SeaWiFS instrument, combined with its wide swath width, permits the observation of sites close enough to the poles more than once per day. For these locations, it is thus possible to compare the original measurements and the derived products from two consecutive orbits, i.e., at about 100 minutes interval. In this period, it is reasonable to expect that the surface has remained essentially the same. Some changes may result from slightly different atmospheric conditions, but the bulk of observed changes must result from variations in the conditions of observation, as the same region is observed from eastward and westward directions (see Figure 3).

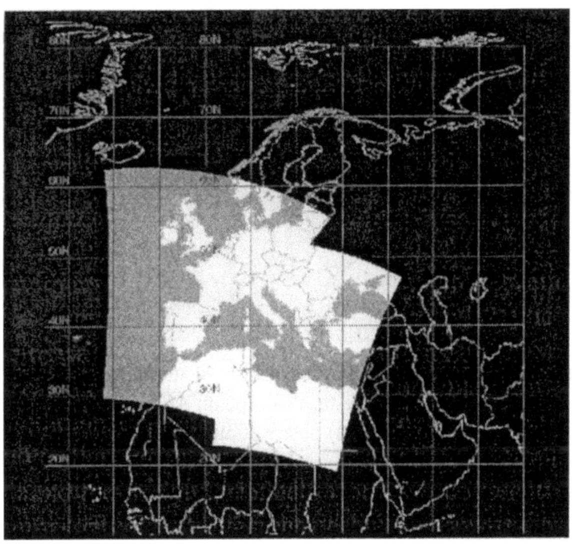

*Figure 3.* Illustration of the trace of two successive orbits of SeaWiFS over Western Europe, for August 7, 1998, at 11:06 and 12:42 UT

For the purpose of this evaluation, we selected data from two consecutive relatively cloud-free SeaWiFS orbits over Northern Europe, acquired on August 7, 1998, at 11:06 and 12:42 UT, respectively. The SeaWiFS-VI values obtained through the procedure described above are displayed in Figure 4 for these two consecutive orbits. The superimposed ellipse on both images delineates the geographical region located approximately between latitude 46° N and 52° N, and longitude 11° E and 15° E, for which further tests are conducted.

Figure 5 (right panel) shows the variations of the SeaWiFS-VI along a particular transect across the mapped data sets for the two consecutive orbits, where the full (dashed) line corresponds to eastward (westward) observation conditions, respectively. Some changes in atmospheric conditions may have occurred in the time period between the two observations, but no significant modifications of the surface properties are expected.

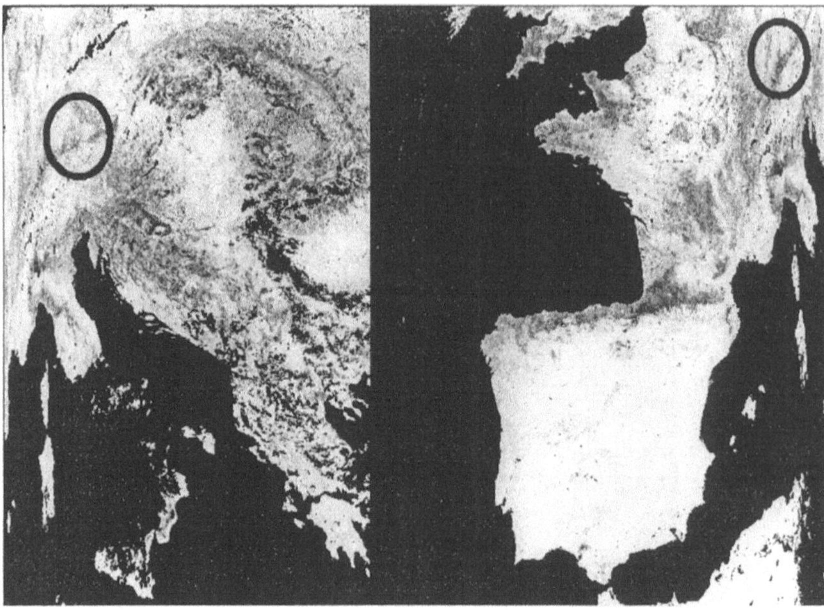

*Figure 4.* Maps of the SeaWiFS-VI estimated for two successive orbits at 11:06 and 12:42 UT, respectively, of SeaWiFS over Western Europe on August 7, 1998. The left (right) image corresponds to an eastward (westward) observation of the region surrounded by an ellipse

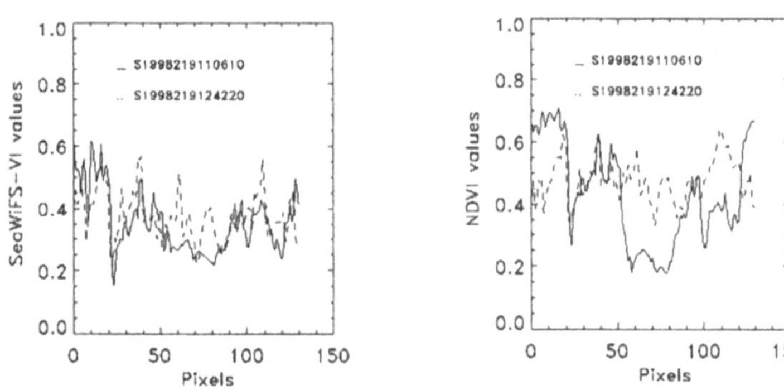


*Figure 5.* Transects of index values extracted from the region identified by the ellipse in Figure 4. The full (dashed) line refers to observations taken at 11:06 (12:42) UT, respectively. The left (right) panel exhibits the SeaWiFS-VI (NDVI) values estimated from data acquired by SeaWiFS for these two successive orbits on August 7, 1998 on the same transect

It can readily be seen that the two SeaWiFS-VI profiles better overlay each other than the corresponding NDVI profiles. Hence, the rectification procedure described earlier has effectively reduced the sensitivity of the optimal index with respect to changes in observation geometry.

The mathematical explanation for these substantial differences in index behavior can be seen in Figure 6, which illustrates the displacements in the spectral space of the data points responsible for an NDVI change of 0.15 between measurements taken from the two successive orbits. The left (right) panel locates these points in the rectified RED-NIR (classical RED-NIR) spectral space and shows the vectors describing the spectral BRF changes during this period. As can be seen, the displacement vectors in spectral space between consecutive orbits occur at significant angles with respect to the NDVI isolines in the original (RED, NIR) space, and are much more parallel to the SeaWiFS-VI isolines in the rectified (RED, NIR) space. This example graphically explains the consistency of the results provided by the SeaWiFS-VI when changing the observation geometry and possibly the atmospheric conditions, and demonstrates the superior performance of SeaWiFS-VI compared to classical indices such as NDVI.

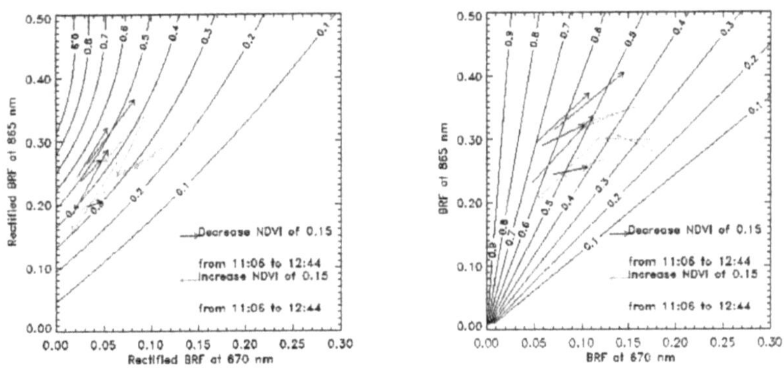


*Figure 6.* Displacements vectors for the data points corresponding to a change in NDVI equal to 0.15 between the two transects shown in Figure 5, acquired at 11:06 and 12:42 UT, respectively. The (left) right panel displays the displacement of the points into the (rectified) red/near-infrared spectral space. Points leading to an increase (decrease) in NDVI values from 11:06 and 12:42 UT are identified in light gray (dark) arrow

# 4. EXPLOITATION OF THE SEAWIFS-VI

The establishment of land cover maps at global and regional scales can be achieved with various tools, techniques and data sets. Currently, this is often done at the global scale, by clustering monthly values of NDVI throughout the year, where each monthly value is itself obtained by selecting, for each given location, the day when the NDVI is maximum during the month (Loveland et al., 1999). Other approaches use directly the spectral dimensions, generally provided by a red and a near-infrared channel (Belward et al., 1990, Ehrlich and Lambin, 1996). Alternative but still exploratory methods exclusively based on a physical interpretation of satellite data sets have been proposed (Gobron et al., 1997 and Gobron et al., 2000b). Whenever vegetation indices are used for this purpose, and independently from the technique applied later for land cover analysis, the use of optimized indicators such as SeaWiFS-VI is fully justified since, at the very least, the latter permit the construction of data sets which are less corrupted by undesirable effects of atmospheric and directional origin. The sensitivity of classical vegetation indices to such perturbations has long been known and extensively documented (see, for instance, Flasse and Verstraete, 1994, Meyer et al., 1995 and Cihlar et al., 1998), and the desire to decrease these sources of noise constituted the original motivation for compositing techniques such as the maximum NDVI described earlier.

To the extent that the SeaWiFS-VI is constrained to fit the FAPAR of the simulated canopies in the simulated data sets, it is likely that the values computed with this index on the basis of actual sensor data will reflect the diversity and spatial distribution of the vegetation and land cover type present in the environment. Of course, this assertion applies only whenever transitions between biome types effectively lead to detectable gradients in FAPAR, either in space or in time. For the sake of the demonstration, we compared, at the regional scale, our SeaWiFS-VI products with a map produced in the context of the Forest Monitoring in Europe with Remote Sensing (FMERS) project which aims at identifying the forests in this region. The FMERS maps are derived from the analysis of one year of data gathered in 1997 by the high resolution IRS 1C WIFS sensor and ancillary ground-based observations (see Hame et al., 1999). This constitutes an independent source of information considered *a priori* adequate to evaluate the suitability of the SeaWiFS-VI for identifying forest patterns. Figure 7 shows that the various forest classes identified by the FMERS project (panel b) are clearly distinguishable on the SeaWiFS-VI derived map (panel a). Note that the latter is produced from SeaWiFS data acquired during a single orbit on August 7, 1998. The map of the corresponding NDVI values (panel c), computed on the basis of the same SeaWiFS data, and using the same color scale, shows patterns that are not immediately related to the FMERS map.

These two applications illustrate the potential benefits of applying optimized indices in general and SeaWiFS-VI in particular to address land cover issues. The major advantage over classical indices is basically a significant increase in the dynamics of the desired information versus the amount of noise due to various perturbing sources. The analysis of spatial and temporal index variations in order to delineate seasonal land cover regions (Loveland and Belward, 1997) is rendered much more reliable and informative due to the noise reduction in the index. It is thus anticipated that the systematic analysis of the global SeaWiFS-VI database will be worthwhile to document land cover changes.

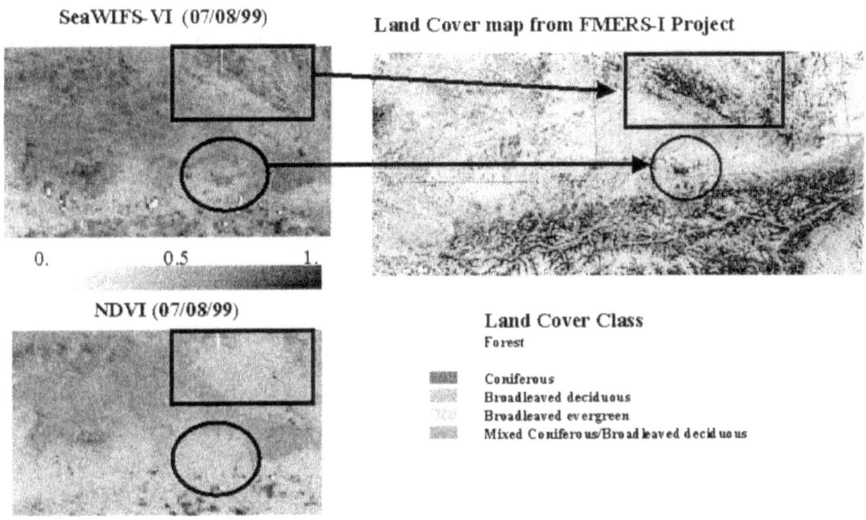


*Figure 7.* Comparison of the SeaWiFS-VI and NDVI values computed with SeaWiFS data taken on August 7, 1998, against a forest map derived from the FMERS project. The specific locations identified on these maps are classified as coniferous forest

# 5. PRODUCTION OF THE SEAWIFS-VI

To support various application projects dealing with the monitoring of land surfaces of SAI, a fast processing system was developed to generate an ensemble of relevant information on the basis of the SeaWiFS data at about 1.5 km resolution.

The system includes a set of algorithms to 1) classify each SeaWiFS pixel on the basis of multispectral BRF measurements into broad categories of geophysical targets such as clouds and bright objects, vegetated surfaces and water bodies and 2) compute the rectified red and near-infrared bands as well as SeaWiFS-VI for those pixels corresponding to vegetated surfaces. A detailed description of the algorithms and of its technical implementation can be found in Gobron et al. (2000c) for terrestrial surfaces and Bulgarelli and Mélin (2000) regarding the retrieval of the optical properties of water pixels. As can be seen from Table 5, the pixel classification is performed on the basis of an ensemble of thresholds using only the values in the bands centered at 443, 670, and 865 nm. These tests were established on the basis of a priori knowledge on the multispectral signatures of each geophysical system. The proposed approach efficiently assigns the vast majority of pixels to these classes without requiring any other ancillary information. A more sophisticated scheme was not deemed necessary or justified given the scientific objectives and computer processing constraints. However, a further screening of undesirable geophysical conditions is imposed such that the values of rectified bands must be within predefined intervals. In practice, for every available individual terrestrial SeaWiFS observation (pixel, date), the algorithm yields either a simple label or, in the case of vegetated surfaces, a string of values including all TOA BRFs, the geometry of illumination and observation, the two rectified bands and the SeaWiFS-VI.

*Table 5.* Pixel classification criteria

| Flag value | Spectral tests | Categories |
|---|---|---|
| 0 | $0 < \rho(443) < 0.3$ and $0 < \rho(670) < 0.5$ and $0 < \rho(865) < 0.7$ and $0 < \rho(443) \leq \rho(865)$ and $\rho(865) \geq 1.25\ \rho(670)$ | Vegetated surface |
| 1 | $\rho(443) \leq 0.0$ or $\rho(670) < 0.0$ or $\rho(865) < 0.0$ | "Bad pixel" |
| 2 | $\rho(443) \geq 0.3$ or $\rho(670) \geq 0.5$ or $\rho(865) \geq 0.7$ | Cloud |
| 3 | $0 < \rho(443) < 0.3$ and $0 < \rho(670) < 0.5$ and $0 < \rho(865) < 0.7$ and $\rho(443) > \rho(865)$ | Water body |
| 4 | $0 < \rho(443) < 0.3$ and $0 < \rho(670) < 0.5$ and $0 < \rho(865) < 0.7$ and $0 < \rho(443) \leq \rho(865)$ and $1.25\rho(670) > \rho(865)$ | Bright surface |
| 5* | $\rho(R670) < 0$ or $\rho(R865) < 0$ | Undefined |
| 6* | VI<0 | No vegetation |
| 7* | VI>1 | Vegetation |

* Generated by the SeaWiFS-VI algorithm internally

For a number of surface applications, it is desirable to ensure a good geographical coverage, which implies the temporal compositing of product time series to fill out the gaps in the daily products created by clouds. Such a procedure is justified to the extent that surface changes occur on a time scale much longer than the one adopted for the compositing. The latter is often performed on the basis of maximum NDVI, over the specified time period, but this procedure has been shown to introduce biases in the resulting data sets due to the preferential selection of measurements collected under specific angular conditions (Holben, 1986 and Meyer et al., 1995). We propose a different scheme that allows the selection of the most representative conditions during the compositing period on the basis of a simple statistical analysis. This analysis, based on the inspection of the daily SeaWiFS-VI values retrieved during each period of ten consecutive days, or monthly period, is implemented as follows. The temporal average and corresponding deviation of the SeaWiFS-VI values over the ten-day (monthly) periods are first estimated:

$$\overline{VI} = \frac{1}{T}\sum_{t=1}^{T} VI(t)$$

$$\Delta_{VI}^{T} = \frac{1}{T}\sum_{t=1}^{T} |VI(t) - \overline{VI}|$$

where $T$ is the number of available clear sky values during the compositing period (10-day or monthly). $\overline{VI}$ is the temporal average index value and $\Delta^{T}_{VI}$ is the average deviation of the distribution. The value selected as the most representative for the given ten day (monthly) period is the actual $VI$ value which minimizes the quantity $|VI(t) - \overline{VI}|$. This procedure thus generates maps of geophysical products for every ten-day period, and monthly period, where each individual value represents the actual measurement or product for the day considered the most representative of that period. The geometry of illumination and observation for the particular day selected is saved as part of the final product, which is thus fully documented and traceable.

The various panels of Figures 8 and 9 provide an example of monthly composite products derived from SeaWiFS measurements for the month of May, 1998, over Western Europe.

Panel a (b) of Figure 8 illustrates the geographical distribution of solar (observation) zenith angles that result from this composition process for the indicated period. In this particular example, the solar zenith angle varies approximately between 12° and 50° from the southern to the northern part of the region considered, while the observation zenith angle varies between 20° and 43°, depending on the outcome of the selection procedure for identifying the most representative day in the entire monthly time series. Frames (c) and (d) of the same figure show the results of the rectification process for the red and near-infrared channels, respectively. Finally, Figure 9 exhibits the composited SeaWiFS-VI itself (left panel) and the associated average deviation of the distribution (right panel), respectively. A detailed inspection of the SeaWiFS-VI map does not reveal any particular bias despite abrupt changes in the satellite observation geometry, nor does it show artifacts that could have been induced by the compositing technique. The average deviation throughout this composite remains less than 0.05, indicating that the processing algorithm leads to rather stationary index values during this monthly period.

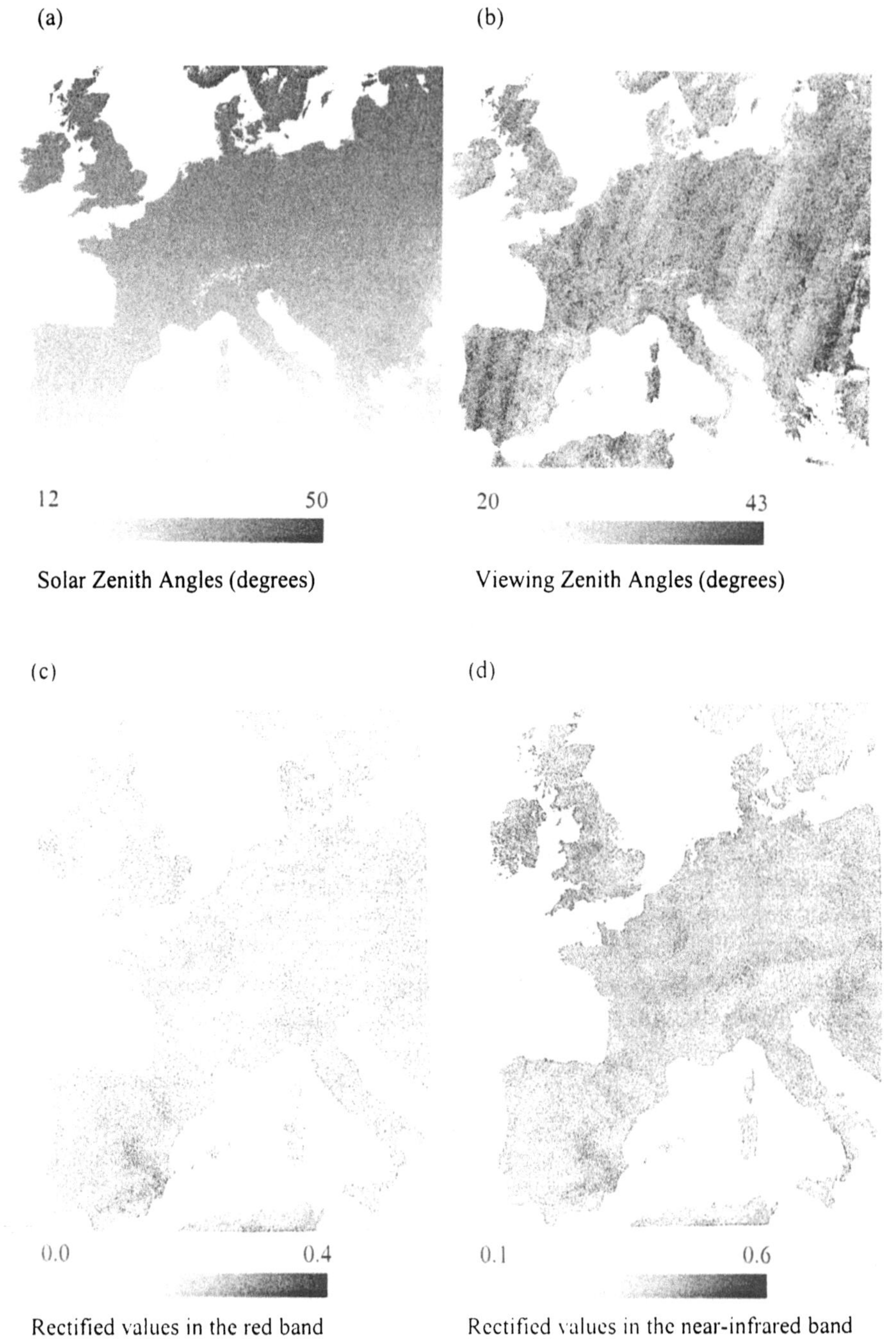


*Figure 8.* Illustration of the retrievals from the SeaWiFS-VI land surface algorithm obtained during the monthly period between May 01 and 31, 1998. Panels (a) and (b) show the Sun and satellite view zenith angles, respectively. The values of the rectified red and near-infrared bands are mapped in panels (c) and (d), respectively

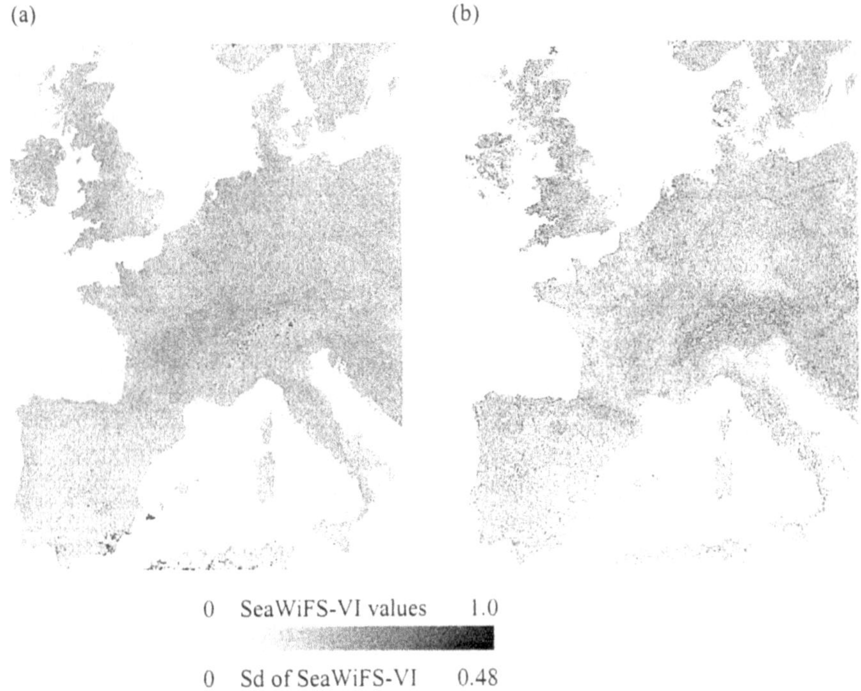


*Figure 9.* Same as Figure 8 except for the SeaWiFS-VI/Fapar (panel a) and the associated average deviation calculated over the monthly composite period (panel b)

# 6. CONCLUSION

This paper documents the implementation and exploitation of a vegetation index whose purpose is to identify and monitor the presence of live green vegetation over a wide diversity of terrestrial surfaces, observed under a variety of atmospheric and angular conditions by the SeaWiFS instrument. The proposed approach is based on a two step procedure. The first step aims at rectifying the red and near-infrared bands from the perturbing effects due to the atmosphere and the changes in the relative geometry of illumination and observation. The second step consists in optimizing the index formula to approximate a one-to-one relationship between the index value and the Fraction of Absorbed Photosynthetically Active Radiation, used as a proxy for detecting the presence of healthy vegetation. The procedure capitalizes on the availability of advanced, coupled, surface-atmosphere radiation transfer models that are exploited to construct the training data set against which the SeaWiFS index optimization is achieved.

A quantitative evaluation of the performance of SeaWiFS-VI has been established through an analysis of actual SeaWiFS observations. This application has shown the capability of the optimized index to be much less sensitive than NDVI to perturbing effects. It also illustrates the ability of this index to distinguish between various biome types on the basis of a single day image. The new products available from this algorithm include the rectified red and near-infrared channels in addition to the final FAPAR estimates. These geophysical products are well suited to address a number of issues related to the documentation and monitoring of land surfaces.

# 7. REFERENCES


Belward, A., J. C. Taylor, M. J. Stuttard, E. Bignal, J. Mathews, and D. Curtis (1990) An unsupervised approach to the classification of semi-natural vegetation from Landsat Thematic Mapper data, *International Journal of Remote Sensing*, **11**, 429–445.

Bulgarelli, B. and F. Mélin (2000) SeaWiFS data processing code REMBRANDT version 1.0: code elements, COASTS Annual Report 2000, Technical Report EUR NC, EC Joint Research Centre.

Cihlar, J., J. Chen, Z. Li, F. Huan, R. Latifovic, and R. Dixon (1998) Can inter-annual land surface signal be discerned in composite AVHRR data? *Journal of Geophysical Research*, **103**, 23,163–23,172.

Ehrlich, D. and E. F. Lambin (1996) Broad scale land-cover classification and inter-annual climatic variability, *International Journal of Remote Sensing*, **17**, 845–862.

Flasse, S. and M. M. Verstraete (1994) Monitoring the environment with vegetation indices: Comparison of NDVI and GEMI using AVHRR data over Africa, in F. Veroustraete and R. Ceulemans (Eds.), *Vegetation, Modelling and Climatic Change Effects*, 107–135. The Hague: SPB Academic Publishing.

Gobron, N., B. Pinty, and M. M. Verstraete (1997) Presentation and application of an advanced model for the scattering of light by vegetation in the solar domain, in *Proceedings of the 7th ISPRS International Symposium on Physical Measurements and Signatures in Remote Sensing*, Courchevel, France, 7–11 April 1997, 267–273, Balkema/Rotterdam/Brookfield.

Gobron, N., B. Pinty, M. M. Verstraete, and Y. Govaerts (1999) The MERIS Global Vegetation Index (MGVI): Description and preliminary application, *International Journal of Remote Sensing*, **20**, 1917–1927.

Gobron, N., B. Pinty, M. M. Verstraete, and J.-L. Widlowski (2000a) Advanced vegetation indices optimized for up-coming sensors: Design, performance and applications, *IEEE Transactions on Geoscience and Remote Sensing*, in print.

Gobron, N., B. Pinty, M. M. Verstraete, J. V. Martonchik, Y. Knyazikhin, and D. J. Diner (2000b) The potential of multi-angular spectral measurements to characterize land surfaces: Conceptual approach and exploratory application, *Journal of Geophysical Research*, in print.

Gobron, N., B. Pinty, M. M. Verstraete, and F. Mélin (2000c) Development of a Vegetation Index Optimized for the SeaWiFS Instrument ATBD, Version 2.0 Technical Report **EUR EN**, Space Applications Institute, In print.

Gobron, N., M. M. Verstraete, and B. Pinty (1998) Development of a spectral index optimized for the GLI Instrument Algorithm Theoretical Basis Document, Technical Report **EUR 18138 EN**, Space Applications Institute.

Govaerts, Y., M. M. Verstraete, B. Pinty, and N. Gobron (1999) Designing optimal spectral indices: A feasibility and proof of concept study, *International Journal of Remote Sensing*, **20**, 1853–1873.

Hame, T., K. Anderson, A. Lohi, M. Kohl, R. Paivinen, E. Carfagna, H. JeanJean, I. Spence, T. leToan, S. Quegan, C. Estreguil, S. Folving, and P. Kennedy (1999) Validated forest variable mapping across Europe using multi-resolution data-results of the FMERS study, in Remote sensing and forest monitoring International IUFRO Conference, Rogow, Poland, June 1-3.

Holben, B. N. (1986) Characteristics of maximum-value composite images from temporal AVHRR data, *International Journal of Remote Sensing*, **7**, 1417–1434.

Huete, A. R., H. Q. Liu, K. Batchily, and W. van Leeuven (1997) A comparison of vegetation indices over a global set of TM images for EOS-MODIS, *Remote Sensing of Environment*, **59**, 440–451.

Kaufman, Y. J. and D. Tanré (1992) Atmospherically resistant vegetation index (ARVI) for EOS-MODIS, *IEEE Transactions on Geoscience and Remote Sensing*, 30, 261–270.

Loveland, T. R. and A. S. Belward (1997) The IGBP-DIS global 1 km land cover data set, DIScover: First results, *International Journal of Remote Sensing*, **18**, 3289–3295.

Loveland, T. R., Z. Zhu, D. O. Ohlen, J. F. Brown, B. C. Reed, and L. Yang (1999) An analysis of the IGBP global land-cover characterization process, *Photogrammetric Engineering and Remote Sensing*, **65**, 1021–1032.

Meyer, D., M. M. Verstraete, and B. Pinty (1995) The effect of surface anisotropy and viewing geometry on the estimation of NDVI from AHVRR, *Remote Sensing Review*, **12**, 3–27.

Rahman, H., B. Pinty, and M. M. Verstraete (1993) Coupled surface-atmosphere reflectance (CSAR) model. 2. Semiempirical surface model usable with NOAA Advanced Very High Resolution Radiometer data, *Journal of Geophysical Research*, **98**, 20,791–20,801.

Verstraete, M. M. and B. Pinty (1996) Designing optimal spectral indices for remote sensing applications, *IEEE Transactions on Geoscience and Remote Sensing*, **34**, 1254–1265.

Verstraete, M. M., B. Pinty, and N. Gobron (1998) Development of a Spectral Index Optimized for the VEGETATION Instrument, Report Phase 1, http://www-vegetation.cst.cnes.fr:8050/vgtprep/verstraete/report2.html.

# Modeling sensible heat flux using estimates of soil and vegetation temperatures: the HEIFE and IMGRASS experiments

Li JIA[1*], Massimo MENENTI[2], Zhongbo SU[3], Zhao-Liang LI[2], Vera DJEPA[4] and Jiemin WANG[1]
[1]*Cold and Arid Regions Environmental and Engineering Research Institute (CAREERI), Chinese Academy of Sciences (CAS), Lanzhou, China*
[2]*Universite Louis Pasteur, Strasbourg, France*
[3]*Alterra Green World Research, Wageningen University and Research Centre, Wageningen, The Netherlands*
[4]*University of Dundee, Dundee UK*
[*]*On leave at Alterra Green World Research, Wageningen, Wageningen, The Netherlands*

Abstract: Heat fluxes at heterogeneous land surfaces are often modeled using single-source resistance-type transport equations, i.e. assuming horizontal homogeneity of the land surface and of the boundary layer. Large deviations from these conditions occur at partial canopies which are geometrically and thermally heterogeneous. Improved models of heat transfer have been proposed in literature to deal with these conditions. Such models require a measure of thermal heterogeneity of the land surface. Directional measurements of the radiance emitted by the land surface have the potential of providing a measure of thermal heterogeneity and improved parameterizations of sensible heat transfer. The paper proposes a methodology, together with two case studies on the use of directional measurements of spectral radiance to estimate the component temperatures of soil and vegetation and their subsequent use to model sensible heat fluxes at length scales of $10^{-1}$m and $10^{3}$m.

The first case study relied on multi-temporal field surface temperature measurements at view angles of $0^0$, $23^0$ and $52^0$ collected at sparse grass covered surface during the Inner-Mongolia Grassland-Atmosphere Surface Study (IMGRASS) experiment in China. This provided useful insights on the applicability of a simple linear mixture model to the analysis of observed directional radiances. Sensible heat fluxes were estimated both at field and regional scales by using The Along-Track Scanning Radiometer (ATSR)-2 observations. The second was done with directional ATSR-1 observations only and was a contribution to the Hei He International Field Experiment (HEIFE) in China. The HEIFE case study was focused on the large oasis of Zhang-Ye and



*M. Beniston and M.M. Verstraete (eds.),*
*Remote Sensing and Climate Modeling: Synergies and Limitations*, 23–49.

led to useful estimate of soil and vegetation temperatures. Sensible heat flux is modeled separately for each component heat source, i.e. soil and vegetation. Heat flux densities were compared with field measurements made with an eddy correlation device and values obtained with vertical profiles of air temperature and horizontal wind speed. Agreement was good for the IMGRASS case study based on field measurements. ATSR-based estimates were also in good agreement with values obtained with observed and modeled through vertical profiles, although few data points were available because of the large spatial scale of the ATSR estimates.

# 1. INTRODUCTION

Remote measurements of spectral directional radiance have been used to estimate heat fluxes at heterogeneous land surfaces (e.g. Menenti, 2000). One active research field is the observation and modeling of sensible heat flux densities at land surfaces using remotely sensed surface temperature and albedo. The basis of this method has been classical one-dimensional resistance-type transport models in which sensible heat flux can be expressed as

$$H=\rho_a C_p(T_{aero}-T_{air})/r_{ah} \tag{1}$$

$\rho_a$ is the air density; $C_p$ is the specific heat of air at constant pressure (=1005 J kg$^{-1}$ K$^{-1}$); $T_{air}$ is the air temperature at a reference level; $T_{aeor}$ is the surface aerodynamic temperature; and $r_{ah}$ is the aerodynamic resistance to heat transfer (s m$^{-1}$) and can be expressed in the near-surface layer (Brutsaert,1982) as :

$$r_{ah}=r_a+r_e \tag{2}$$

where $r_a$ is aerodynamic resistance for momentum, and $r_e$ is a so-called 'excess resistance' which originally arises from the different transfer mechanisms for heat and momentum at the surface so that resistance to heat transport is higher than that to momentum transport. Transport resistances are parameterized as functions of a roughness length for momentum and a roughness length for heat transport. The 'excess resistance' may therefore be expressed in terms of $kB^{-1}$ (Chamberlain, 1968):

$$kB^{-1} = \ln(z_{0m}/z_{0h}) \tag{3}$$

where k is the von Karman's constant, $z_{0m}$ and $z_{0h}$ are the roughness lengths for momentum and heat transfer respectively. The roughness lengths are typically measured with moderate accuracy and the experimental error on $kB^{-1}$ is large.

In classical single-source resistance-type models, $T_{aero}$ is derived from the extrapolation of the air temperature profile to the apparent canopy height (the displacement height + the roughness length) and may not actually exist or be measurable except for smooth surfaces (Norman et al, 1995a). For practical purposes radiometric surface temperature $T_{rad}$ is used in place of $T_{aero}$ in such single-source heat transfer models. Radiometric surface temperature can be measured by a radiometer and is more appropriate to the application of remote sensing at various spatial scales. However, when using $T_{rad}$ instead of $T_{aero}$ in Eq.(1), an empirical adjustment must be made because $T_{rad}$ is not equal to $T_{aero}$, which results in an additional resistance added to the resistance term in a single-source model. The moderate accuracy of $kB^{-1}$ values makes it very difficult to determine the two terms of the correction separately. Therefore, when using $T_{rad}$ in single-source models, one can consider the 'excess resistance' in terms of $kB^{-1}$ as a combination of adjustments which account for the difference between $z_{0m}$ and $z_{0h}$ and the difference between $T_{rad}$ and $T_{aero}$ even though these two additional resistances are different conceptually. Most of the studies on 'excess resistance' has focused on the determination of $kB^{-1}$ (Table 1) and the values of $kB^{-1}$ (or $z_{0h}$) are always related to the 'surface' temperature.

For most homogeneous 'permeable-rough' surfaces such as uniform and full canopy cover, $kB^{-1}$ is approximately 2 to 3 (Brusaert, 1982) and the single-source resistance methods have been applied successfully (Deardorff, 1978; Kustas, 1990). Over heterogeneous sparse canopies, however, widely varying values for $kB^{-1}$ are found in literature (Kustas et al, 1989; Beljaars and Holtslag, 1991; Stewart et al, 1994) (see Table 1). This implies that the value of $kB^{-1}$ cannot be approximated by a constant in case of sparse cover and it must be determined through calibration. A fixed value of $kB^{-1}$ (or $z_{0h}$) will introduce errors into the estimation of heat flux (Kohsiek et al., 1993, Stewart et al, 1994). Some authors related $kB^{-1}$ to surface wind speed and the difference between surface temperature and air temperature (Kustas et al., 1989). It seems that regressing $kB^{-1}$ with wind speed and difference of surface and air temperature does not provide a general formula for any sparse canopy. Consequently, it is difficult to develop a simple method to relate $kB^{-1}$ to surface properties.

Recently, efforts have been made to develop dual or multi-source models to estimate sensible heat flux and evaporation from partial canopies (Choudhury and Monteith, 1988; Kustas, 1990; Lhomme et al, 1994; Norman et al, 1995) so that the empirical adjustment of resistance in single-source models can be avoided. Vegetation and the substrate (i.e. the soil), in fact, interact separately with the air in the canopy space hereby affecting the sensible and latent heat flux densities in sparsely covered canopies especially when the temperatures of the cooler vegetation and the warmer soil surface are significantly different. Consequently, resistances between vegetation (foliage) and the air in the canopy space and between the soil and the air in the canopy space have to be parameterized in dual–source models. The difficulty with this approach to account for the mechanisms determining heat transfer in the vicinity of leaves and soil is that in the canopy space there is no defined surface layer, i.e. there is no defined vertical structure and no horizontal homogeneity. We propose a different conceptual model of heat transfer in the canopy space to describe separately heat exchanges between leaves, soil and air.

In our dual-source model, component temperatures have to be known. Multi-angle and multi-channel remote sensing technology such as The Along-Track Scanning Radiometer (ATSR)-1/2 on board the European Remote Sensing Satellites (ERS)-1/2 provides an opportunity to extract component temperatures from directional measurements of existance (Menenti et al 1999). A new dual-source model is developed in our study and used to estimate sensible heat fluxes based on component temperatures for incomplete canopy cover both at field scale and at regional scale. As mentioned above, our model is different from other authors' in the resistance scheme and is simplified. It is applicable at regional scale where meteorological information near surface is not always available for each pixel.

*Table 1.* Values of $kB^{-1}$ at various (sparse) surfaces

| Project(or site) | Surface type | $kB^{-1}$(avg) | $kB^{-1}$(std) | Reference author(s) |
|---|---|---|---|---|
| | Homogeneous and full vegetated canopy | 2 or 3 | | Brusaert(1982) |
| Cabauw | Grassland | 8.8 | 0.24 | Beljaars&Holtslag(1991) |
| SEBEX | Savanna | 5.8 | 2.9 | Stewart(1994) |
| SEBEX | Open forest | 8.3 | 3.3 | as above |
| MONSOON 90 | Grass | 3.8 | 2.8 | as above |
| MONSOON 90 | Shrubs | 5.6 | 2.8 | as above |
| Owens Valley | Shrubs | 8.0 | 3.8 | as above |
| Smith Creek Valley | Shrubs | 12.4 | 5.9 | as above |
| Smoke Creek Desert | Shrubs | 8.4 | 4.9 | as above |
| La Crau | Grass/Stones | 4.5 | 2.1 | as above |
| Owens Valley | Bushes | 5.6 | 3.2 | Kustas et al.(1989) |
| HEIFE Gobi | Gobi with Shrubs | 5.5 | 4.1 | this study |
| AECMP'95(HEIFE) Desert site | Shrubs | 12.3 | 6.0 | this study |

# 2. THEORY

## 2.1 Basic equations of the dual-source model for estimation of sensible heat flux density from a composite surface.

Our dual-source model parameterizes heat exchange between vegetation, soil and air using the resistance scheme in Fig.1. Sensible heat flux from the (vegetation + soil) mixture is considered as the sum of contributions from vegetation and soil and can be expressed as

$$H=H_v+H_s \tag{4}$$

where $H_v$ and $H_s$ are the sensible heat flux densities from vegetation and the soil (W $m^{-2}$) respectively,

$$H_v=\rho_a C_p(T_v- T_{ac})/r_{ah,v} \tag{5}$$

$$H_s=\rho_a C_p(T_s- T_{ac})/r_{ah,s} \tag{6}$$

where $T_v$ and $T_s$ are vegetation and soil temperature respectively, $T_{ac}$ is aerodynamic temperature in the canopy space, $r_{ah,v}$ and $r_{ah,s}$ are resistances for heat transfer from vegetation and soil to air in the canopy space having temperature $T_{ac}$. The total heat flux H from the canopy space to the surface layer may also be expressed as

$$H= \rho_a C_p(T_{ac} -T_{ref})/r_{ah} \tag{7}$$

where $T_{ref}$ is air temperature at the reference height, z, above the canopy, $r_{ah}$ is the classical aerodynamic resistance for heat transfer between the reference source height, $z_0$, in the canopy and a reference level above the canopy.

To use this heat transfer model, parameterizations have to be developed to estimate the three resistances $r_{ah,v}$ , $r_{ah,s}$ and $r_{ah}$.

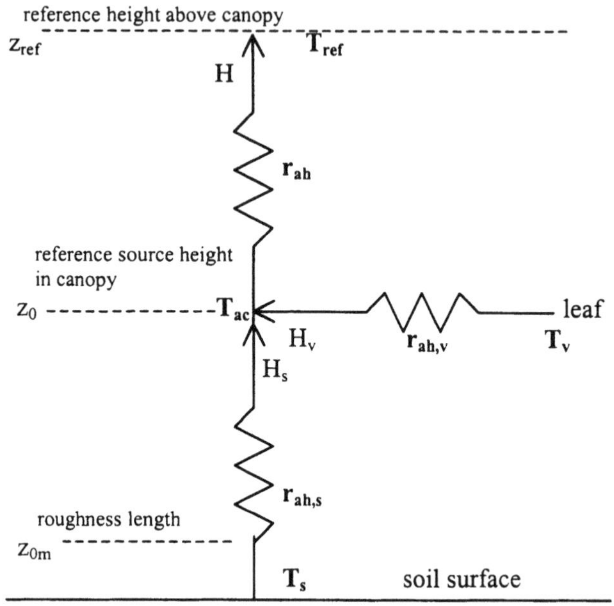


*Figure 1.* Resistance scheme of dual-source model to estimate sensible heat flux using component temperatures

## 2.2 Resistances scheme

According to Stanghellini (1987) the leaf resistance for sensible heat transfer $r_{ah,e}$ can be expressed as

$$r_{ah,e}=\rho_a C_p l/(\lambda_a Nu) \tag{8}$$

where Nu is a mixed convection Nusselt number given by:

$$Nu=0.37(Gr+6.92R_e^{\,2})^{0.25} \tag{9}$$

where Gr the Grashof number is:

$$Gr=g\beta l^3(T_{surf}-T_{c\text{-}ref})/\upsilon^2 \tag{10}$$

and Reynolds number Re:

$$Re=ul/\upsilon \tag{11}$$

where :

| | | |
|---|---|---|
| $l$ | - | characteristic dimension of the surface (m) |
| $\lambda_a$ | - | thermal conductivity of air (W $m^{-1}$ $K^{-1}$) |
| u | - | wind speed (m $s^{-1}$) |
| $\upsilon$ | - | kinematic viscosity of air ($m^2$ $s^{-1}$) |
| g | - | acceleration due to gravity (m $s^{-2}$) |
| $\beta$ | - | coefficient of thermal expansion of air ($K^{-1}$) |
| $T_{surf}$ | - | leaf surface temperature (K) |
| $T_{c\text{-}ref}$ | - | air temperature at a reference height in the canopy (K) |

After substitutions and using the numeric values of air properties, $r_{ah,e}$ can be written as (see Stanghellini 1987 for details)

$$r_{ah,e} = 1174 l_v^{0.5}/(l_v|T_v - T_{ac}|+207u_0^2)^{0.25} \tag{12}$$

where $T_v$ and $T_{ac}$ as defined above and $u_0$ is wind speed at a level in the canopy, $l_v$ is the mean leaf size.

Using a parallel resistance scheme for leaf resistances, the total vegetation resistance is:

$$r_{ah,v}=r_{ah,e}/(2LAI) \tag{13}$$

Namely:

$$r_{ah,v}=1174l_v^{0.5}/(l_v|T_v - T_{ac}|+207u_0^2)^{0.25}/(2LAI) \tag{14}$$

Experimental validation of this parameterization was given by Stanghellini (1987).

To parameterize the soil resistance in a similar way, a suitable linear dimension of the soil surface for the (vegetation + soil) mixture must be identified and estimated. We propose to take typical linear dimension of the soil surface as the square root of the fractional soil cover, i.e. the fraction of horizontal unit area occupied by soil:

$$l_s = (1-f_v)^{0.5} \tag{15}$$

where $f_v$ is the fractional vegetation cover, $(1-f_v)$ is then the fractional soil cover.

The parameterization for the soil resistance is then given by:

$$r_{ah,s}=1174\, l_s^{0.5}/(l_s|T_s - T_{ac}|+207u_0^2)^{0.25} \tag{16}$$

In neutral conditions, aerodynamic resistance for heat transfer between a level in the canopy and the reference height above the canopy is expressed as

$$r_{ah0}=ln[(z_{ref}-d)/z_0]/(k^2 u_{ref}) \tag{17}$$

where $u_{ref}$ is the wind speed (m $s^{-1}$) at the same height as $T_{ref}$, $k$ is the von Karman's constant taken as 0.4, d is the displacement (m) and $z_0$ is the reference source height (m) in the canopy. Following the expressions given by Choudhury et al (1986) and Kalma and Jupp (1990) for incompletely covered surface, the stability corrected aerodynamic resistance $r_{ah}$ is:

$$r_{ah}=r_{ah0}/(1+\eta)^p \tag{18}$$

with $p$=0.75 in unstable conditions and $p$=2 in stable conditions, and

$$\eta=5g(z_{ref}-d)(T_{ac}-T_{ref})/(T_{ref}u_{ref}^2) \tag{19}$$

Aerodynamic surface temperature of the canopy, $T_{ac}$, is not measurable directly as discussed above. In our dual-source model, $T_{ac}$ is an ancillary variable. Combining Eq.(4), (5), (6) and (7), one can get:

$$T_{ac}=\frac{\frac{T_{ref}}{r_{ah}}+\frac{T_v}{r_{ah,v}}+\frac{T_s}{r_{ah,s}}}{\frac{1}{r_{ah}}+\frac{1}{r_{ah,v}}+\frac{1}{r_{ah,s}}} \tag{20}$$

Iterations are made between Eq.(14), Eq.(16), Eq.(18) and Eq.(20) to determine the values of the variables $T_{ac}$ first, then $r_{ah}$, $r_{ah,v}$ and $r_{ah,s}$, $H_v$, $H_s$ and H are determined finally.

## 2.3 Retrieval of component temperatures and vegetation information from directional measurements

In a dual-source model, component surface temperature $T_v$ and $T_s$ are needed instead of one radiometric surface temperature used in a single-source model. Vegetation information such as leaf area index or fractional cover of vegetation is also needed. Neglecting the cavity effect in the canopy, the radiometric surface temperature $T_{rad}$ can be related to component temperatures by a simple linear mixture model as the following (Norman et al, 1995a):

$$\varepsilon(\lambda)T_{rad}^{n}(\theta,\lambda) = f_v(\theta)\varepsilon_v(\lambda)\ T_v^n + f_s(\theta)\varepsilon_s(\lambda)\ T_s^n \qquad (21)$$

where n 4 for spectral bands in λ=8~14μm and λ=10~12μm ( Becker and Li,1990), ε, $\varepsilon_v$ and $\varepsilon_s$ are the emissivity of the (vegetation + soil) mixture, vegetation and the soil respectively, θ is the zenith view angle of the sensor, $f_v(\theta)$ and $f_s(\theta)$ are the fractions of vegetation and the soil in the field of view of radiometer when looking at the surface at zenith view angle (θ), $f_s(\theta)=1-f_v(\theta)$, is the soil fraction. $f_v(\theta)$ depends on the typc of vegetation and the architecture of the canopy. Assuming a random canopy with a spherical leaf angle distribution (Norman et al, 1995b),

$$f_v(\theta) = 1- exp[-0.5LAI/cos(\theta)] \qquad (22)$$

where LAI is leaf area index. For nadir view, $f_v(\theta)$ is the fractional vegetation cover, $f_c$. Usually, a radiometer measure surface brightness temperature $T_{B0}(\lambda,\theta)$ and Eq. (21) can be rewritten as

$$T_{B0}^{n}(\theta,\lambda) = f_v(\theta)\varepsilon_v(\lambda)\ T_v^n + f_s(\theta)\varepsilon_s(\lambda)\ T_s^n \qquad (23)$$

When the surface brightness temperature $T_{B0}(\lambda,\theta)$ at two or more view angles can be obtained from the measurements of radiance, it is possible to derive $T_v$ and $T_s$ from $T_{B0}(\lambda,\theta)$ through Eq.(23).

## 2.4 Atmospheric correction for ATSR thermal channels – single channel method

Space-borne radiometers measure brightness temperature at the top of the atmosphere, $T_B(\lambda,\theta)$, not the surface brightness temperature, $T_{B0}(\lambda,\theta)$. At wavelength λ and zenith view angle θ, the radiances measured by the radiometers on the satellite are from three contributions: (1) emittance from the land surface that is attenuated by the atmosphere between the surface and the sensor, (2) the downwelling atmospheric emittance to the surface and then reflected by the surface to the sensor, (3) upwelling atmospheric emittance. With this concept, the radiative transfer equation in thermal bands can be written as

$$B(T_B(\lambda,\theta)) = \varepsilon(\lambda,\theta)B(T_{rad}(\lambda,\theta))\tau(\lambda,\theta) + I_{atm}\uparrow(\lambda,\theta) + I_{ref}(\lambda)\tau(\lambda,\theta) \qquad (24)$$

where B is Planck function, $\varepsilon(\lambda,\theta)$ is the surface emissivity, $\tau(\lambda,\theta)$ is the total atmospheric path transmittance, $I_{atm}\uparrow(\lambda,\theta)$ is the upwelling atmospheric emittance, $I_{ref}(\lambda)$ is reflected downwelling atmospheric emittance by the surface. Neglecting the reflected downwelling atmospheric emittance, Eq.(24) can be rewritten in terms of brightness temperature

$$B(T_B(\lambda,\theta)) = B(T_{B0}(\lambda,\theta))\tau(\lambda,\theta) + I_{atm}\uparrow(\lambda,\theta) \quad (25)$$

One can get the corrected brightness temperature at the surface, $T_{B0}(\lambda,\theta)$, by inverting the Planck's function $B(T_{B0}(\lambda,\theta))$ in Eq.(25), knowing the total atmospheric path transmittance and the atmospheric emittance at different wavelengths and zenith view angles which can be computed using a radiative transfer model such as MODTRAN.

# 3. DATA

## 3.1 Field measurements

Data used in this study are from two experiments on land surface processes carried out in China, namely the Hei He International Field Experiment (HEIFE) (Mitsuta, 1993) and the Inner-Mongolia Grassland-Atmosphere Surface Study (IMGRASS) (Su et al 1999).

### 3.1.1 HEIFE

The large-scale field experiment HEIFE has been carried out in the arid zone of north-west China during several years (1989-1995). In the area, long-term measurements (Table 2) were made by means of towers, radiometers, automatic weather stations, and by means of additional eddy correlation and Bowen ratio devices during several short-term intensive observation periods. One of the basic sites, named **Zhang-Ye**, is selected for present study. **Zhang-Ye** site is located in the central part of a large oasis with crops such as bean, corn and a smaller fraction of orchard. Windbreaks are widely used to protect crops. At the 1 km (ATSR pixel size) scale the surface is relatively homogeneous in a statistical sense.

Surface radiometric temperatures on the ground were measured using a radiometer (EKO Thermo-Hunter) operating in the spectral range 7-20 μm with a radiometric resolution 0.1 $^{0}$C and mounted at a 1.5 m height with zenith view angle $2^{0}$ at HEIFE sites. Measurements of heat fluxes from the surface to the atmosphere were made at 2.9 m height by means of eddy correlation systems which consists of three-dimensional sonic anemometer-thermometer (SAT: Kaijo, DAT-300 with TR-61A Probe), infrared hygrometer (Kaijo, AH-300), Clinometer (Kaijo, CM-100) and rotator (Kaijo,

502MSA). Profiles of wind speed, air temperature and humidity were measured on a 20 m high tower between the 0.5 and 20 m levels.

### 3.1.2 IMGRASS

IMGRASS's field experiment was carried out in Inner Mongolia grassland in north-east of China during May to August in 1998. Its main aim is to understand the effects of changing vegetation on the hydrologic and heat cycle of Inner Mongolia grassland and to provide surface measurements of water, heat and trace gases over various scales for development and validation of remote sensing algorithms. The land cover in the experiment area is grassland with various species of grass and fractional cover. The site in the present study, **Baiyinsumu**, has sparse grass cover and is a so-called degraded prairie.

Heat flux measurements near the surface were made by using the same method and instruments as those in HEIFE campaign but at 4.9m height. A 10m high tower was set up to measure wind speed, air temperature and humidity profiles with 5 levels at 0.5m, 1m, 2m, 4m and 8m height respectively. The baseline measurements, which include vertical profiles of wind speed, air temperature and humidity can be used to estimate sensible and latent fluxes. This method has been used successfully to estimate sensible and latent heat fluxes in early HEIFE studies (see Zuo et al, 1993). An Eppley pyrgeometer PIR with spectral range 4-50 μm was used to measure radiant flux density at IMGRASS site 4 from which surface temperature was obtained.

During the period of 26-31 July 1998, directional surface brightness temperature was measured using an IR-AH portable digital radiation thermometer operating in the spectral range 8-13 μm, footprint diameter = [distance/50] (m). Observations at nadir, $23^0$ and $52^0$ zenith view angle were done. The measurement height at nadir was 1.5 m and corresponds to a footprint diameter of about 3 cm. The field of view of the sensor is therefore small enough so that unobstructed bare soil can be seen at nadir view angle. Leaf area index of 0.5 was determined by counting grass leaf area in a meter square (Su et al,1999).

At both sites, standard meteorological radiosounding data closest to the satellite overpass time were collected to perform atmospheric correction. At HEIFE site, lower level radiosounding and tethered-balloon measurements were also used for atmospheric correction of ATSR data.

## 3.2 Remote sensing data

ATSR-1 observations on 19 August 1991 were collected for HEIFE area, and ATSR-2 on 14 June 1998 for IMGRASS area. Dual view angle data (approximately 0° and 53° at surface) are acquired in four channels by ATSR-1 and seven channels by ATSR-2. The two thermal channels (TIR) in ATSR1 and 2 (the central wavelengths are 11 and 12μm) are employed in our study. The TIR sensors are saturated at about 311 K. The standard ATSR-1/2 gridded brightness temperature image (512x512 $km^2$) was produced from nadir and forward view instrument pixels which are collocated and gridded into a 1 km grid resolution (see the World Wide Web site at http://www.atsr.rl.ac.uk/software.html for details).

The satellite overpass time is around 11h at local solar time for HEIFE and IMGRASS respectively. Subsets of 20 x 20 $km^2$ for HEIFE and 40 x 40 $km^2$ for IMGRASS were extracted from the ATSR-1 and - 2 images. The standard meteorological radiosounding station in HEIFE is located in the subset. The one for IMGRASS is about 40 km away from the central point of the subset where the Baiyinsumu site is located.

*Table 2.* Data used in the study at the field sites in HEIFE and IMGRASS

| Measurements | Measurements Height | Date |
|---|---|---|
| *HEIFE* | | |
| Flux (eddy correlation system) | 2.9m | 19/08/1991 |
| Tower profile | 0.5-20m | 08/1991 |
| Tethered balloon sounding | 0-1000m | 19/08/1991 |
| Low-level sounding | 1000-2000m | 19/08/1991 |
| Standard meteorological Radiosounding | 2000-16000m | 19/08/1991 |
| Surface temperature | 1.5m | 19/08/1991 |
| ATSR-1 | 780km | 19/08/1991 |
| ***IMGRASS*** | | |
| Directional radiometric surface temperature | 1.5m | 26-31/07/1998 |
| Flux (eddy correlation system) | 4.9m | 26-31/07/1998 |
| Tower profile | 0.5-8m | 14/06, 07/1998 |
| Standard meteorological Radiosounding | Surface-16000m | 14/06/1998 |
| Surface temperature | 1.5m | 14/06/1998 |
| ATSR-2 | 780km | 14/06/1998 |

# 4. APPROACH

## 4.1 Atmospheric correction

Atmospheric correction is made using MODTRAN code combining atmospheric temperature and humidity profiles. Lower-level sounding, tethered balloon measurements, standard meteorological radiosounding and middle latitude summer model atmosphere are used for HEIFE subset of ATSR image, while only standard meteorological radiosounding and middle latitude summer model atmosphere are used for IMGRASS subset because of lack of the other measurements. No sounding measurements were available at exactly the same time as the satellite overpass, therefore the closest ones are employed which were acquired at 07:30h and 07h in the morning for HEIFE and IMGRASS respectivcly. Table 3 gives the results obtained with MODTRAN and atmospheric profiles.

*Table 3.* MODTRAN results for atmospheric transmission and the path radiance for each channel at nadir and forward view angle( n=nadir,f=forward;11=11μm,12=12μm)

| Site | Water vapor ($g/cm^3$) | Transmission 11n | 11f | 12n | 12f | Path Radiance ($W\ cm^{-2}\ sr^{-1}\mu m^{-1}$) $x10^{-4}$ 11n | 11f | 12n | 12f |
|---|---|---|---|---|---|---|---|---|---|
| HEIFE | 1.77 | 0.8603 | 0.7832 | 0.7755 | 0.6702 | 0.776 | 1.29 | 1.31 | 1.90 |
| IMGRASS | 1.52 | 0.8290 | 0.7307 | 0.7346 | 0.6110 | 0.942 | 1.49 | 1.60 | 2.36 |

## 4.2 Data screening

The inversion of $T_v$ and $T_s$ from radiometric temperature is based on the assumption that the change of radiometric surface temperature with view angles is only caused by the changing fraction of vegetation cover in the field of view of the radiometer. Therefore, a pre-analysis is made to evaluate the quality of ATSR directional surface temperature prior to retrieve $T_v$ and $T_s$ from the brightness measurements. Three cases can be distinguished which will not be used in the inversion of $T_v$ and $T_s$, namely:

### 4.2.1 $T_{B0}(\lambda, nadir) - T_{B0}(\lambda, forward) < 0$

Fig. 2 shows there are some clouds dispersed in the IMGRASS ATSR subset. In the case that clouds fall in the field of view of the radiometer with the sensor looking at the surface at nadir view angle, while less or no clouds exist in the forward view, $T_{B0}(\lambda,nadir) - T_{B0}(\lambda,forward) < 0$ could be observed (see Fig. 3a). This situation may also be observed because of heterogeneity, i.e. a large fraction of vegetation is observed at nadir while it is mixed with a large fraction of bare soil in the forward view. Observations meeting this criterion, i.e. $T_{B0}(\lambda,nadir) - T_{B0}(\lambda,forward) < 0$, are not considered in our study.

### 4.2.2 $T_{B0}(\lambda, nadir) - T_{B0}(\lambda, forward) < 0.5K$

This happens when the surface is rather homogeneous with either bare soil or full canopy, under which directional effects in the surface temperature are not significant. Considering the surface types in these two experiments area, full canopies exists most probably in the HEIFE subset and bare soil surface exists in IMGRASS subset for the pixels with $0 < T_{B0}(\lambda,nadir) - T_{B0}(\lambda,forward) < 0.5K$. The threshold T = 0.5 K is the nominal accuracy of surface temperature estimates based on ATSR- radiances. This criterion implies that directional changes smaller than the accuracy of observed brightness temperature are neglected.

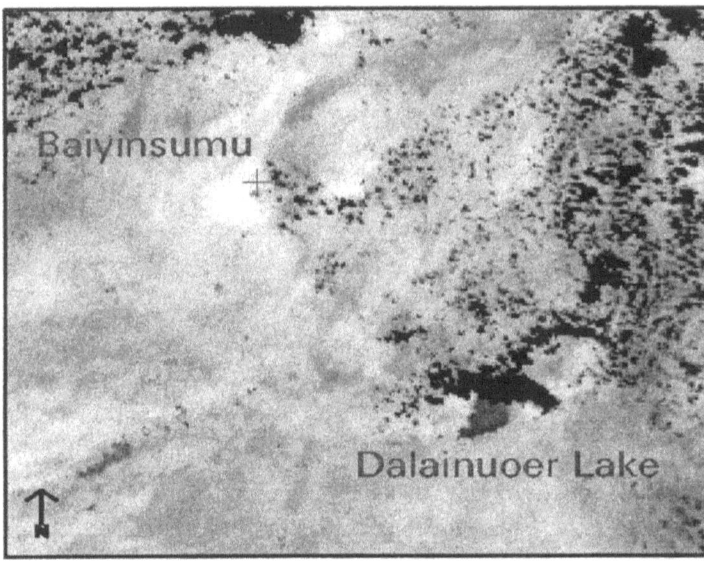


*Figure 2.* ATSR image at nadir view of 14-06-1998 in IMGRASS area, showing the brightness temperature in the 11μm channel; clouds and water are black (Baiyinsumu site location : 44°16′44′′N,115°56′32′′E)

### 4.2.3 $T_{B0}(\lambda,nadir) - T_{B0}(\lambda,forward) >> 0$

The directional brightness temperature measurements collected duirng IMGRASS gave a mean difference of $T_{B0}$ between nadir and forward views ($52^0$ was taken in the field measurements) of 2.6 K. Kimes and Kirchner (1983) found 16.2 K differential between the 0 and 80 zenith view angles at noon and 0.9 K differential in the early morning on a cotton canopy with mean height of 44 cm and mean row spacing of 1 m. In their case, the large difference at high solar zenith is due to the large change in the portion of sunlit soil or shaded vegetation with view angle for a relatively higher canopy with row structure. A maximum difference up to 3.5 K for a corn canopy and 1.5 K for grass (with 20 cm height) between 0 and $60^0$ were observed around solar noon by Lagouarde and Kerr (1993). For the HEIFE subset with corn and bean surface, and for IMGRASS subset with sparse short grass the difference in $T_{B0}$ between nadir and forward view should not be significantly larger than 10K for instance. Much larger difference in $T_{B0}$ between nadir and forward view are probably caused by the fact that there are clouds in the forward view but no or less in the nadir view (see Fig. 3b).

The pixels in the three categories A, B and C described above are not considered in retrieving $T_v$ and $T_s$ from ATSR directional brightness surface temperature measurements in our study.

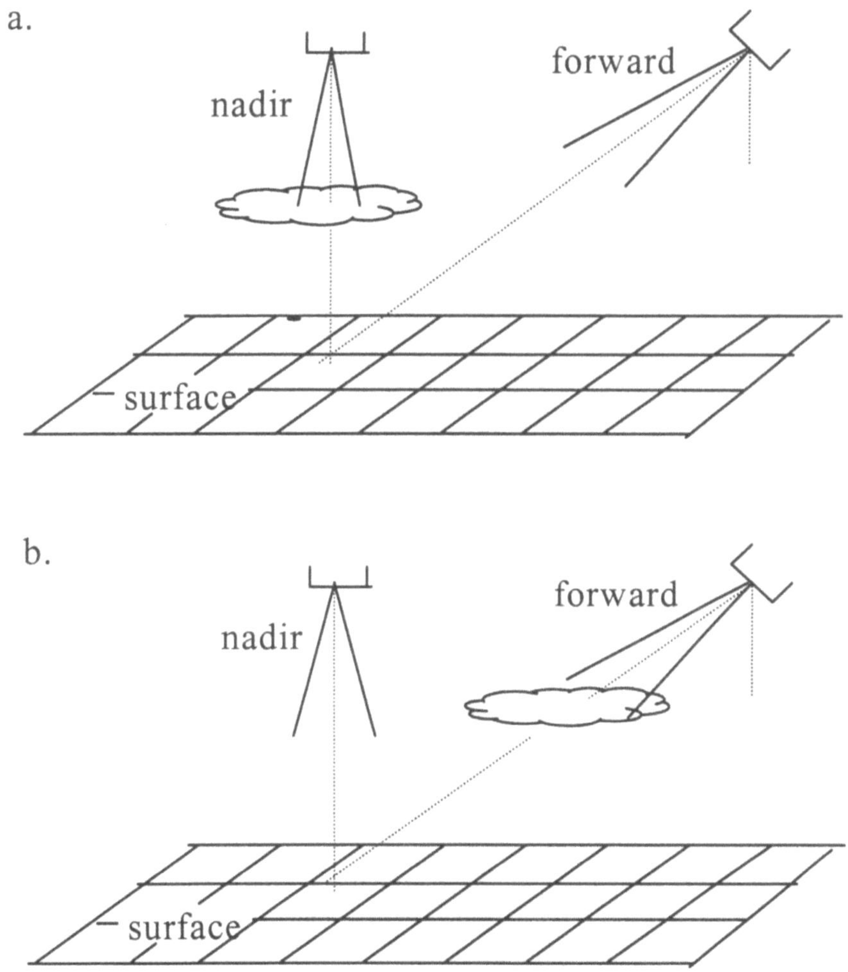


*Figure 3.* Illustration for nadir and forward view of ATSR sensor in cloudy condition

## 4.3 Inversion of directional brightness surface temperature measurements to retrieve $T_v$ and $T_s$

With two angles measurements of brightness surface temperature, $T_{B0}(\lambda, \theta)$, we can rewrite Eq.(23) as:

$$T_{B0}(\lambda, \theta_1) = [\, f_v(\theta_1)\, \varepsilon_v\, T_v^{n(\lambda)} + f_s(\theta_1)\, \varepsilon_s\, T_s^{n(\lambda)}\, ]^{1/n(\lambda)} \quad (26)$$

$$T_{B0}(\lambda, \theta_2) = [\, f_v(\theta_2)\, \varepsilon_v\, T_v^{n(\lambda)} + f_s(\theta_2)\, \varepsilon_s\, T_s^{n(\lambda)}\, ]^{1/n(\lambda)} \quad (27)$$

where n(11)=4.5, n(12)=4.2. At wavelengths in the range 11 μm to 12 μm the spectral emissivity of foliage is rather high and constant. Reliable estimates of $\varepsilon_v$ can be found in e.g. Rubio et al., (1997).

After obtaining the atmospherically corrected surface brightness temperature at two angles and two channels, $T_{B0}(\lambda,\theta)$, from ATSR image, one can derive $T_s$, $T_v$ and LAI by rewriting Eq.(23) as four equations with four unknowns $T_v$, $T_s$, LAI (or $f_v$) and $\delta\varepsilon$ (= $\varepsilon_v$ -$\varepsilon_s$). In this study we have used the same $\varepsilon_v$ at both 11μm to 12μm, although different values might have been used. We have also neglected directional changes in $\varepsilon_v$ and $\varepsilon_s$, although this may be easily taken into account if the explicit dependence of $\varepsilon_v$ and $\varepsilon_s$ is known.

## 4.4 Surface characteristics and meteorological variables

To calculate the resistances, several surface characteristics are needed.

Local roughness length for momentum is determined by eddy correlation measurements at lower height in IMGRASS site. The reference height is taken as 2 meters. The regional effective values of $z_{0m}$ are estimated by fitting wind-speed measurements at different levels to the logarithmic velocity profiles using least-square method and taking $d=2/3h_c$ for higher and denser canopy and d=0 for the lower and sparse covered canopy. Blending height is considered as a suitable reference level (Brutsaert and Sugita, 1992) to estimate regional heat flux. In the HEIFE area, the available lower-level sounding was measured two hour earlier (09:00h in the morning) than the satellite overpass time in another site of HEIFE which is about 30km away from Zhang-Ye site. This lower-level sounding was used to determine blending height and wind speed and potential temperature at this height. Unfortunately, we do not have the same measurements during the day of ATSR-2 acquisition in the IMGRASS campaign. Reference height and meteorological variables at the reference height are simply taken as those at the lowest level of standard meteorological radiosounding in the early morning (at 07:00h).

The reference source height in the canopy, $z_0$, is considered as the order of magnitude of roughness length for momentum, $z_{0m}$ or the proportion of canopy height. In our study we have taken $z_0 = 2/3h_c$ for both areas. For partial canopies, $u_0$ can be obtained by extrapolating the logarithmic profiles of wind speed to $z_0$ with the assumption that the logarithmic form of wind speed is valid down to the canopy space. The surface and meteorological variables used in the study are listed in Table 4.

*Table 4.* Surface and meteorological data used in the study

| Site | | $h_c$(m) | $Z_{0m}$(m) | Reference height(m) | $u_{ref}$(m/s) | $T_{ref}$($^0$C) | LAI |
|---|---|---|---|---|---|---|---|
| IMGRASS | field | 0.1 | 0.02 | 2 | * | * | 0.5 |
| | regional | 0.2 | 0.02 | 445 | 10 | 20.2 | * |
| HEIFE | regional | 1.2 | 0.5 | 500 | 1.8 | 25.4 | * |

Note: * values are from measurements (for $u_{ref}$ and $T_{ref}$) or modeling (for LAI). For regional scale, $T_{ref}$ is the potential temperature.

# 5. RESULTS

## 5.1 Retrieval of $T_v$ and $T_s$ – case study based on field measurements in IMGRASS

With the measurements of directional surface brightness temperature at three view angles, e.g. $0^0$, $23^0$ and $52^0$, $T_v$ and $T_s$ are retrieved using Eq.(26) and Eq.(27). For the studied area, LAI was measured as 0.5 corresponding to $f_c$ 30%; $\varepsilon_v$ and $\varepsilon_s$ were 0.98 and 0.95 respectively.

Theoretically, $T_v$ and $T_s$ can be derived using measurements of brightness at any pair of view angles, such as $T_{B0}(0)$ and $T_{B0}(23)$, $T_{B0}(0)$ and $T_{B0}(52)$, and $T_{B0}(23)$ and $T_{B0}(52)$. However, in our study, due to the small field of view of the radiometer used, only bare soil was seen in the field of view when measurements were done at nadir, so that the $T_{B0}(0)$ is the soil brightness temperature. To obtain $T_s$ from $T_{B0}$ (0), $\varepsilon_s = 0.95$ was used. This was taken as the reference to evaluate the $T_s$ retrieved from $T_{B0}(23)$ and $T_{B0}(52)$. The same measurements of $T_{B0}(23)$ and $T_{B0}(52)$ give $T_v$. Agreement of $T_s$ between retrievals and measurements was good, with a root mean square difference (RMSD) of 0.8 K (Fig.4).

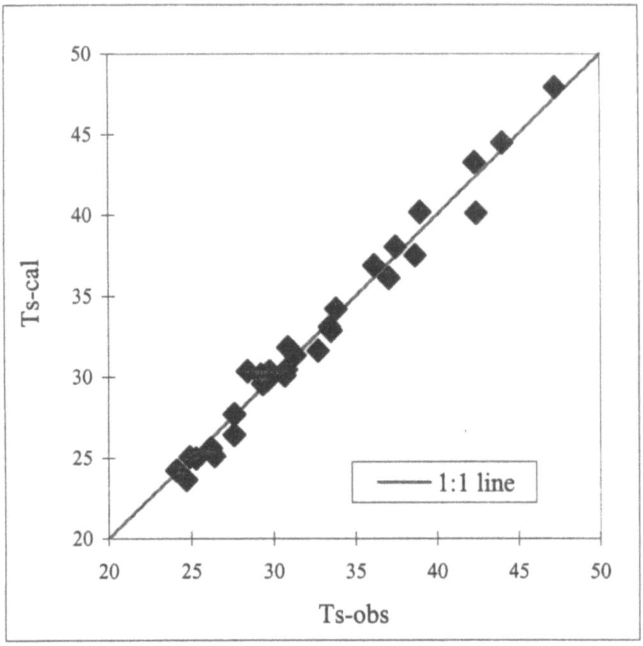


*Figure 4.* Comparison of the observed soil surface temperature with estimates obtained using eqs. 26 and 27 (RMSD=0.8K)

## 5.2 Impact of errors in the atmospheric correction.

The proposed approach to estimate foliage and soil temperatures requires the determination of Bottom-Of-Atmosphere (BOA) spectral directional radiance. Uncertainty in the knowledge of the atmospheric state affects the accuracy of the retrieved radiance. On the other hand there is a simple relation between the impact of atmospheric state at two view angles, when the atmospheric state is known. We have analyzed the impact of uncertainty on the atmospheric state by comparing the frequency distribution of $[T_{B0}(\lambda,0) - T_{B0}(\lambda,53)]$ as obtained past the data screening procedure described above in three different cases:

a) Atmospheric transmittance and path radiance calculated (MODTRAN) with actual radio-soundings;
b) Same as A, but atmospheric profile modified to give a 20% increase in the column water content;
c) Same as B, but for a 20% decrease.

The frequency distributions of $[T_{B0}(\lambda,0) - T_{B0}(\lambda,53)]$ for cases A, B and C are given in Fig. 5.

The range of the $T_{B0}$ difference [nadir minus forward] is limited by the data screening bounds described above. The data screening does not modify significantly the distributions for the 11 μm case while it has a clear impact on the 12 μm case, where it cuts out part of the lower tail. On the other hand, the impact of errors in the atmospheric correction appears limited, since the distributions are similar in the three cases. We note that the distributions relate to all valid (i.e. past the screening) observations used to retrieve soil and vegetation temperatures.

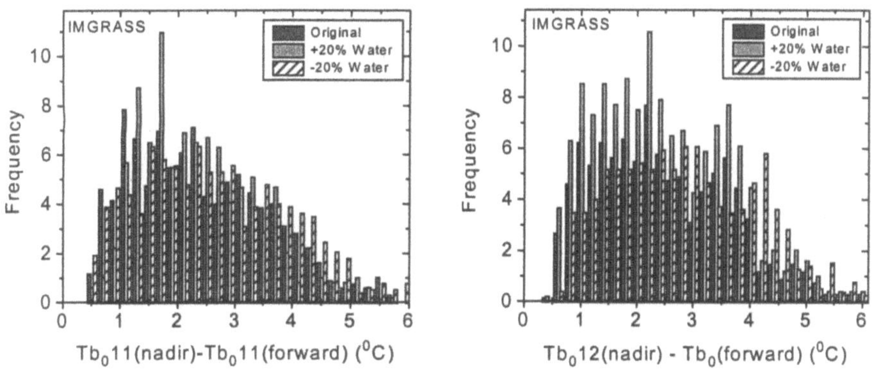


*Figure 5.* Frequency distribution of the difference in BOA brightness temperature in the nadir and forward view direction: A: 11μm and B: 12μm

## 5.3 Retrieval of $T_v$ and $T_s$ - case studies based on ATSR data

After screening the dual view ATSR observations as described in the previous section, $T_v$ and $T_s$ are derived using the inversion method described in sections 3 and 4. Fig.6 gives the histograms of derived $T_v$ and $T_s$ for each subset. In the HEIFE area, the peak of $T_v$ appears around 36$^0$C, while there is no obvious peak for $T_s$ which vary between 35-50$^0$C. On the contrary, in the IMGRASS area, $T_s$ has a peak around 46$^0$C, $T_v$ are scattered between 20-40$^0$C. This can be explained by the different surface types and fractional vegetation cover in these two areas described above. Table 5 shows the comparison of surface temperature between field and ATSR observations for each area.

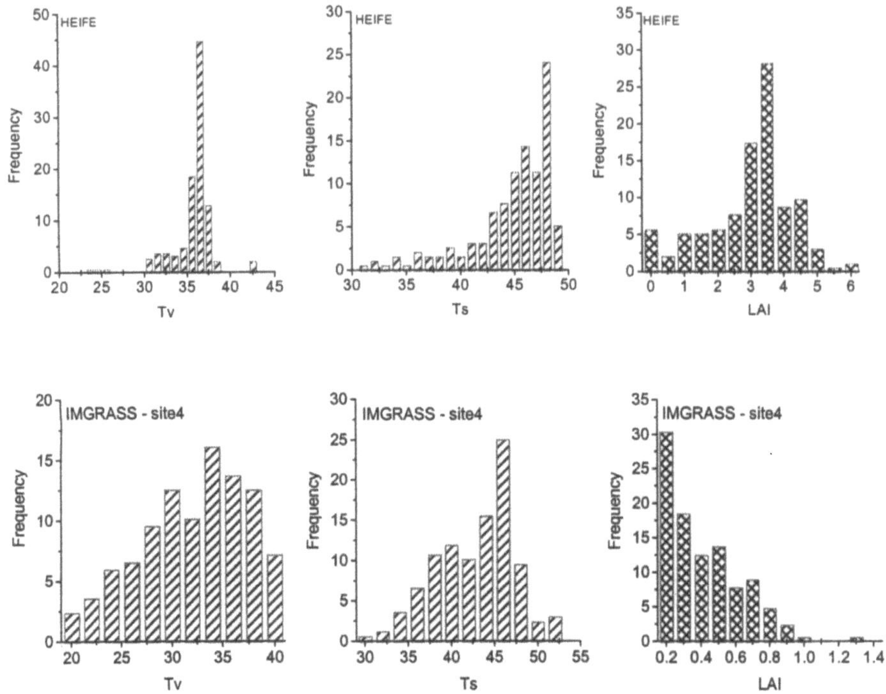


*Figure 6.* Vegetation and the soil temperature derived from ATSR for HEIFE and IMGRASS subset

*Table 5.* Comparison between brightness (nadir and forward observations), vegetation and soil temperature from ATSR and the observed surface temperature in the field; the range indicated is ± σ (standard deviation)

| | $T_v \pm \sigma$ (K) | $T_s \pm \sigma$ (K) | $T_{B0\text{-}N} \pm \sigma$ (K) | $T_{B0\text{-}F} \pm \sigma$ (K) | $T_B$-obs (K) |
|---|---|---|---|---|---|
| HEIFE | 306.2 ± 1.05 | 317.2 ± 3.97 | 307.2 ± 0.33 | 305.2 ± 0.37 | 305.7 |
| IMGRASS | 307.2 ± 7.3 | 321.2 ± 0.44 | 316.9 ± 0.45 | 315.1 ± 0.74 | 311.3 |

## 5.4 Estimates of sensible heat fluxes using the developed dual-source model

The dual-source model developed in our study is used to estimate the sensible heat flux density. The model is tested first on the field measurements in IMGRASS site (Fig.7): the mean RMSD was 27.1 $Wm^{-2}$. The measurements of directional brightness surface temperature and sensible heat flux density were not exactly simultaneous. Moreover, a series of observations of directional radiometric surface temperature at three view angles and four azimuth angles required a few minutes, while the measurements of sensible heat flux density were averaged over the thirty minutes centered at each hour and half-hour. This may contribute to the observed scatter (Fig. 7).

However, we may conclude that the sensible heat flux density calculated by our dual-source model agree reasonably well with the observed values, particularly after taking into account that 30 $Wm^{-2}$ is the order of magnitude of the accuracy of eddy correlation systems like the one used during the IMGRASS experiment.

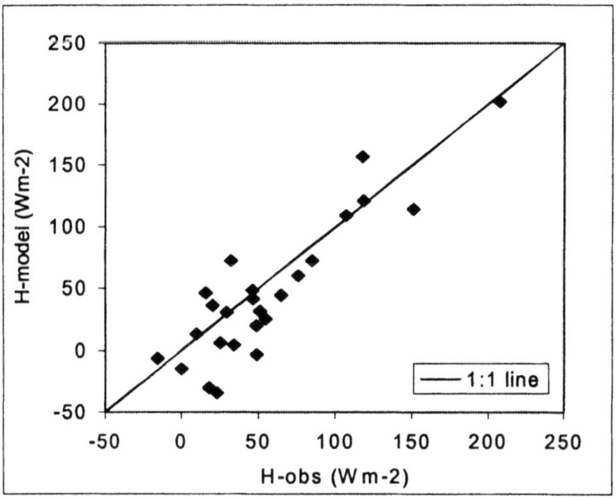


*Figure 7.* Comparison of modeled sensible heat flux with measurements in IMGRASS site

The ATSR-1 and -2 aboard the ERS-1 and -2 provide the opportunity to derive $T_v$ and $T_s$ for estimating heat flux density using our dual-source model at regional scale.

The calculations at the regional scale are done first for the HEIFE area using $T_v$ and $T_s$ derived from ATSR-1 image and the meteorological observations at blending height (Table 6). Fig.8 gives the histograms of the sensible heat flux density obtained in this way. At the satellite overpass time, sensible heat fluxes measured in the field were **46.6 $Wm^{-2}$**. The estimated mean values of H for 9 pixels close to the site was **52.5** $Wm^{-2}$ with a standard deviation **4.2** $Wm^{-2}$. The agreement between measured H and modeled H seems fairly good although we have used the atmospheric sounding 3 hours earlier than satellite overpass time to obtain blending height information. Table 6 also gives the modeled sensible heat flux using RAMS (Regional Atmospheric Modeling System) (Yan et al 1999) in the HEIFE area with 4km x 4km grid resolution. The value of H modeled by RAMS in Table 6 was taken from the model grid of RAMS where Zhang-Ye site is located. It appears that the RAMS H-values were significantly larger than the value observed (relative errors larger than 60%) and the value obtained with our dual-source model.

Sensible heat flux density was also estimated for the IMGRASS case using $T_v$ and $T_s$ retrieved pixel by pixel from ATSR-2. Atmospheric variables were estimated by assuming the lowest level of sounding as the reference height. The relative error between H modeled and H observed is smaller than 15%(see Table 6).

Fig. 8 gives the histograms of modeled sensible heat fluxes for both the HEIFE and IMGRASS areas. The modus is 60 $Wm^{-2}$ for HEIFE, a reasonable value in comparison with the field measurements. The limited range of H-values indicates that the land surface is relatively homogeneous. On the contrary, in the IMGRASS area, the values of H vary in a wide range with two peaks (one is around 50 $Wm^{-2}$, the other one is around 180 $Wm^{-2}$) which is the consequence of the sparse grass cover and more heterogeneous surfaces.

*Table 6.* Comparison of sensible heat flux density estimated with our model and RAMS (Yan et al.,1999) with ground measurements

| Site | time | Dual-source Mean($Wm^{-2}$) (3x3 pixels) | Dual-source $\sigma_H$($Wm^{-2}$) (3x3 pixels) | Dual-source RE | RAMS model ($Wm^{-2}$) | RAMS RE | Observation ($Wm^{-2}$) |
|---|---|---|---|---|---|---|---|
| IMGRASS | 11h,14/06/1998 | 174.6 | 30.1 | 8% | - | | 190.0 |
| HEIFE | 11h,19/08/1991 | 52.5 | 4.2 | 13% | - | | 46.6 |
| | 10h,14/08/1991 | - | - | - | 120.0 | 62% | 74.0 |
| | 10h,15/08/1991 | - | - | - | 160.0 | 64% | 97.6 |

mean: mean value of H modeled in 3x3 pixels closest to the site;
$\sigma_H$: standard deviation of modeled H in 3x3 pixels closest to the site;
RE: relative error between the value of modeled H and observed H.

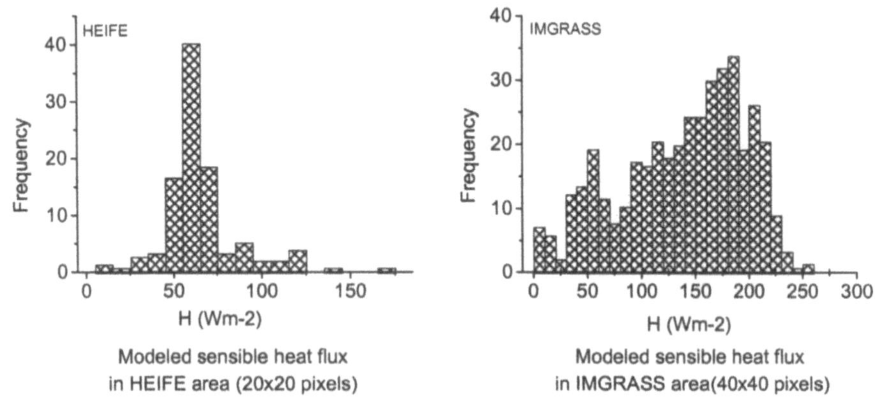


*Figure 8.* Histograms of estimated sensible heat flux density for the HEIFE and IMGRASS areas

## 6. DISCUSSION

*Reference air temperature.* We have taken the blending height as the reference height to estimate regional sensible heat flux density with our model. Atmospheric variables such as wind speed and potential temperature at blending height can be treated as horizontally uniform over rather large areas. The turbulent boundary layer under unstable conditions has typical ratios of vertical to horizontal scale around 1/10 to 1/100 and the structure of a well-developed ABL is the result of boundary conditions and surface exchange processes over the upwind region (Brutsaert and Parlange 1992).

In the HEIFE case study, a lower-level sounding is used to obtain the blending height and the variables at this height for the area. Though the launching site is about 30km away from the Zhang-Ye site, it is still in the appropriate fetch and should not lead to large errors. Mahrt and Sun (1996) noted that using the value of potential temperature at the top of the surface layer leads to small errors without considering spatial variations of flow at this height. In the IMGRASS case the interval between standard meteorological radiosounding and satellite overpassi time was more than 3 hours in which larger variation in surface layer structure may happen. Moreover, in this area, the reference height has been taken simply as the lowest level of the standard meteorological radiosounding. The impact of the chosen reference air temperature on the accuracy of estimated sensible heat flux density should be evaluated more precisely in a separate study.

*Validation of surface temperatures.* Validation of vegetation and soil temperatures is only feasible with field measurements. We have shown that the retrieved soil temperatures are in good agreement with observations and concluded that the simple linear mixture model describes correctly the change in brightness temperature with the view angle. Comparison of satellite with field observations is complex, especially at the relatively low spatial resolution of ATSR. Brightness temperature compared well with observations in the case of HEIFE, less so for IMGRASS (Table 5). A first difference is the type of measurements. The HEIFE observations were done with a thermal infrared thermometer operating in the 7–20 μm spectral region, while an Eppley pyranometer (wavelength 4 –50μm; facing – down) was used during IMGRASS. Differences in emissivity in the non- – overlapping spectral ranges (4–7 μm and 20–50 μm) may contribute to differences between the IMGRASS and HEIFE cases as regards to the satellite vs. field comparison. Moreover the ATSR brightness temperatures are narrow-band values in the range 10.5–12.5 μm. where spectral emissivity is higher than in the other two spectral intervals. Another, more likely explanation of the large differences observed in the case of IMGRASS and small differences observed in the case of HEIFE is spatial variability. In the HEIFE Zhang-Ye oasis fractional vegetation cover is rather high, the surface brightness and vegetation temperature have limited spatial variability (Fig.6) and field measurements were done by positioning the radiometer over a well developed canopy. It is therefore not surprising that field and satellite measurements agree reasonably well, even though the spatial scale of the observations is very different. In the IMGRASS case fractional vegetation cover is low and spatial variability of surface brightness and soil temperature is large. This made rather difficult to position the Eppley pyranometer to obtain a representative sample of the spatial distribution of surface temperature.

Proper validation of satellite observations of heterogeneous land surfaces at resolutions of 1 km or lower remains challenging. Reliable sampling of the spatial variability of surface temperature using field measurements only is feasible when dealing with rather homogeneous land surfaces. Airborne or satellite observations at higher spatial resolution would be actually needed to evaluate more precisely the proposed inversion of directional measurements of brightness temperature over heterogeneous land.

*Validation of sensible heat flux.* As in the case of field measurements of surface brightness temperature, agreement of H-values obtained with our dual-source parameterization and observations is good (Fig.7), given the moderate accuracy of eddy correlation systems. Estimates of H obtained with ATSR data were close to the observations in both the HEIFE and IMGRASS cases, notwithstanding the high spatial variability of retrieved soil temperature (HEIFE) and vegetation temperature (IMGRASS). We note the larger spatial scale of turbulent flux measurements as compared with thermal infrared radiometers. The eddy correlation system in the Zhang-Ye oasis was mounted at a 3 m elevation and in the IMGRASS site at 5 m. Assuming a ratio of vertical to horizontal scales of 1 /10 to 1/100, these elevations imply that the HEIFE system had a footprint of 30 to 300 m, while the IMGRASS one had 50 to 500 m. Although still smaller than the ATSR pixel size, such footprints provide a significant better sampling than the radiometers used to measure surface brightness temperature. The large deviation of RAMS-values of sensible heat fluxes compared with either measurements or our model calculations seem to confirm the inadequacy of single-source parameterizations (as used in RAMS) to describe heat transfer at heterogeneous land surfaces.

## 7. CONCLUSIONS

This paper describes a new dual-source model of heat transfer at heterogeneous land surfaces. This model avoids assumptions on the vertical and horizontal structure of the surface layer by dealing separately with heat transfer in the canopy air space and in the surface layer above the canopy. The model requires observations of surface brightness temperature at two view angles: nadir and a large off-nadir angle. These observations are currently provided at low spatial resolution by the ATSR instruments on-board ERS-1, ERS-2 and ENVISAT in the near future. The Land Surface Processes and Interactions Mission (LSPIM) under consideration at ESA would provide observations at much higher spatial resolution and at additional view angles. The model does also require fractional vegetation cover and a characteristic linear dimension of plant leaves. Fractional vegetation cover can be estimated with a variety of algorithms and observations of the spectro- directional reflectance in the visible and near infrared spectral region. Estimation of leaf size is obviously more difficult, although it may be retrieved by inverting radiative transfer modeling of spectro-directional radiometric measurements. Regional representative leaf size can also be estimated for each canopy given a correct vegetation classification. We note that the alternative single-source model requires the determination of the roughness length for heat transport. Several studies have demonstrated the difficulties involved in the determination of generally applicable values of this land surface property.

Values of sensible heat flux obtained with our model and ATSR data were compared with field measurements collected during two field experiments in China (HEIFE and IMGRASS). Agreement was good in both cases. Detailed field directional measurements of brightness temperature were collected during IMGRASS. This made feasible the comparison with measurements of H throughout the entire campaign. Agreement was good, taking into account the moderate accuracy of eddy correlation systems.

The analysis of field measurements indicates that the dual-source model proposed in this paper describes correctly heat transfer in the canopy air space and to the surface layer above the canopy. The scope of the validation of estimates based on ATSR data was limited since only one ATSR data set was analyzed in each experiment. Future work will address this aspect taking advantage of the algorithms developed for this study and of easier access to ATSR data.

# 8. ACKNOWLEDGMENTS

This study was performed with support of the Netherlands Board of Remote Sensing (BCRS), the European Space Agency (ESA), the Royal Netherlands Academy of Arts and Sciences (KNAW) and the Dutch Ministry of Agricultural, Fishery and Nature (LNV). The senior author (L. Jia) is grateful to the BCRS and ALTERRA (formerly the Winand Staring

Centre) for the opportunity to spend a period of research in Wageningen,The Netherlands. The authors would like to acknowledge Ir. Claire Jacobs at ALTERRA Green world Research, Wageningen University and Research for her help with preparing ATSR image.

# 9. REFERENCES

Becker,F. and Z.-L.,Li,1990, Temperature-independent spectral indices in thermal infrared bands, *Remote Sens. Environ.*, 35:161-173.

Beljaars,A.C.M.,and Holtslag,A.A.M.,1991,Flux parameterization over land surface for atmospheric models, *J.Appl.Meteorol.*, 30:327-341.

Brusaert,W.H.,1982, *Evaporation into the Atmosphere*, Reidel, Dordrecht, The Netherlands.

Brutsaert,W., and M.B., Parlange, 1992, The unstable surface layer above forest: regional evaporation and heat flux, *Water Resources Res.*, 28(12):3129-3134.

Brutsaert,W., and M. Sugita,1992, Regional surface fluxes from satellite-derived surface temperatures(AVHRR) and radiosonde profiles, *Boundary-Layer Meteorology*, 58:355-366.

Chamberlain,A.C.,1968, Transport of gases to and from surfaces with bluff and wave-like roughness elements, *Quart. J. Roy. Meteor. Soc.*, 94: 318-332.

Choudhury,B.J., and J.L.,Monteith, 1988, A four-layer model for the heat budget of homogeneous land surfaces, *Quart. J. Roy. Meteor. Soc.*, 114:373-398.

Choudhury,B.J., R.J.,Reginato, S.B.,Iso, 1986, An analysis of infrared temperature observations over wheat and calculation of latent heat flux, *Agr. and Forest Meteorol., 37:75-88.*

Deardorff,J.W.,1978,Effective prediction of ground surface temperature and moisture, with inclusion of a layer of vegetation, *J. Geophys. Res.*, 83:1889-1903.

Garratt,J.R.,1992,*The atmospheric boundary layer*, Cambridge University, New York,USA.

Kalma,J.D., and D.L.B.,Jupp, 1990, Estimating evaporation from pasture using infrared thermometry: evaluation of a one-layer resistance model, *Agr. and Forest Meteorol.*,51:223-246.

Kimes,D.S., and J.A.Kirchner, 1983, Directional radiometric measurements of row-crop temperatures, *Int. J. Remote Sensing*, 4(2):299-311.

Kohsiek,W.,H.A.R.De Bruin,H.The and B.Van Den Hurk, 1993, Estimation of the sensible heat flux of a semi-arid area using surface radiometric temperarure measurements, *Boudary-Layer Meteorol.*, 63:213-230.

Kustas,W.P.,B.J.Choudhury, M.S.Moran,R.J.Reginato and R.D.Jackson,L.W.Gay and H.L.Weaver, 1989, Determination of sensible heat flux over sparse canopy using thermal infrared data, *Agr. and Forest Meteorol., 44:197-216.*

Kustas,W.P., 1990, Estimates of evapotranspiration with a one- and two-layer model of heat transfer over partial canopy cover, *J.Appl. Meteorol.*, 29:704-715.

Lagouarde,J.P., and Y.,Kerr,1993, Experimental study of angular effects on brightness surface temperature for various types of surfaces, *Workshop on Thermal Remote Sensing of the Energy and Water Balance over Vegetation in Conjunction with Other Sensors*, La Londe Les Maures, France.

Lhomme,J.P.,B.Monteny,M.Amadou,1994, Estimating sensible heat flux from radiometric temperature over sparse millet, *Agr. and Forest Meteorol.*, 68:77-91.

Mahrt,L., and J.,Sun,1996, Dependence of surface exchange coefficients on averaging scale and grid size, *Quart.J.Roy. Meteor. Soc.*, 121:1835-1852.

Menenti,M., Z.-L.,Li, V.,Djiepa, J.,Wang, M.P.,Stoll, L.,Jia, Z.B.,Su, and M.,Rast, 1999, Estimation of soil and vegetation temperatures with directional thermal infrared observations: The HEIFE,SGP97 and IMGRASS experiments, Second International Workshop on Multiangular measurements and Models, 15-17 Sept. 1999, Ispra; Italy.

Menenti,M., 2000. Evaporation. Chapter 8 in: G.A. Schultz and E.T. Engman (eds.). Remote Sensing in Hydrology and Water Management. Spinger Verlag, Heidelberg : (in press)

Mitsuta Y.(Ed.),1993, *Proc. Int. Symp. on HEIFE*, Disaster Prevention Research Institute, Kyoto University, Kyoto.

Norman,J.M., M.,Divakarla, and N.S.,Goel, 1995a, Algorithms for Extracting Information from Remote Thermal-IR Observations of the Earth's Surface, *Remote Sens. Environ.*, 51:157-168.

Norman,J.M., W.P.,Kustas, K.S.,Humes,1995b, Tow-source approach for estimating soil and vegetation energy fluxes in observations of directional radiometric surface temperature, *Agr. and Forest Meteorol*, 77:263-293.

Rubio,E., V.,Caselles and C.,Badenas,1997, Emissivity measurements of several soils and vegetation types in the 8-14μm wave bands: analysis of two field methods, *Remote Sensing Environ.*, 59:490-521.

Sobrino,J.A., Z-L.,Li, M.P.Stoll and F.Becker,1996, Multi-channel and multi-angle algorithms for estimating sea and land surface temperature with ATSR data, *Int. J. Remote Sensing*, 17(11):2089-2114.

Stanghellini,C., 1987, Transpiration of greenhouse crops – an aid to climate management, *Ph.D thesis*, Agriculture University,Wageningen, The Netherlands.

Stanghellini,C., 1993, Mixed convection above greenhouse crop canopies, *Agr. and Forest Meteorol.*, 66:111-117.

Stewart,J.B.,Kustas,W.P.,Humes,K.S.,Nichols,W.D.,Moran,M.S., and de Bruin,A.A.R., 1994, 'Sensible heat flux-radiometric surface temperature relationship for eight semiarid areas', *J.Appl. Meteorol.*, 33:1110-1117.

Su,Z., J.,Wang, J.,Wen, L.,Jia, M.,Menenti,1999, Field observations during IMGRASS - An examination on possibilities of using AATSR data to estimate soil and vegetation temperature, Proc.Int. Geosci. And Remote Sens. Symp., 1999, p.634-645. Also in: Mesoscale climate hydrology: the contribution of the new observing systems, Report USP-2(Editors: Z.,Su and M.Menenti), Winand Staring Centre, Wageningen UR, The Netherlands, pp141.

Yan, Y.-P., J.M.,Wang, M.,Menenti, R.,Hutjes, Z.,Su,,1999, Heterogeneous land surfaces and meso-scale atmospheric boundary layer processes: a case study on the HEIFE/HeiHe basin with the model RAMS, in: Mesoscale climate hydrology: the contribution of the new observing systems (Editors: Z.Su and M.Menenti), Winnand Staring Centre, Wageningen UR, The Netherlands, pp141.

Zuo,H.-C. and Y.-Q.,Hu, 1993, the comparison and seasonal variation of microclimatic characteristics between oasis and Gobi in HEIFE, *Proc. Int. Syp. on HEIFE*, Disaster Prevention Research Institute, Kyoto University, Kyoto.

# Exploitation of Surface Albedo Derived From the Meteosat Data to Characterize Land Surface Changes

Bernard PINTY[1], Michel M. VERSTRAETE[1], Nadine GOBRON[1], Fausto ROVEDA[2], Yves GOVAERTS[2], John V. MARTONCHIK[3], David J. DINER[3] and Ralph A. KAHN[3]

[1]*Space Applications Institute, Ispra, Italy*
[2]*EUMETSAT, Darmstadt, Germany*
[3]*Jet Propulsion Laboratory, Pasadena (CA), USA.*

Abstract: Land surface albedo constitutes a critical climatic variable, since it largely controls the actual amount of solar energy available to the Earth system. From a mathematical point of view, the determination of the surface albedo corresponds to the estimation of a boundary condition for the radiation transfer problem in the coupled surface-atmosphere system. A relatively large database of 10 years or more of Meteosat data has been accumulated by EUMETSAT. These data, collected at half-hourly intervals over the entire Earth disk visible from longitude 0 degree, constitute a unique resource to describe the anisotropy of the coupled surface-atmosphere system, and provide the opportunity to document changes in surface albedo which may have occurred in these regions over that period. An advanced algorithm to retrieve the radiative properties of terrestrial surfaces sampled by the Meteosat visible instrument has been derived and a preliminary analysis of a one-year (1996) set of Meteosat data was performed. The accumulation of results in 10-day periods permits evaluating the seasonal albedo changes occurring at a continental scale. These first results, supported by additional radiation transfer simulations, suggest that anthropogenic fire activities induce significant perturbations of the surface albedo values in the inter-tropical zones at that scale.

## 1. INTRODUCTION

Satellite-borne instruments constitute, *a priori*, a unique tool for monitoring surface albedo values at the global scale and at spatial and temporal resolutions adequate for meteorological and climate studies. However, the



*M. Beniston and M.M. Verstraete (eds.),*
*Remote Sensing and Climate Modeling: Synergies and Limitations,* 51–67.

effectiveness of this approach hinges on the availability of tools and models that can accurately account for the radiative contributions to the measured radiances from the atmospheric and surface components of the observed system, including the spectral and directional variations resulting from the anisotropy of terrestrial surfaces. The accurate representation of the upward radiance field at the land surface (bottom of the atmosphere), taking into account the convolution of surface and atmospheric scattering properties, is a major scientific problem to be solved.

Atmospheric aerosols of diverse origins exhibit significant spatial and temporal variations and strongly impact radiation transfer processes at solar wavelengths, but their properties have never been made available as an operational product. This lack of reliable information on aerosol load and properties reinforces the need to invert coupled surface-atmosphere radiation transfer models against space remote sensing data. As is usual with inverse problems, a minimum number of input data of sufficient quality and a small set of critical state variables are required to guarantee a reliable assessment of the retrieved properties. Martonchik et al. (1998a) demonstrated the possibility of retrieving surface radiative properties from an analysis of quasi-instantaneous multi-angular spectral measurements of the radiance fields emerging at the Top Of the Atmosphere (TOA). Martonchik et al. (1998b), Kahn et al. (1997) and Kahn et al. (1998) showed that aerosol properties can similarly be estimated. The design of the Meteosat VIS band does not yield a comprehensive spectral and directional sampling of the radiance fields scattered by the Earth. However, thanks to its geostationary orbit, this sensor is able to sample the radiance field emerging at TOA every thirty minutes during the course of the day, i.e., for different solar illumination conditions. In other words, assuming that the geophysical system under observation does not change drastically during the daily period of solar illumination, Meteosat data provide a useful angular sampling of the radiance field scattered by the Earth system. Whenever and wherever this assumption is acceptable, the Meteosat temporal sampling of the radiance field for a given location can thus be interpreted as an angular sampling; this approach constitutes the cornerstone of our strategy to estimate surface albedo values.

This paper summarizes the methodology developed to address various issues related to the actual application of a multi-angular approach for estimating surface properties from the Meteosat data set. These issues include

1. the optimal modeling of the radiation transfer for clear sky conditions as measured by the Meteosat instrument for finding solutions to an inverse problem in an operational context,
2. the selection, for each pixel (location), of those time observations during the day which are not contaminated by cloud radiative effects, and

3. the identification of the optimal solution, i.e., the joint characterization of the surface and the atmosphere for each pixel and each day through the set of potential solutions.

This paper also discusses the results of this approach when applied to one year (1996) of Meteosat data. It will be seen that our results point to anthropogenic effects, and in particular biomass burning, as the likely process of surface albedo changes in savannas and woodlands.

## 2. OUTLINE OF THE RETRIEVAL PROCEDURE

Fundamentally, the basic physical quantity required by any kind of surface application of remote sensing in the optical domain is the Bidirectional Reflectance Factor (BRF). Indeed, this quantity expresses the probability of radiation coming from one specific direction, $\Omega'$ ($\Omega_0$ for the particular solar direction), to be scattered into another specific direction, $\Omega$, normalized by the reflectance of a Lambertian target illuminated and observed under identical conditions. Accordingly, the upwelling radiance field, $I^{\uparrow}$ $(z_0, \Omega, \Omega_0)$ at the surface level $z_0$ in the direction $\Omega$, can be expressed as follows:

$$I^{\uparrow}(z_0,\Omega,\Omega_0)=\frac{1}{\pi}\int_{2\pi}\rho_{sfc}(z_0,\Omega'\rightarrow\Omega)I^{\downarrow}(z_0,\Omega',\Omega_0)\mu' d\Omega' \qquad (1)$$

where $\mu'$ is the cosine of the radiation incident from direction $\Omega'$, $\rho_{sfc}$ $(z_0, \Omega'\rightarrow\Omega)$ represents the BRF of the surface, and $I^{\downarrow}(z_0,\Omega',\Omega_0)$ is the downwelling radiance in the direction $\Omega'$ at the bottom of the atmosphere which is generated when the Sun is illuminating from the direction $\Omega_0$. All physical quantities in Equation 1 are monochromatic spectral quantities.

The surface BRF is used to estimate various angularly integrated quantities or albedos, including Directional Hemispherical Reflectances (DHRs):

$$DHR(z_0,\mu_0)=\frac{1}{\pi}\int_{2\pi}\rho_{sfc}(z_0,\Omega_0\rightarrow\Omega)\mu d\Omega \qquad (2)$$

The estimation of surface BRF values from satellite measurements requires solving an inverse problem in the atmosphere to determine the lowest boundary condition. However, the radiance field emerging at the top of the atmosphere depends on a large number of state variables characterizing the absorption and scattering properties of both the atmosphere and the surface. The inverse problem can therefore be solved in a reliable manner only for the most sensitive state variables, and the radiation transfer model simulating the radiance fields measured by a space-borne instrument must be constrained by a sufficient number of independent observations. The cornerstone of the surface albedo algorithm relies on the exploitation of the temporal sampling of Meteosat VIS channel (individual observations for any given location are acquired every 30 minutes) as if it were an instantaneous angular sampling (such observations are accumulated from sunrise to sunset). The VIS channel of the Meteosat sensor series extends from approximately 0.4 μm to 1.1 μm with a maximum response around 0.65 μm. As such, it is affected by all radiation transfer processes involving the ozone and water vapor contents of the atmospheric column. Since this algorithm is implemented in the EUMETSAT re-processing environment, it benefits from estimates of the total vertical content of ozone and water vapor provided by observations from the Total Ozone Mapping Spectrometer (TOMS) and analyses from the European Centre for Medium-range Weather Forecasts (ECMWF). This reliability permits reducing the full radiation transfer problem to a surface-aerosol absorption-scattering problem. An exhaustive description of the algorithm is given in Pinty et al. (2000a). It is assumed that only a finite set of pre-defined types of atmospheres can be considered and that atmospheric functions and radiance fields can be pre-computed for discrete values of the aerosol optical depth and black surface conditions. This was done for a US-62 type of standard atmosphere implementing a continental aerosol model which includes dust-like, water soluble and soot components (see Vermote et al., 1997, for complete information about this aerosol model). To limit the number of entries in the look-up tables (LUTs), the approach implements a simplified atmospheric model where the gas absorbing layers are located on top of the scattering layers. This scheme is similar to the one adopted in the Second Simulation of the Satellite Signal in the Solar Spectrum (6S) code (Vermote et al., 1997).

The surface BRF, $\rho_{sfc}(z_0,\Omega' \rightarrow \Omega)$ is represented by the RPV parametric BRF model proposed by Rahman et al. (1993):

$$\rho_{sfc}(z_0,\Omega_0 \rightarrow \Omega;\rho_0,\rho_c,\Theta,k) = \rho_0\,\breve{\rho}_{sfc}(z_0,\Omega_0 \rightarrow \Omega;\rho_c,\Theta,k) \quad (3)$$

where $\rho_0$ and $\breve{\rho}_{sfc}(z_0,\Omega_0 \rightarrow \Omega;\ \rho_c,\ \Theta,\ k)$ describe the amplitude and the angular variability of the surface BRF, respectively. The solution of the coupled surface-aerosol absorption-scattering problem is obtained dynamically during the retrieval, given the pre-computation of (1) the function $\breve{\rho}_{sfc}(z_0,\Omega_0 \rightarrow \Omega;\ \rho_c,\ \Theta,\ k)$ for a set of pre-defined $\Theta$ and $k$ parameter values and, (2) all the atmospheric functions required to solve the atmospheric radiation transfer problem for a black surface condition and a set of pre-defined aerosol models.

After some mathematical manipulations, the modeled Meteosat spectral response to the total BRF emerging at the top of the atmosphere in direction $\Omega(\mu,\phi)$ when the Sun is illuminating the system from direction $\Omega_0(\mu_0,\phi_0)$, can be approximated by:

$$R_M(z_{sat},-\mu,\mu_0,\phi-\phi_0;\tau,\tilde{\rho}_0,\tilde{k},\tilde{\Theta}_{HG}) = T^{gas}_{Msat}(-\mu,\mu_0;U_{O3},U_{H2O})$$
$$[\tilde{\rho}_{atm}(z_{soa},-\mu,\mu_0,\phi-\phi_0;\tau) + \tag{4}$$
$$\tilde{\rho}_0\,\breve{\tilde{\rho}}_{sfc}(z_{soa},-\mu,\mu_0,\phi-\phi_0;\tau,\tilde{\rho}_0,\tilde{\rho}_c,\tilde{k},\tilde{\Theta}_{HG})]$$

where

$$R_M(z_{sat},-\mu,\mu_0,\phi-\phi_0;U_{O3},U_{H2O},\tau,\tilde{\rho}_0,\tilde{k},\tilde{\Theta}_{HG}) = \frac{I^{\Uparrow}_M(z_{sat},\Omega,\Omega_0)\pi}{\mu_0\int_{\lambda 1}^{\lambda 2} E_0(\theta_0)S(\lambda)d\lambda} \tag{5}$$

and

$$\tilde{\rho}_{atm}(z_{soa},-\mu,\mu_0,\phi-\phi_0;\tau) = \frac{\tilde{I}^{\uparrow}_{atm}(z_{soa},\Omega,\Omega_0)\pi}{\mu_0\int_{\lambda 1}^{\lambda 2} E_0(\theta_0)S(\lambda)d\lambda} \tag{6}$$

In Equations 4, 5 and 6, $\tilde{\rho}_{atm}$ ($z_{soa}$, $-\mu$, $\mu_0$, $\phi-\phi_0$; $\tau$) represents the contribution of the intrinsic reflectance of the scattering-only-atmosphere (soa) to the total BRF, weighted by the Meteosat spectral response $S(\lambda)$; $T_{Msat}{}^{gas}(-\mu, \mu_0; U_{O3}, U_{H2O})$ denotes the transmission factor due to gaseous absorption ($U_{O3}$ and $U_{H2O}$ are the total content in ozone and water vapor, respectively), weighted by $S(\lambda)$; $E_0(\theta_0)$ is the spectral extra-terrestrial solar irradiance; $I_M{}^{\Uparrow}(z_{sat}, \Omega, \Omega_0)$ is the radiance measured by the Meteosat sensor and $I_{atm}{}^{\uparrow}(z_{soa}, \Omega, \Omega_0)$ is the scattered radiance field emerging at the top of the scattering-only-atmosphere, i.e., without considering the gaseous absorption effects, bounded by a black surface and weighted by the Meteosat spectral response.

This formulation summarizes the set of dependent and independent variables required to simulate the Meteosat observations under a variety of geophysical situations. The four most critical mathematical manipulations concern

1. the decoupling of the gaseous absorption and aerosol absorption and scattering processes,
2. the linearization of the TOA BRF with respect to the parameter describing the amplitude of the surface BRF ($\tilde{\rho}_0$),
3. the expansion of scattered radiation as a Fourier series in relative azimuth angles and,
4. the explicit contribution of atmospheric functions related to the radiation transfer regime for a black surface condition.

This strategy allows a straightforward implementation of the forward radiation transfer model since only sums and products of functions are required during the retrieval process. The Fourier expansion in $d\phi$ values also avoids creating LUTs with an entry for this coordinate and, therefore, significantly reduces the memory size required by the processing.

For similar reasons, and as suggested by a sensitivity study, the $\rho_c$ value controlling the hot spot function in $\rho^{\cup}_{sfc}(z_0,\Omega'\rightarrow\Omega;\rho_c,\Theta_{HG},k)$ was fixed at a value equal to 0.15. This strategy follows the approach applied to the MISR instrument for the retrieval of aerosol over dark surfaces (Martonchik et al., 1998).

The estimated values of $\rho_0$, denoted by $\tilde{\rho}_0$, for all the pre-defined conditions of the surface-atmosphere scattering model, described by the pre-defined values of the aerosol optical depth $\tau$, $k$ and $\Theta_{HG}$ parameters, are determined by the expression:

$$\tilde{\rho}_0 = \frac{\sum_i W_{inv}(i)[R_{Msat}(z_{sat},i)/T^{gas}_{Msat}(i;U_{O3},U_{H2O}) - \tilde{\rho}_{atm}(z_{soa},i;\tau)]}{\sum_i W_{inv}(i)\,\tilde{\rho}^{\cup}_{sfc}(z_{soa},i;\tau,\tilde{\rho}_0,\tilde{\rho}_c,\tilde{k},\tilde{\Theta}_{HG})} \qquad (7)$$

where the index $i$ designates the slot (image) number in the daily sequence, and $W_{inv}(i)$ is a weighting function. Since the angular variability function $\tilde{\rho}^{\cup}_{sfc}(z_{soa}, -\mu, \mu_0, \phi-\phi_0;\ \tau, \tilde{\rho}_0, \tilde{\rho}_c, \tilde{k}, \tilde{\Theta}_{HG})$ in Equation 4 is a function of $\tilde{\rho}_0$, an iteration procedure can be applied to solve Equation 7 until the convergence criterion $|\tilde{\rho}_0(n) - \tilde{\rho}_0(n+1)| \leq 10^{-3}$ is satisfied. This convergence is generally achieved in as few as 3 iterations.

The selection of acceptable solutions from the ensemble of retrievals, obtained using the pre-defined models, depends on a comparison of a cost function for each retrieval to a threshold value. This $\chi^2$ metrics (Kahn et al., 1997) is described by:

$$\chi^2 = \frac{\sum_i W_{cost}(i)[R_{Msat}(z_{sat}, i) - R_M(z_{sat}, i; U_{O3}, U_{H2O}, \tau, \tilde{\rho}_0, \tilde{k}, \tilde{\Theta}_{HG})]^2}{\sum_i W_{cost}(i)\sigma^2_{data}(i)} \quad (8)$$

where $W_{cost}(i)$ is a weighting function, $R_{Msat}(z_{sat}, i)$ is the TOA BRF value measured by Meteosat at the current slot $i$, and $\sigma_{data}(i)$ is the assumed uncertainty in both the observation simulations and the actual data $R_{Msat}(z_{sat}, i)$. The uncertainty $\sigma_{data}(i)$ is difficult to assess precisely using a theoretical approach, since it takes into account the limitations of the instrument, the uncertainties in the calibration, the stability of the instrument and geometrical rectification, as well as the inaccuracies inherent to the modeling of daily series of Meteosat BRF observations. The value of $\sigma_{data}(i)$ impacts the number of combinations of surface and atmospheric variables which represent acceptable solutions of the inverse problem obtained daily for all the processed pixels: the larger its value, the larger the number of solutions that are considered acceptable from the radiative point of view. The weighting functions, namely $W_{inv}(i)$ and $W_{cost}(i)$, can be chosen such as to maximize the impact of the large solar angles and the corresponding increased atmospheric paths on the retrieval; this, in turn, should lead to a better accuracy in the estimation of the downwelling radiance fields and surface BRFs values.

This approach allows us to identify, for each pixel and on a daily basis, a set of radiatively consistent atmospheric and surface conditions, leading to $\chi^2$ values less than unity. Furthermore, anyone of these sets of conditions is considered accurate enough to interpret the Meteosat "clear-sky" daily time series with an accuracy at least equal to the value of the denominator of Equation 8. This inversion procedure yields the simultaneous estimate of parameter values characterizing (1) the amplitude ($\tilde{\rho}_0$) and the shape ($\tilde{k}$ and $\tilde{\Theta}_{HG}$) of the surface scattering function and, (2) an indication of the aerosol load provided as an effective aerosol optical thickness at 550 nm. Since more than one solution can be retrieved for every single day, the selection of the "Likely" solution is based on inspection of the distribution of retrieved values for $\tilde{\rho}_0$, their mean, $\overline{\rho}_0$, and their average deviation, $\Delta_{\rho 0}$:

$$\overline{\rho_0} = \frac{1}{N}\sum_{j=1}^{N}\tilde{\rho}_0(j)$$
$$\Delta_{\rho 0} = \frac{1}{N}\sum_{j=1}^{N}|\tilde{\rho}_0(j) - \overline{\rho_0}| \quad (9)$$

where $N$ is the number of retrieved solutions.

The solution selected as being the "Likely" solution, $\hat{\rho}_0$, is the one minimizing first the quantity $|\tilde{\rho}_0(j)-\bar{\rho}_0|$ from among those that are not further away from $\bar{\rho}_0$ than $\Delta_{\rho 0}$, and, second, the associated $\chi^2[\tilde{\rho}_0(j)]$ values. This criterion selects the solution giving the lowest $\chi^2$ value in the range $\bar{\rho}_0 \pm \Delta_{\rho 0}$. Once the "Likely" solution for $\tilde{\rho}_0$ is identified, the associated values of aerosol optical depth, $\tau$, and surface anisotropy parameters, $\tilde{k}$ and $\tilde{\Theta}_{HG}$ are extracted. This procedure gives explicitly more weight to the control of the "Likely" solution by the value of the amplitude factor of the surface BRF field.

The various experiments conducted with synthetic Meteosat data have shown that the ensemble of solutions to the inverse problem that characterize the surface radiative state can be sampled in an appropriate manner with respect to the envisaged applications (Pinty et al., 2000a). The documentation of the state of the atmosphere is currently tentative, due to the intrinsic nature of the radiative effects and the specific spectral sampling of the Meteosat instrument. The potential to extract an indication of the probable aerosol load over relatively dark surfaces exists, however. Since there is no guarantee that the proper aerosol type is applied at any given time and location, the retrieved aerosol optical depth values must be considered as "effective" in the sense that it permits the interpretation of Meteosat observations at the accuracy prescribed in the inversion scheme. However, this "effective" value allows the accurate reconstruction of the downwelling atmospheric radiance fields at the surface level.

## 3. APPLICATION

In the context of an operational application where the containment of computational expenses is a significant driver, it is essential to ensure that the inversion procedure is restricted to daily sequences showing a high level of temporal consistency that conforms to the physical expectations expressed by the Meteosat data simulator. These expectations are such that, for all Meteosat pixels, the intrinsic variations of the BRF data strings built from the accumulation of half-hourly measurements from sunrise to sunset for all Meteosat pixels can be fully explained by Equation 4. Based on the classical plane-parallel approach, this equation is only valid for stationary clear-sky systems and any BRF measurement corrupted by clouds and/or cloud shadows and/or rapid change in aerosol load and radiative properties must be excluded before entering the inversion procedure. In addition, artificial BRF changes in a daily data string may occur due to an inaccurate pointing of the same region during the daily sequence of BRF data accumulation. The latter is a particularly sensitive issue for those pixels close to sharp geophysical boundaries such as lake shores, rivers, coastlines and mountains. In other words, for every pixel, variations in the time series during the day due to undesired geophysical and instrumental effects should be rejected. A first level of screening is performed by setting a threshold value equal to 0.6 on the TOA BRF measurements in order to eliminate obviously cloudy conditions. A second level, yielding a finer screening of undesired conditions, is achieved by implementing a Data Consistency Procedure (DCP) to produce an angularly smooth but coherent TOA BRF series which accounts for hot spot conditions. This procedure checks the consistency of the pre-screened TOA BRF values by attempting to fit the data series against a generic parametric BRF model, namely the Modified version of the RPV (MRPV) model (Engelsen et al., 1996). The MRPV model permits to fit angularly consistent BRF data strings, including the effects due to hot spot conditions, in the case of daily "clear-sky" situations (Pinty et al., 2000b). This constitutes a novel approach to cloud screening conditions since it does not require any additional information from thermal bands, as is usually the case for cloud identification techniques. This novel approach is entirely based on the analysis of the angular coherence of the bi-directional shapes emulated by the daily accumulation of TOA BRF measurements.

The procedure compares the values of the standard deviation of the fit, $\sigma_{act}$, against a pre-defined threshold value, $\sigma_{DCP}$, which represents the maximum value of the standard deviation of the fit that is considered acceptable for successful interpretability. When the condition $\sigma_{act} \leq \sigma_{DCP}$ is fulfilled, the procedure ends and the daily data time series is interpreted by the algorithm described in Section 2. Otherwise, the observed BRF value exhibiting the largest absolute departure with respect to the model prediction is eliminated and the series of observed BRF values is screened again. This iterative procedure is pursued until an acceptable fit is obtained, or the number of BRF data points remaining in the time series becomes too low to ensure a reliable retrieval of the geophysical parameters. In practice, the value of the following $\chi_{DCP}^2$ function is estimated:

$$\chi^2_{DCP}(z_{sat};r_0,k_m,b_m) = \frac{\sum_i [R_{Msat}(z_{sat},i) - R_M(z_{sat},i;r_0,k_m,b_m)]^2}{\sum_i \sigma^2_{DCP}(i)} \qquad (10)$$

where $R_{Msat}(z_{sat}, i)$ is the TOA BRF value measured at level $z_{sat}$ by Meteosat for the current slot $i$, $R_M(z_{sat}, i; r_0, k_m, b_m)$ is the TOA BRF value simulated with the MRPV model for the same image $i$ using the optimal parameter values retrieved as indicated above, and $\sigma_{DCP}(i)$ is the maximum acceptable standard deviation value to guarantee an appropriate smoothness and angular consistency between the reflectances in the various images of the same day for a given pixel. This smoothness condition is deemed verified when the $\chi_{DCP}^2$ value is equal to or less than one.

In summary, the data consistency procedure guarantees the selection of samples of the Meteosat BRF fields (at the full pixel resolution, as well as for each and every pixel of the image) which can be interpreted at a pre-specified quality level given by the value of the $\sigma_{DCP}$ parameter controlling the cost function. It should be underscored that the procedure does produce valuable geophysical information concerning

1. the characterization of the fields required to estimate TOA albedos,
2. the identification of clouds and cloud-shadows every thirty minutes in the daily sequence, and
3. the detection of potential error sources due to the inaccuracy in the geo-rectification process of the raw data.

For all practical purposes, the choice of the numerical values for the $\sigma_{DCP}$ and $\sigma_{data}$ parameters results from the compromise between generating accurate products and retrieving the desired information over a maximum number of pixels. Too small a value for the $\sigma_{DCP}$ parameter translates into the rejection of a high number of slots for all pixels. Although this would ensure that an angularly consistent string of BRF values is retained, a too small number of slots may not provide sufficient angular constraints on the inversion procedure which, in turn, may affect the reliability of the final products since too many acceptable solutions would be identified. In order to ensure that these constraints remain strong enough, it was decided to impose that a total of at least 9 solar angles would be required for performing the inversion. Although the $\sigma_{data}$ parameter value should be as small as possible to limit the number of acceptable solutions, too small a value, corresponding to a high accuracy in the data fitting exercise, may not permit us to identify even a single solution. In the present application based on Meteosat-5 data, $\sigma_{DCP}$, $\sigma_{data}$ and the minimum number of solar angles were set to 5%, 8% and 9, respectively, regardless of the pixel location and period of the year.

## 4. SURFACE ALBEDO CHANGES

The algorithm described above was implemented in the operational processing chain of EUMETSAT and then applied to a full year (1996) of Meteosat-5 data. The operational version of the algorithm permitted the retrieval, on a daily basis and for every pixel, of the surface parameters characterizing the BRF shape and amplitude. On this basis, it is then possible to estimate the associated DHR values for any particular location of the Sun.

To provide spatially comparable values, the surface albedo has been computed everywhere for a standard zenith angle of 30°.
For a number of surface applications, it is desirable to ensure a good geographical coverage, which implies the temporal compositing of these products in time. Such procedures are justified to the extent that surface changes occur over time scales longer than the period of composition. Traditional analyses based on vegetation index products, such as NDVI, recommend the application of a simple algorithm, for instance selecting the maximum value during the compositing period, but it has been shown that this procedure biases the composite data sets by selecting results towards measurements collected under specific angular conditions (see for instance, Holben, 1986; Meyer et al., 1995). Here, we propose a different scheme, which allows the selection of the most representative conditions during a compositing period on the basis of a simple statistical analysis. This analysis is based on the inspection of the daily $\hat{\rho}_0$ retrieved values for every period of ten consecutive days. The daily $\hat{\rho}_0$ likely values have been analyzed for every period of ten consecutive days in order to select the most representative value. This latter step was implemented by estimating the temporal average and corresponding deviation of the $\hat{\rho}_0$ values over the 10-day periods:

$$\begin{aligned} \overline{\hat{P}} &= \frac{1}{T}\sum_{t=1}^{T}\hat{P}(t) \\ \Delta^{T}_{\hat{P}} &= \frac{1}{T}\sum_{t=1}^{T}|\hat{P}(t)-\overline{\hat{P}}| \end{aligned} \tag{11}$$

where $T$ is the number of available values during the 10-days period of temporal accumulation, $\overline{\hat{P}}$ is the temporal averaged value estimated for parameter $\hat{P}$ and $\Delta^{T}_{\hat{P}}$ is the average deviation of the distribution.

The 10-days representative value for the $\tilde{\rho}_0$ parameter is the actual $\hat{\rho}_0$ value minimizing the quantity $|\hat{\rho}_0 - \overline{\hat{\rho}}_0|$. Since this solution corresponds to one of the daily "Likely" solutions selected in the complete 10-day time series, the associated discrete values for the $\tilde{k}$ and $\Theta_{HG}$ parameters are easily assessed. This procedure defines the most representative 10-day values of the three surface parameters characterizing the surface radiative properties, namely, $\{\hat{\rho}_0\}$, $\{\hat{k}\}$ and $\{\hat{\Theta}_{HG}\}$, as well as the corresponding DHR (30°) values. It also ensures that these selected values are sufficient to generate a radiation field consistent with at least one of the radiation fields actually measured during one of the 10-days period by the Meteosat instrument.

In order to deliver the most complete possible maps of geophysical products we implemented an accumulation procedure for every period of ten consecutive days during the year 1996. The accumulation procedure simply consists in the sequential filling of the remaining gaps in the results available for any $\{P^{\hat{}}\}$ parameter, starting from day 1 and ending on day 10 of the time series. In other words, we produced maps of geophysical products for every ten-day period, which are made up of the most representative retrievals selected over these ten days. According to this procedure, the actual conditions of observations, the performances and the results of the inversion procedure are always fully documented for these successful retrievals composing the maps. Figure 1 displays a sample of the maps of the DHR (30°) values obtained over the Sahelian region of North Africa, for the first ten days during the months of November, January and May, on the basis of this accumulation procedure. A very large North-South gradient (absolute DHR values of about 0.55 and 0.08 are observed over the Sahara and the Equatorial forest, respectively) with values decreasing with latitude is shown on these maps. The most striking feature is the relative decrease of the DHR values over the entire continent in quite a broad band of latitudes from November to January and, conversely, a relative increase from January to May.

The seasonal migration of the Inter-Tropical Convergence Zone (ITCZ) is the most important meteorological process over the western part of these African regions. The increase of rainfall associated with the northward displacement of the ITCZ over the continent, between April-May and August-September, translates into a corresponding growth of vegetation in these bands of latitude. Conversely, the southward migration of the ITCZ, which generally occurs from September to March-April, is associated with onset of the dry season and vegetation, mainly savanna, suffers from curing (see Cheney and Sullivan, 1997), i.e., plants are basically drying out and dying. The DHR (30°) values, as retrieved from the Meteosat-5 instrument, were simulated for a variety of leaves and underlying soil properties (Pinty et al., 2000b). These simulations have revealed that the Meteosat-5 DHR (30°) values should increase with a decrease in the chlorophyll content of the leaves. However, Figure 1 indicates that, instead, a significant decrease of roughly 0.1 is occurring during the onset of the dry season, while, on the contrary, a relative increase of about the same amplitude is observed from January to May.

These results cannot be interpreted solely on the basis of natural phenomena controlled by the tropical meteorology. As a mater of fact, these bands of latitude are also subject to major anthropogenic activities related to biomass burning. Interestingly, Figure 2 displays the location of the major fires which have been identified from AVHRR data in these African regions, accumulated during the months of December and April 1993 (Arino and Melinotte, 1998). Though similar results for 1996 are not yet available, the seasonality of fire activities is very well established (see for instance, Cooke et al., 1996 and Koffi et al., 1996). On this basis, it is reasonable to consider the results obtained in 1993 data as representative of usual conditions for the sake of the present discussion. This figure illustrates the intense biomass burning activities occurring during the onset of the dry season, typically in December, with some definite slowing down of these activities as the dry season goes on (during April, for instance). Comparing Figures 1 and 2 strongly suggests that fire activities constitute a major environmental land cover change able to significantly impact the surface albedo values at a continental scale. The co-location of the detected fires in December and the regions affected by a decrease in surface albedo between November and January is indeed quite obvious. The relative increase in surface albedo values from January to May may result from various phenomena including a slight re-growth of vegetation and also a change in soil cover due to the removal of the dark burnt material by winds. The simulated impacts of these processes on the variations of the surface albedo values (see Figure 17 in Pinty et al., 2000b) are in agreement with results from radiative transfer simulations. This provides some evidence that these processes are *a priori* good candidates to interpret these fast changes in surface albedo values at the continental scale.

## 5. CONCLUSIONS

An advanced algorithm for characterizing the radiative state of the surface and the atmosphere over the Meteosat visible band has been designed and tested against actual data. The proposed algorithm capitalizes on the capability of the Meteosat instrument to acquire radiance data every 1/2 hour, suggesting that, for a given geophysical system, the successive relative locations of the Sun during the same day (or even half-day) provide a good angular sampling of the radiation field emerging at the top of the atmosphere. The main products delivered by this algorithm are

1. the quantitative characterization of surface radiative properties which can be used to document the state and monitor the evolution of the land surface,
2. an indication on the probable aerosol load provided as an effective optical thickness,
3. the description of the "clear sky" reflectance field at the top of the atmosphere, and
4. the detection of clouds and their associated shadows during the day.

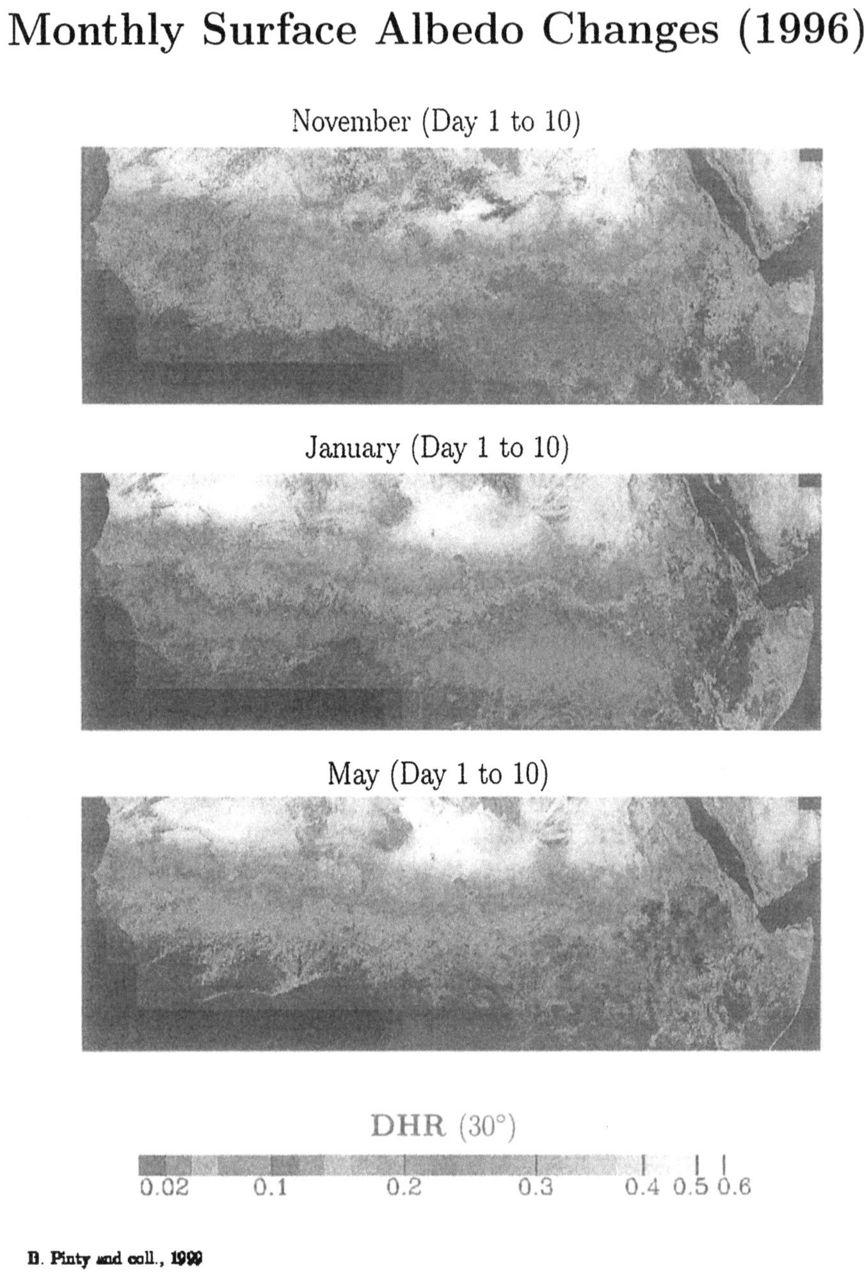


*Figure 1.* Maps of the DHR (30°) values obtained over the Sahelian regions of North Africa, for the first ten days during the months of November, January and May on the basis of the accumulation procedure. Each value corresponds to the selection of the most representative "Likely" solution retrieved each day. The black color identifies the pixels for which no reliable solution was delivered by the algorithm. These DHR values are weighted by the spectral response of the Meteosat VIS channel which extends approximately from 0.4μm to 1.1μm with a maximum response at 0.65μm

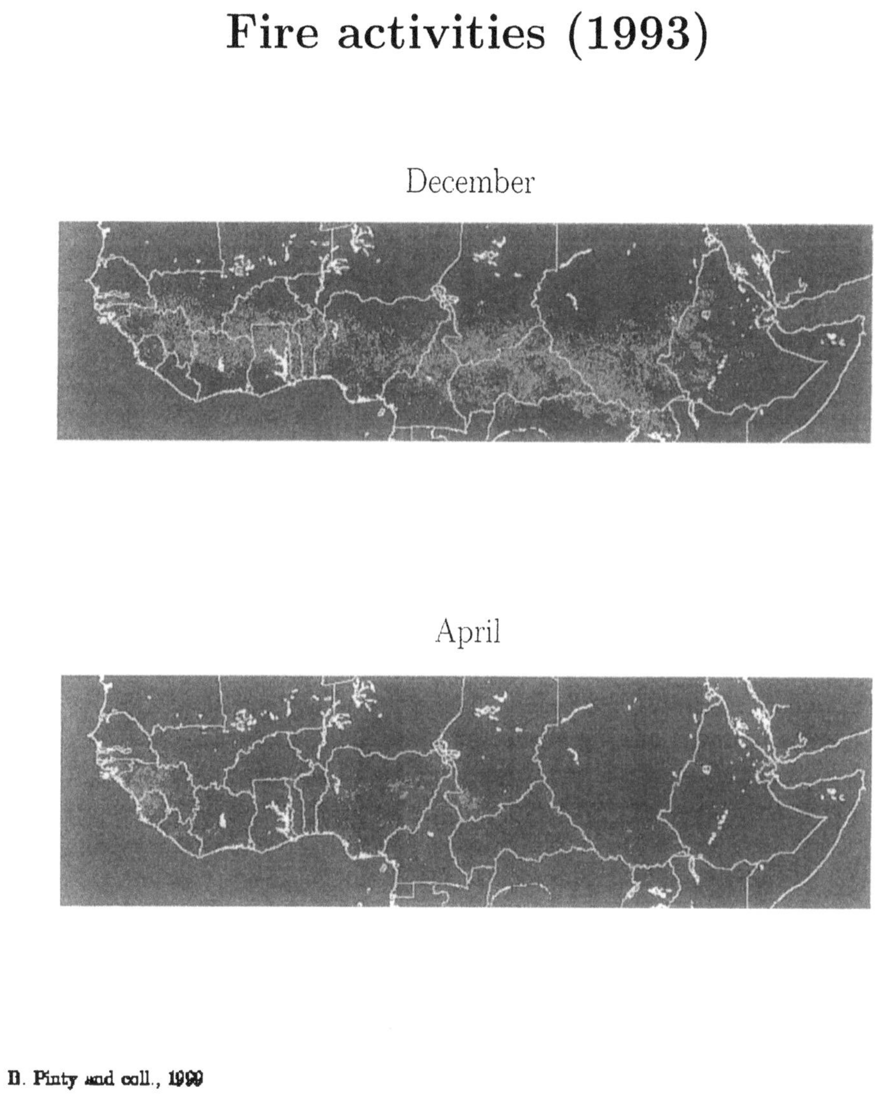


*Figure 2.* Monthly maps of the active fire occurrence detected on the basis of AVHRR data for year 1993. The detection follows from a series of tests on the values measured in the various AVHRR channels (see Arino and Melinotte, 1998, for more details)

The analysis of the results obtained for the year 1996 shows that accurate surface albedo maps may help assessing large land cover changes at the continental scale. As a matter of fact, the interpretation of the monthly surface albedo changes strongly suggests that biomass burning activities may be the dominant environmental factor over large African regions, even masking the natural changes that would be induced by the North-South migration of the monsoon.

The results presented here are based on a small portion of the entire set of products generated by EUMETSAT for year 1996. An assessment of the quality of these products will require further work and is under way. However, these results already illustrate, on the one hand, the great potential of a radiation transfer algorithm based on multi-angular measurements and, on the other hand, the type and value of the information that may be retrieved from the geostationary satellites in general and Meteosat in particular.

## 6. ACKNOWLEDGEMENTS

This study has been motivated by the wish of the Space Applications Institute (SAI) and EUMETSAT to join their efforts to re-process the existing Meteosat archive and to propose new climate products derived from Meteosat data. Informal discussions about fire activities in Africa with Pietro Ceccato and Jean-Marie Grégoire, from the SAI/GVM of the Joint Research Centre have been very helpful in the analysis of the fire activities.

## 7. REFERENCES

Arino, O. and J.-M. Melinotte (1998) The 1993 Africa fire map, *International Journal of Remote Sensing*, **19**, 2019–2023.

Cheney, P. and A. Sullivan (1997) Grassfires: fuel, weather and fire behaviour, Collingwood 3066, Australia: CSIRO Publishing.

Cooke, W. F., B. Koffi, and J.-M. Grégoire (1996) Seasonality of vegetation fires in Africa from remote sensing data and application to a global chemistry model, *Journal of Geophysical Research*, **101**, 21,051–21,065.

Engelsen, O., B. Pinty, M. M. Verstraete, and J. V. Martonchik (1996) Parametric bidirectional reflectance factor models: Evaluation, improvements and applications, Technical Report **EUR 16426 EN**, EC Joint Research Centre.

Holben, B. N. (1986) Characteristics of maximum-value composite images from temporal AVHRR data, *International Journal of Remote Sensing*, **7**, 1417–1434.

Kahn, R., P. Banerjee, D. McDonald, and D. J. Diner (1998) Sensitivity of multiangle imaging to aerosol optical depth and to pure-particle size distribution and composition over ocean, *Journal of Geophysical Research*, **103**, 32,195–32,213.

Kahn, R., R. West, D. McDonald, and B. Rheingans (1997) Sensitivity of multiangle remote sensing observations to aerosol sphericity, *Journal of Geophysical Research*, **102**, 16,861–16,870.

Koffi, B., E. Koffi, and J.-M. Grégoire (1996) Atlas of fire seasonality and its interannual variability for the African continent, Technical Report **EUR 16407 EN**, EC Joint Research Centre.

Martonchik, J. V., D. J. Diner, R. A. Kahn, T. P. Ackerman, M. M. Verstraete, B. Pinty, and H. R. Gordon (1998) Techniques for the retrieval of aerosol properties over land and ocean using multi-angle imaging, *IEEE, Transactions on Geoscience and Remote Sensing*, **36**, 1212–1227.

Martonchik, J. V., D. J. Diner, B. Pinty, M. M. Verstraete, R. B. Myneni, Y. Knyazikhin, and H. R. Gordon (1998) Determination of land and ocean reflective, radiative, and biophysical properties using multiangle imaging, *IEEE, Transactions on Geoscience and Remote Sensing*, **36**, 1266–1281.

Meyer, D., M. M. Verstraete and B. Pinty (1995) The effect of surface anisotropy and viewing geometry on the estimation of NDVI from AVHRR, *Remote Sensing Reviews*, **12**, 3–27.

Pinty, B., F. Roveda, M. M. Verstraete, N. Gobron, Y. Govaerts, J. Martonchik, D. Diner, and R. Kahn (2000a) Surface albedo retrieval from METEOSAT - Part 1: Theory, *Journal of Geophysical Research*, in print.

Pinty, B., F. Roveda, M. M. Verstraete, N. Gobron, Y. Govaerts, J. Martonchik, D. Diner, and R. Kahn (2000b) Surface albedo retrieval from METEOSAT - Part 2: Application, *Journal of Geophysical Research*, in print.

Rahman, H., B. Pinty, and M. M. Verstraete (1993) Coupled surface-atmosphere reflectance (CSAR) model. 2. Semiempirical surface model usable with NOAA Advanced Very High Resolution Radiometer data, *Journal of Geophysical Research*, **98**, 20,791–20,801.

Vermote, E., D. Tanré, J. L. Deuzé, M. Herman, and J. J. Morcrette (1997) Second simulation of the satellite signal in the solar spectrum: An overview, *IEEE Transactions on Geoscience Remote Sensing*, **35**, 675–686.

# Towards a Climatology of Australian Land Surface Albedo for use in Climate Models

IAN F. GRANT
*CSIRO Atmospheric Research, Australia*

Abstract: This paper describes the motivation and an approach for deriving a time series of albedo maps of Australia from historical Advanced Very High Resolution Radiometer (AVHRR) data. Polarization and Directionality of the Earth's Reflectances (POLDER) measurements will be used to test the angular correction algorithm. Some initial results from a survey of POLDER directional reflectance signatures of Australian land cover are presented. Those results show that, while there is much correspondence between the spatial patterns of directional signatures and land cover types, there is a large spread of signatures within each land cover type. However, the similarity of two of the kernels of the bidirectional reflectance distribution function (BRDF) model used to parameterise the directional signatures can produce spurious variations in the model parameters. Finally, some field measurements of grassland albedo are used to make the point that for the greatest accuracy in the estimation of land surface albedo from satellites, it is necessary to account for the detailed shape of the diurnal variation and the effect of the cloudiness on albedo.

## 1. INTRODUCTION

This paper describes the motivation and approach behind work to derive time series of maps of albedo of the Australian land surface from historical Advanced Very High Resolution Radiometer (AVHRR) data. The need for better albedo maps in Australian climate modelling is pointed out, and current Australian efforts to develop uniform best practice processing of AVHRR data are outlined. Then the approach adopted to develop a treatment of view angle effects in the AVHRR data is described, and some initial results of an investigation of the angular reflectance signatures of Australian land surfaces are presented. Lastly, some ground-based albedo observations



*M. Beniston and M.M. Verstraete (eds.),*
*Remote Sensing and Climate Modeling: Synergies and Limitations*, 69–84.

are used to illustrate the importance of sun direction and sky cloudiness in controlling land surface albedo.

## 2. MOTIVATION FOR AN AVHRR ALBEDO ALGORITHM

The Commonwealth Scientific and Industrial Research Organisation (CSIRO) is Australia's primary public organisation for scientific research. The Atmospheric Research division of CSIRO has developed a general circulation model and regional climate models, and these models require as input the surface broadband albedo. The albedo fields used currently are those of the Simple Biosphere (SiB) model developed by Dorman and Sellers (1989), in which a single albedo value is assigned to each land cover class. The albedo for each class has a fixed seasonal cycle and no interannual variation, whereas large regions of Australia experience significant inter-annual rainfall variations that could be expected to produce significant inter-annual variations in the land cover and hence in the albedo. Also, while the SiB albedo of the desert class is assigned the value appropriate for African deserts, there is evidence that a different value would be more appropriate for Australian deserts, which cover much of the Australian continent. A multiyear time series of albedo maps derived from satellite observations would provide more realistic estimates of the mean and range of variation of albedo. The time series of albedo maps would be of even greater value if it was analysed in conjunction with datasets of those parameters that influence albedo, such as soil moisture and vegetation properties. This would help to refine model parameterisations of the dependence of albedo on those parameters.

The CSIRO has some twenty divisions, which conduct research for the benefit of Australia in diverse fields including the ocean, forestry, agriculture, inland water resources, biodiversity and mining exploration. Many divisions have a small complement of specialists in satellite-based earth observation. In 1996 the CSIRO Earth Observation Centre (EOC) was formed to coordinate remote sensing research across CSIRO. One of the EOC's tasks has been to standardise CSIRO's processing of AVHRR data. To this end, the EOC has established and funded teams to develop CSIRO best practice algorithms for navigation, calibration, and atmospheric and angular corrections, either by comparing methods already in use, or by conducting research where no method has been developed. Furthermore, these algorithms are being implemented in a new efficient and flexible software package called the Common AVHRR Processing System (CAPS), which is based on the Tcl/Tk environment. The completion of this effort is expected to see all Australian providers and processors of AVHRR data using the same software and the same algorithms.

The EOC has also embarked on taking the historical data from several AVHRR reception sites around Australia and merging them into a single archive with a uniform format. A feature of this effort is that acquisitions of a single pass from several stations will be stitched into a superpass, resulting in a reduction of data volume and allowing the correction of reception errors in cases of overlap. The application of the best practice algorithms to the stitched archive will enable the generation of uniformly processed, higher level, continental scale AVHRR datasets, extending back fifteen years for some parts of the country.

The availability of the stitched archive and good processing algorithms and software makes the generation of a multiyear time series of continental albedo maps from AVHRR feasible. Two steps in the processing need development: the allowance for changes in surface reflectance due to the large variations in the direction from which AVHRR views any particular site; and the conversion from albedo in the narrow AVHRR spectral bands to albedo in the solar broadband. In the next section an approach to deriving narrowband albedo and an angular correction from AVHRR is described. Narrowband to broadband conversion for AVHRR will not be touched on here but is discussed, for example, in Li and Leighton (1992) and references therein.

## 3. APPROACH TO DERIVATION OF AVHRR ALBEDO ALGORITHM

The albedo of a surface is the ratio of upwelling radiative flux density to downwelling radiative flux density at the surface, and radiation propagating in all directions over the upward and downward hemispheres, respectively, is included. While the surface strongly controls the albedo, it also depends on the angular distribution of the downwelling radiation, which is different, for instance, for clear and cloudy skies. These comments apply to monochromatic radiation. They also apply to a broad spectral band such as the visible, near infrared, or solar (shortwave) bands, if the radiative fluxes are integrated over the spectral band. Surface broadband albedos are affected not only by the spectral reflectance properties of the surface but by the spectral distribution of the downwelling radiation.

In general the reflectance of a surface depends on the view direction, sun direction, and wavelength, and is formally described by the spectral bidirectional reflectance distribution function (BRDF). The direction from which AVHRR views a particular surface point varies with a cycle of about ten days (Figure 1). At mid-latitudes, at the start of the cycle the views are from roughly west at a zenith angle at the surface of approximately 60°, progress through nadir, and finish being from roughly east at 60° zenith angle. For the mid-afternoon satellite passes the variation of view direction corresponds to a shift from reflection in the backward direction to reflection in the forward direction, and commonly introduces a variation in reflectance of as much as 30 to 50%. Figure 2 shows a time series of AVHRR channel 1 reflectance at Tinga Tingana in the Strzelecki Desert in South Australia. The target is clear for the entire 40-day series. The site is very sparsely vegetated and has very little rainfall and can be presumed to be unchanging through the period shown. Cyclic variations in reflectance of relative amplitude 20% are clear and can be ascribed to the varying view geometry.

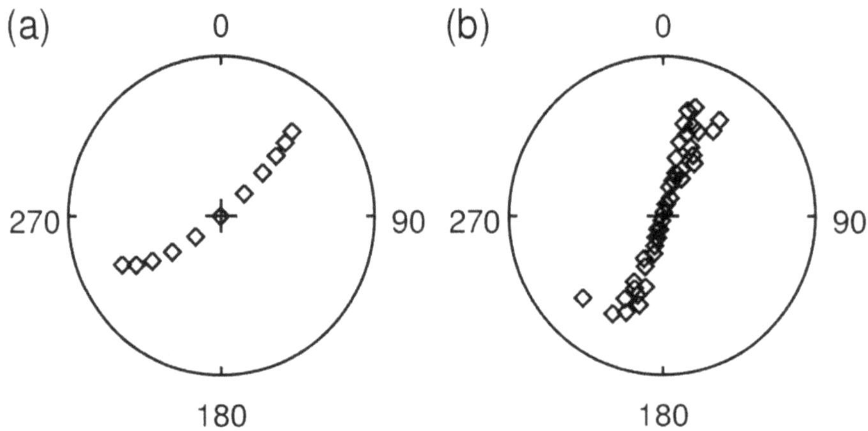


*Figure 1.* View directions for AVHRR. Each polar plot represents the view zenith angle from 0° to 90° as the radius and the sun-view relative azimuth as the polar angle. (a) View directions over eleven consecutive days. (b) View directions over forty days

For many users of AVHRR data, a desirable consistency of reflectance time series can be achieved by correcting the daily observations to a standard view and illumination geometry, nadir being the obvious choice for the view direction. This is essentially an interpolation problem, since the AVHRR does indeed sometimes view the target at nadir (although cloud-free conditions will not always coincide with this ideal viewing geometry). For albedo estimation, the sampling of view directions would ideally cover the whole hemisphere represented in Figure 1. While Figure 1 illustrates the fact that AVHRR samples part of the hemisphere of view directions, albedo estimation will effectively require extrapolation to view directions that are never sampled by AVHRR.

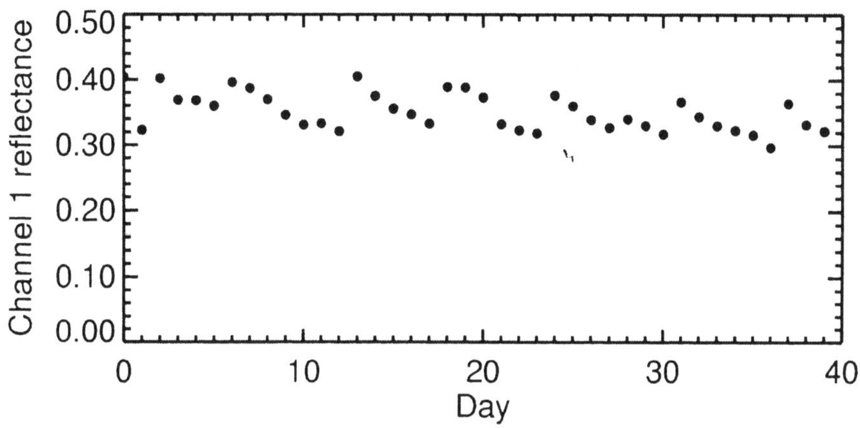


*Figure 2.* Time series of AVHRR channel 1 reflectances at Tinga Tingana

O'Brien et al. (1998) developed an algorithm to correct top-of-the-atmosphere AVHRR time series to nadir viewing by fitting a parametric BRDF model to the time series within a sliding window. The algorithm was tested at semi-arid sites in Australia and performed well, judging by the consistency of the time series and the rapid response of the resulting NDVI time series to rainfall. The algorithm was also successful at predicting the radiance measured by the Japanese Geostationary Meteorological Satellite, which views the site from a direction unsampled by AVHRR. This suggests that the technique may be capable of the extrapolation in view angle required for albedo estimation.

Work is underway to further develop this BRDF fitting technique to apply to the whole of Australia, for the correction of AVHRR time series to nadir viewing and for the estimation of albedo in the AVHRR shortwave bands. The technique must be demonstrated to work for a wide variety of Australian land cover types and to work in the presence of cloud, which was largely absent for the tests at semi-arid sites described above. A variety of BRDF models will be assessed with the method, which has only been attempted so far with the model of Staylor and Suttles (1986) and Li's model for sparse canopies (Wanner et al., 1995). The criteria for ranking models, following the approach used by Hautecœur and Leroy (1996), will include robustness for different land cover types, robustness in the presence of data gaps caused by cloudiness, and the ability to extrapolate to unmeasured view directions. It will also be of interest to determine whether the model must be fitted separately to individual pixels, or whether single model fits can be satisfactorily applied to regions of similar land cover type, giving savings in computation.

The interpolation of the AVHRR view to nadir and the angular extrapolation required for albedo will be tested with data from the Polarization and Directionality of the Earth's Reflectances (POLDER) sensor that operated on Japan's Advanced Earth Observing Satellite (ADEOS) satellite from November 1996 until June 1997. This sensor was the first to acquire global scale observations of reflectance with good angular coverage. Figure 3 shows that over a few days POLDER densely samples the hemisphere of view directions out to zenith angles of about 60°. The comparison between AVHRR and POLDER will be done in three steps:

1. for cases where the AVHRR and POLDER view directions and solar zenith angles are similar, to gauge the effect of the spectral mismatch between the bands of the two sensors;
2. for cases where the POLDER view direction is at nadir, to test the BRDF-correction of AVHRR to this standard view direction;
3. for all POLDER view directions, to test, over the largest range of view directions, the BRDF that has been fitted to AVHRR data.

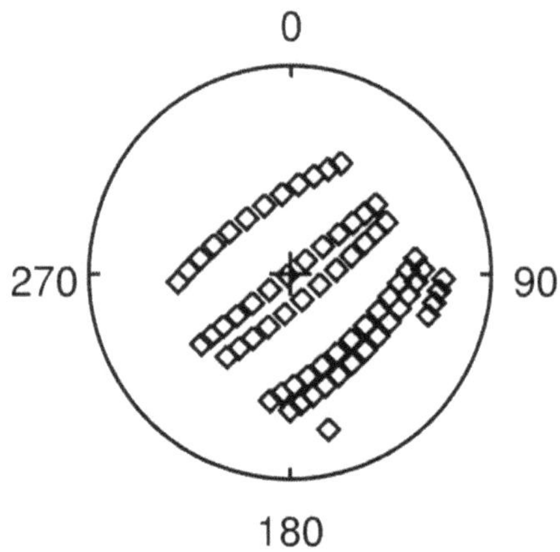


*Figure 3.* View directions for POLDER over six consecutive days. The plot format is the same as Figure 1

These comparisons will be made at the top of the atmosphere, to avoid the uncertainty associated with atmospheric correction of the satellite observations in the absence of good knowledge of the aerosol and water vapour content of the atmosphere. It is assumed that if the BRDF fitting technique is shown to be robust at the top of the atmosphere then it will also perform well for surface BRDFs.

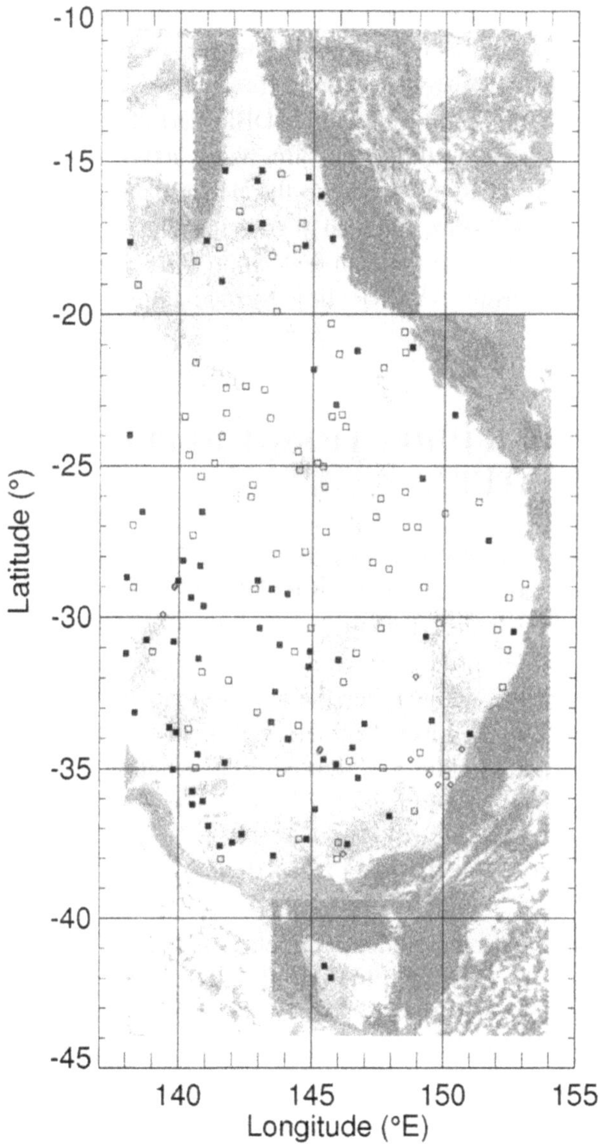


*Figure 4.* Locations of sites in eastern Australia for testing AVHRR algorithms for angular correction and albedo, plotted on a mosaic of AVHRR channel 2 images. Filled squares are uniform sites chosen to span a wide range of vegetation types, open squares are randomly selected sites, and diamonds are field data sites

About seventy sites in eastern Australia have been selected at which to compare the AVHRR and POLDER data. This region has been chosen because it has good angular coverage by the AVHRR data in the archive held at CSIRO Atmospheric Research. The sites represent every class in a thirty-two class land cover map based on AUSLIG (1990), and have been chosen to be among the most uniform within their respective classes to reduce the effect of misregistration errors. Time series of AVHRR data covering at least the eight-month POLDER period are being extracted at each site. At the same time, data are being extracted at about seventy randomly selected locations to gauge the performance of the BRDF fitting algorithm at "typical" sites, and also at sites for which field measurements or airborne scanner data exist that can be used to further verify the BRDFs and albedos retrieved from AVHRR time series. Figure 4 shows all of the selected sites.

## 4. POLDER DIRECTIONAL REFLECTANCE SIGNATURES

In order to gain insight into how the surface directional reflectance varies spatially and temporally, the POLDER data are also being used to explore the directional reflectance signatures of Australian land cover types. The POLDER Level 3 product *Land surfaces and atmosphere: surface directional signatures* summarises the atmospherically corrected multiangular measurements by the three parameters of the Roujean BRDF model fitted to the measurements within a 30-day sliding window. The Roujean BRDF model (Roujean et al., 1992) is:

$$\rho(\theta, \theta_0, \phi) = k_0 + k_1 f_1(\theta, \theta_0, \phi) + k_2 f_2(\theta, \theta_0, \phi), \qquad (1)$$

where $k_0$, $k_1$ and $k_2$ are the model parameters and $f_1$ and $f_2$ are purely geometric functions of the view zenith angle $\theta$, the solar zenith angle $\theta_0$ and the view-solar relative azimuth $\phi$. The kernels $f_1$ and $f_2$ are derived from simple physical models of the interaction of light with the structured surface. The "geometric" kernel $f_1$ represents the effects of shading by protrusions on the surface. The "volume scattering" kernel $f_2$ captures the effects of multiple scattering by reflecting facets (leaves or soil particles) spread through a thick layer.

For the POLDER surface directional signature product, the Roujean model is fitted separately for each $7 \times 7$ km pixel and for each of the four spectral bands. Thus for each 30-day windowing period, there are twelve maps at 7-km resolution: the three model parameters at each of 443, 670, 765 and 865 nm. This dataset is being compared with a 32-type classification of Australian land cover in terms of growth form and fractional foliage cover that is based on AUSLIG (1990). Some initial results from this analysis are now presented. Grant (2000) gives more detail.

The top panel of Figure 5 is a map over Australia of the $k_2$ parameter at 670 nm for the thirty days centred on 15 June 1997. The bottom panel of Figure 5 shows the distribution of sparse grassland with a projective foliage cover of 10–30% (class code G2). The regions of $k_2 > 0.36$ in northern Australia correspond strikingly to a subset of the grassland regions. Similar correspondences also appear for other vegetation types, and in other parameters and in other bands. This indicates that the vegetation type influences the directional signature, as is expected from observations and modelling reported in the literature. If the directional signatures are similar within a region of uniform land cover type, then it might be feasible to fit a single BRDF model or model shape to a region.

In order to examine how uniquely the model parameters are predicted from the vegetation class, two-dimensional histograms such as those in Figure 6 have been examined. In Figure 6 the two-dimensional histogram of parameters $k_2$ and $k_0$ is plotted for each of several vegetation classes. Some trends in the distributions' central tendency with growth form and density are evident. However, the spread of parameters within classes is at least as large as the differences between classes. This large spread also appears for $k_1$ and other bands, and suggests that factors other than the coarse description of vegetation structure represented in the land cover map used control the directional reflectance. These could include details of the vegetation canopy structure, leaf area index, soil type, rainfall history and topography.

Caution is needed in interpreting parameter triplets ($k_0$, $k_1$, $k_2$) as uniquely labelling a particular BRDF. The kernel functions $f_1$ and $f_2$ can be similar in shape, particularly for small solar zenith angles. Thus when fitting the Roujean model to a set of data, an error in parameter $k_1$ can be compensated by the error in parameter $k_2$. Figure 7 shows a region of the map of the parameter $k_1$ at 670 nm for one particular vegetation class, sparse tussock grassland. The area is in tropical Australia in December, so the sun is high (i.e. the solar zenith angle is small) at the time of the ADEOS overpass around 1030 local time. The parameter $k_1$ shows strong spatial structure on all scales down to the pixel level (7 km $\cong$ 0.07°). Maps of $k_1$ and $k_2$ for the whole continent show similar fine structure in December, but not in June when the sun is relatively low (not shown). However, when the Roujean model is evaluated with the mapped parameters ($k_1$ as in Figure 7, and the corresponding $k_0$ and $k_2$) for one particular set of sun and view angles, yielding a directional reflectance, the spatial structure is much smoother (Figure 8) than for the parameter $k_1$ in Figure 7. Thus $k_1$ and $k_2$ are correlated, and some of the spread in two-dimensional histograms such as those in Figure 6 may give a misleading indication of the spread of directional signatures within each class. Grant (2000) discusses this in more detail. Eliminating the spurious spread in the Roujean parameters is the next step to be taken in this survey of Australian directional reflectance signatures.

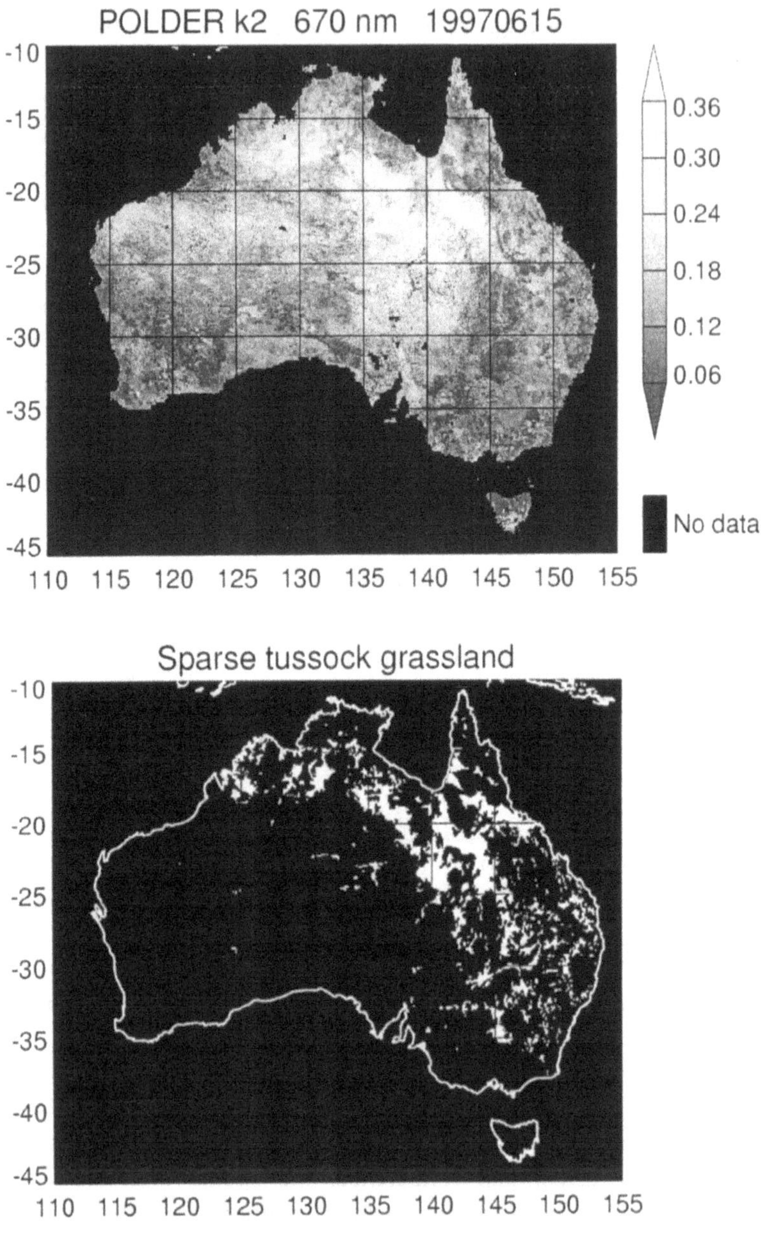


*Figure 5.* Top: POLDER parameter k2 at 670 nm in June 1997.
Bottom: Distribution of sparse tussock grassland

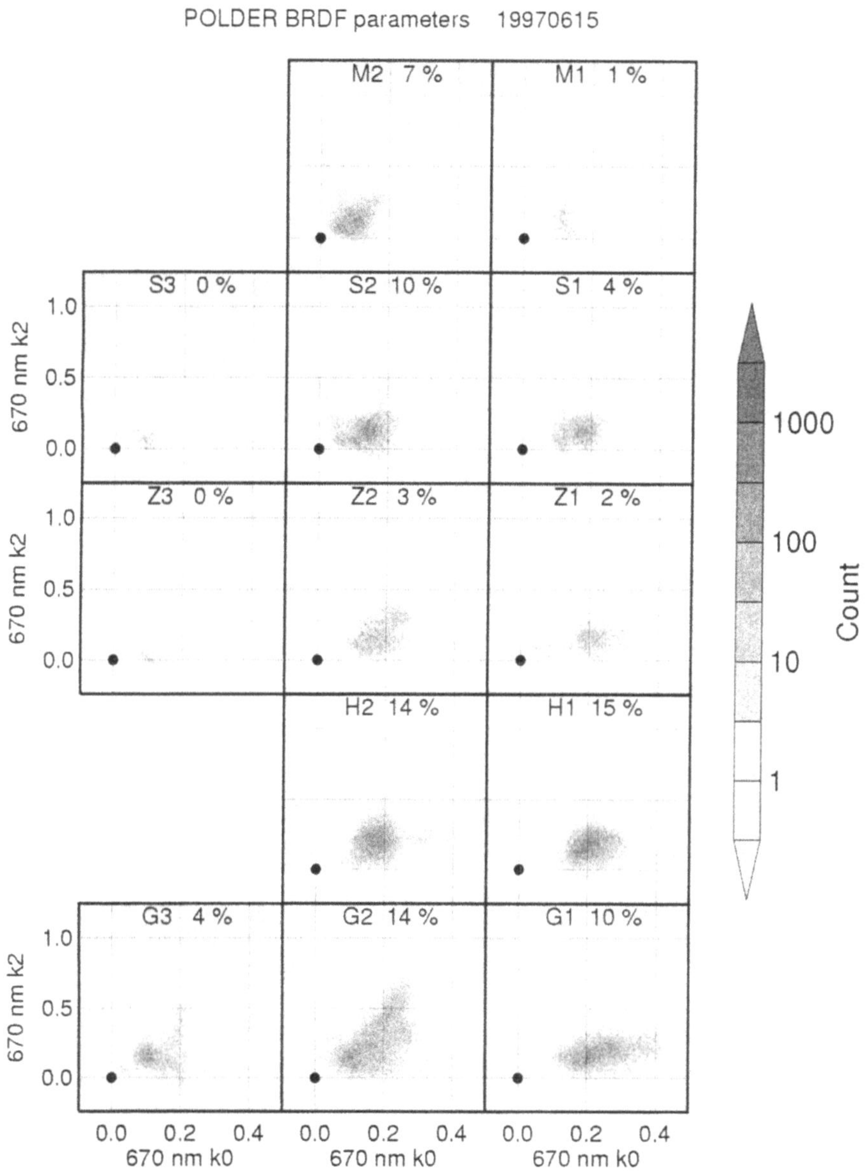


*Figure 6.* Two-dimensional histograms of ($k_0$, $k_2$) at 670 nm for several vegetation types. Following the coding in AUSLIG (1990), each type is labelled by a letter indicating the vegetation growth form (M = medium trees, S = shrubs over 2 metres, Z = shrubs under 2 metres, H = hummock grasses, G = tussock grasses) and a number indicating the projective foliage density (1 = under 10%, 2 = 10–30%, 3 = 30–70%)

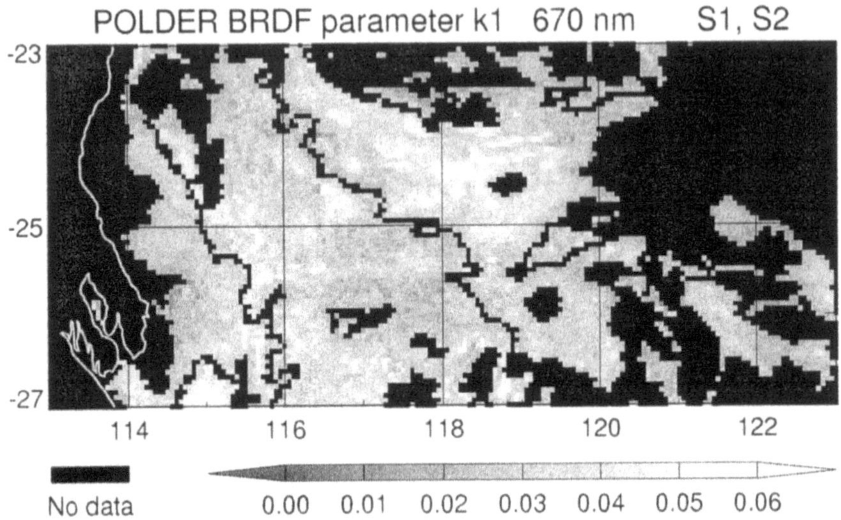


*Figure 7.* POLDER parameter k1 at 670 nm in December 1996 for two vegetation types in a region of Western Australia: tall shrubs with density of under 10% and 10–30%. The mixed pixels on the boundary between the types are not plotted here

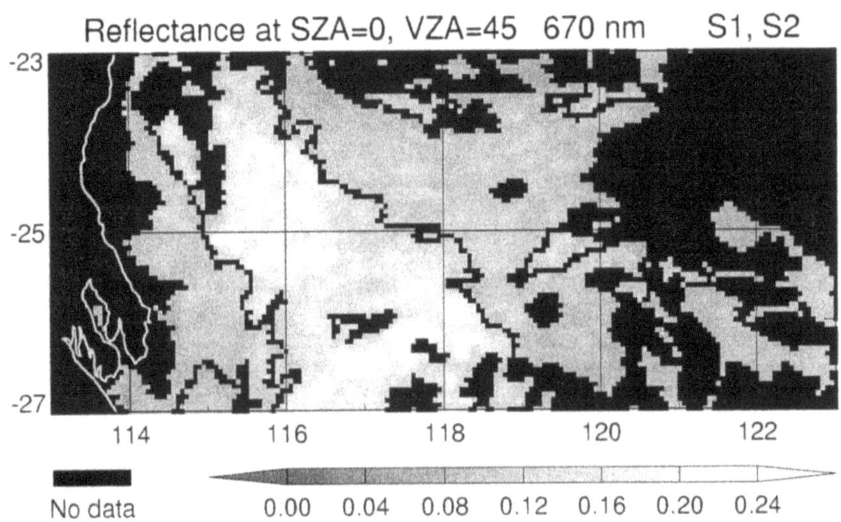


*Figure 8.* Reflectance at 670 nm for overhead sun and view zenith angle 45° for the region and month shown in Figure 7, evaluated from the POLDER Roujean coefficients

# 5. DEPENDENCE OF ALBEDO ON SUN DIRECTION AND CLOUDINESS

This section presents some observations to emphasise that albedo is not only controlled by the surface but also by the illumination, in particular the solar zenith angle and cloudiness. The measurements were collected at a grazed grassland site at Uardry in southeastern Australia and have been processed into the 30-minute mean albedos presented here. Prata et al. (1998) and Grant et al. (2000) give details on the site, and the measurements and their analysis.

Figure 9 shows the diurnal variation of the albedo on three clear days in three different months. The albedo has a minimum near noon on each day, as has previously been well reported from observations, and as is expected from theoretical considerations. However, the amplitude of the diurnal variations reduces markedly over several months, there is sometimes a difference between the morning and afternoon albedo at corresponding solar zenith angles, and departures on the time scale of a few hours from a smooth variation are sometimes apparent. These three features have been seldom, if ever, noted, but if neglected will introduce errors in any estimation of the diurnal cycle of albedo, or daily or monthly mean albedo, from an estimate of the instantaneous albedo at one local time such as would be made with a polar orbiting satellite. Grant et al. (2000) show that for the Uardry site an estimate of the daily mean albedo from the 1030 local time albedo could be in error by 0.01–0.03 for an albedo of 0.20. That is, a relative error of 5 to 15%, depending upon the available level of detail of knowledge of the diurnal variation of albedo.

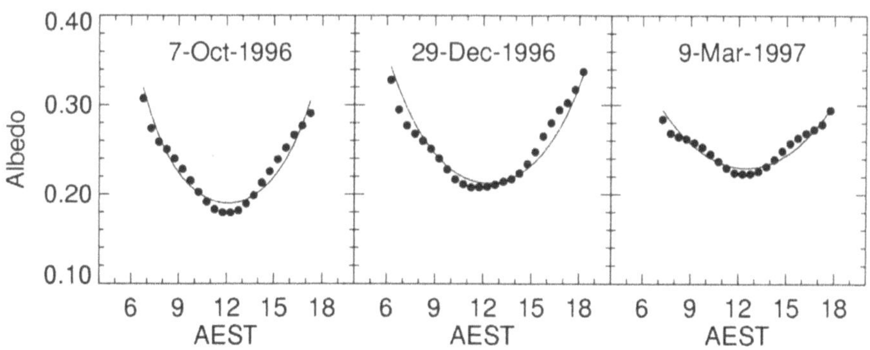


*Figure 9.* The diurnal cycle of albedo at Uardry on three clear days over several months. The dots are the observed 30-minute mean albedos. The smooth curves are a fitted model and are included as an aid to judging the symmetry and smoothness of the observed variations

Figure 10 shows the diurnal cycle of downwelling solar irradiance and albedo for three days of an eight-day period. The first and third days are clear throughout and the albedos are similar. Presumably the surface changed only slightly over the period, perhaps in greenness or soil wetness. The second day was cloudy throughout, reducing the downwelling irradiance to a roughly constant fraction of its clear-sky value. The albedo on the cloudy day was significantly below that on the two clear days, and had a flatter diurnal variation. Satellite-based estimates of land surface albedo rely on measurements in clear-sky conditions. A likely approach to the estimation of cloudy-sky albedo from satellite observations is to measure the surface BRDF under clear conditions and integrate it with a cloudy-sky irradiance field. For sufficiently stringent demands on the accuracy of the albedos input to climate models, the dependence of the albedo on the sky condition will have to be taken into account.

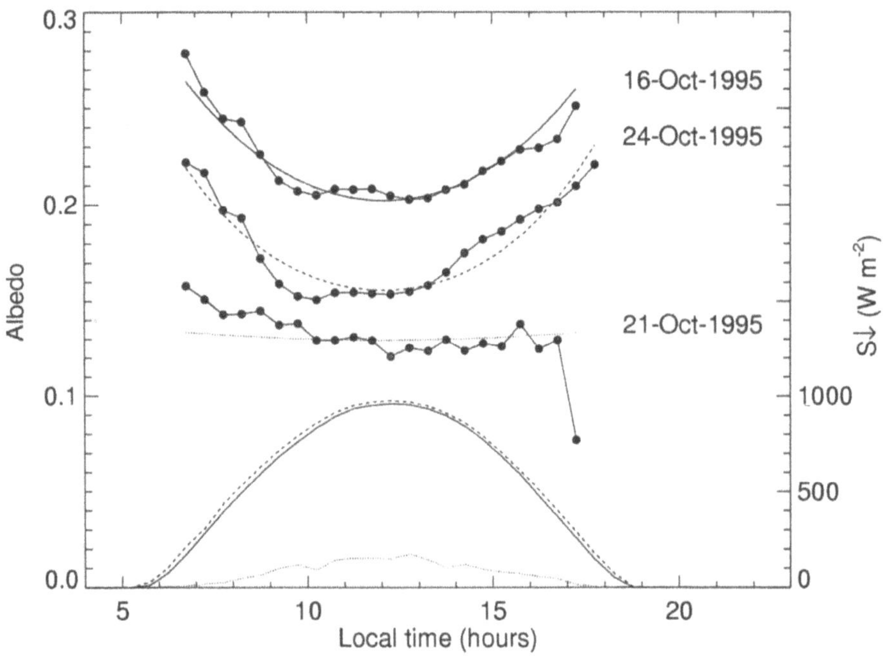


*Figure 10.* The diurnal cycle of downwelling solar irradiance S↓ (lower three curves) and albedo (upper three pairs of curves) at Uardry on three days, two of which are clear and one cloudy. For the albedo the joined dots are the 30-minute observations and the curves through them are fitted models. The irradiance and model albedo curves for any particular day have the same line style

## 6. CONCLUSIONS

This paper has outlined an approach to developing a robust algorithm to derive albedo and time series of reflectances that are corrected for view direction effects from AVHRR time series. POLDER top-of-the-atmosphere radiance measurements will be used to test the algorithm. Some initial results from a survey of POLDER directional reflectance signatures of Australian land cover were presented, which showed that while there was much correspondence between the spatial patterns of directional signatures and land cover types, there was a large spread of signatures within each land cover type. However, the similarity of two of the kernels of the BRDF model used to parameterise the directional signatures can produce spurious variations in the model parameters. Finally, some field measurements of grassland albedo were used to make the point that for the greatest accuracy in the estimation of land surface albedo from satellites, it is necessary to account for the detailed shape of the diurnal variation and the effect of the cloudiness on albedo.

## 7. ACKNOWLEDGMENTS

Dean Graetz is thanked for supplying the vegetation classification map. The POLDER data were obtained from CNES's POLDER on board NASDA's ADEOS. This paper was presented at the International Workshop on Satellite Remote Sensing and Climate Simulations: Synergies and Limitations, which was held in Les Diablerets, Switzerland, 20–24 September 1999. The CSIRO Earth Observation Centre supported Ian Grant and his attendance at the Workshop.

## 8. REFERENCES

Australian Surveying and Land Information Group (AUSLIG), Atlas of Australian Resources: third series, volume 6 *Vegetation*, Dep. of Admin. Serv., Canberra, ACT, Australia (1990).

Dorman, J. L. and P. J. Sellers, A global climatology of albedo, roughness length and stomatal resistance for atmospheric general circulation models as represented by the Simple Biosphere Model (SiB), J. Appl. Meteorol., 28: 833–855 (1989).

Grant, I. F., Investigation of the variability of the directional reflectance of Australian land cover types, in Proceedings of the Second International Workshop on Multiangular Measurements and Models, Ispra, Italy, 17–19 September 1999, submitted to Remote Sens. Rev. (2000).

Grant, I. F., A. J. Prata and R. P. Cechet, The impact of the diurnal variation of albedo on the remote sensing of the daily mean albedo of grassland, J. Appl. Meteorol., 39: 231–244 (2000).

Hautecœur, O., and M. Leroy, Intercomparison of several BRDF models for the compositing of POLDER data over land surfaces, in Proceedings of the IGARSS'96 Conference, edited by T. I. Stein, pp. 204–208, Lincoln, Nebraska, IEEE Publications, Picataway, NJ (1996).

Li, Z., and H. G. Leighton, Narrowband to broadband conversion with spatially autocorrelated reflectance measurements, J. Appl. Meteorol., 31: 421–432 (1992).

O'Brien, D. M., R. M. Mitchell, M. Edwards and C. C. Elsum, Estimation of BRDF from AVHRR short-wave channels: tests over semiarid Australian sites, Remote Sens. Environ., 66: 71–86 (1998).

Prata, A. J., I. F. Grant, R. P. Cechet and G. F. Rutter, Five years of shortwave radiation budget measurements at a continental land site in southeastern Australia, J. Geophys. Res., 103: 26 093–26 106 (1998).

Roujean, J.-L., M. Leroy and P.-Y. Deschamps, A bidirectional reflectance model of the Earth's surface for the correction of remote sensing data, J. Geophys. Res., 97: 20 455–20 468 (1992).

Staylor, W. F., and J. T. Suttles, Reflection and emission models derived from Nimbus 7 ERB scanner measurements, J. Clim. Appl. Meteorol., 25: 196–202 (1986).

Wanner, W., X. Li and A. H. Strahler, On the derivation of kernels for kernel-driven models of bidirectional reflectance, J. Geophys. Res., 100: 21 077–21 089 (1995).

# Collocated surface and satellite observations as constraints for Earth radiation budget simulations with global climate models

Martin WILD
*Swiss Federal Institute of Technology, Department of Geography, Zurich, Switzerland*

Abstract: Satellite measurements show that the exchange of solar energy between the global climate system and outer space is well simulated by the current generation of General Circulation Models (GCM). However, this alone does not ensure that these models also reproduce the distribution of solar energy *within* the simulated climate system correctly. Thus, the present study uses in addition to the satellite data a collocated set of surface observations for a more vigorous assessment of the solar energy in the climate system than could ever be achieved using satellite data alone. It is shown that GCMs typically underestimate the absorption of solar energy in the atmosphere, by 10 -20 $Wm^{-2}$. In other words, the present study suggests that the global mean shortwave atmospheric absorption, a highly debated quantity, should rather be between 80-90 $Wm^{-2}$, than around 70 $Wm^{-2}$ as found in many current GCMs. This leads to excessive insolation at the GCM surface compared to more than 700 globally distributed observation sites. In a case study based on data from observation sites in Germany, the relative portion of solar energy absorbed in the cloud-free atmosphere and its cloudy counterpart is investigated. No indications are found that the absorption of solar radiation in the GCM atmospheres should be significantly enhanced when clouds are present, which has been postulated in other studies. Rather, the underestimation in the atmospheric absorption in many GCMs seems to be caused by a lack of absorption in the cloud-free atmosphere, related to an underestimated water-vapor and aerosol absorption.

## 1. INTRODUCTION

The radiation balance of the Earth plays a fundamental role in the global climate system and in the radiatively-induced climate change. It is therefore essential that General Circulation Models (GCM) which attempt to re-



*M. Beniston and M.M. Verstraete (eds.),*
*Remote Sensing and Climate Modeling: Synergies and Limitations*, 85–102.

produce the Earth system should be able to simulated the radiative processes with high confidence. Due to recent satellite programs such as ERBE, the total amount of solar energy absorbed by the global climate system is well established (Barkstrom et al. 1990). To date the current generation of GCMs simulates this net exchange of solar energy between outer space and the global climate system realistically when compared to the satellite observation. However, the good agreement between the satellite-observed and simulated fluxes at the top of atmosphere (TOA) only implies that the *total* absorption of solar energy in the climate system is quantitatively correctly captured in the GCMs. It does not ensure that the solar energy is absorbed at the proper places *within* the climate system.

Therefore, rather than relying on a validation of the TOA budgets using satellite data only, validation studies trying to assess radiation in GCMs should make use of the additional information available from ground observations.

The present study outlines how such combined surface/satellite data sets can improve our knowledge on the distribution of solar energy in the climate system and its representation in GCMs.

## 2. OBSERVATIONAL DATA

The satellite data used in the present study are ensemble averages of radiative fluxes at the Top of Atmosphere (TOA) from the Earth Radiation Budget Experiment (ERBE, Barkstrom 1990). These are broad band measurements representative for the period 1985 - 1989, with a resolution of 2.5° x 2.5°. The uncertainties in the monthly averaged scanner data are estimated within +/- 5 $Wm^{-2}$.

The observational data at the surface are retrieved from a database containing the world-wide instrumentally measured surface energy fluxes, the Global Energy Balance Archive (GEBA, Ohmura et al. 1989, Gilgen and Ohmura 1999). This database currently possesses 220,000 monthly mean fluxes for approximately 1600 sites and has been used in a number of studies to assess model and satellite derived estimates of surface energy fluxes (e.g., Garratt 1994, Li et al. 1995, Wild et al. 1995, 1997, 1998, Wild 1999, Arking 1996, Konzelmann et al. 1996, Rossow and Zhang 1995, Cusack et al. 1998). Gilgen et al. (1998) estimated the relative random error (root mean square error / mean) of the incoming shortwave radiation values in the GEBA at 5% for the monthly means and 2% for annual means.

For the assessment of the all-sky shortwave radiation budgets in GCMs, long-term surface observations from 720 GEBA sites together with their collocated TOA fluxes from ERBE are used. The global distribution of these sites is given in Fig. 1. Their temporal and spatial representativity has been investigated in Wild (1997).

For the assessment of the GCM-simulated solar radiation specifically under clear-sky conditions, we are currently preparing an observational data-set of surface clear-sky climatologies at selected sites worldwide. So far, observed clear-sky climatologies of surface insolation have been determined for a number of sites in Germany (Wild and Liepert 1998). The clear-sky insolation climatologies were obtained from composites of cloud-free episodes which were identified on an hourly basis using additional information on cloudiness and sunshine duration. Monthly all-sky climatologies of surface insolation for the same sites and period were available from the Global Energy Balance Archive.

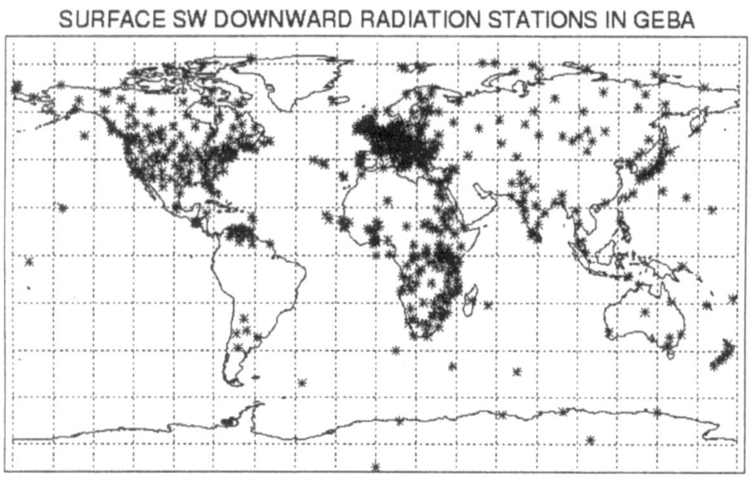


*Figure 1.* Global distribution of the 720 sites with long-term incoming shortwave radiation measurements selected for this analysis from the Global Energy Balance Archive

To determine the amount of shortwave radiation absorbed at the surface, all above mentioned insolation climatologies were combined with the collocated values of a surface albedo climatology provided by the Surface Radiation Budget Project (SRB, Darnell et al. 1992) representative for the period 1985 - 1989. In an attempt to estimate potential errors introduced by the surface albedo, the measured insolation was additionally combined with two alternative sets of albedo climatologies. They did not alter the surface absorption significantly. The clear- and all-sky radiative fluxes at the TOA collocated with the surface measurements at the German sites are again taken from ERBE. Finally, estimates of clear- and all-sky absorption within the atmosphere were obtained from the respective differences between the absorbed radiation implied by satellite observations at the TOA and the absorbed radiation at the surface from the ground measurements.

## 3. MODELS

Several GCMs are considered in this study, namely ECHAM3 (Roeckner et al. 1992) from the Max Planck Institute for Meteorology, Hamburg, ARPEGE (Déqué et al. 1994) from Meteo-France, Toulouse, and model versions HadAM2b (Stratton 1999) and HadAM3 (Cusack et al. 1998) from the Hadley Centre, Bracknell. All model data stem from AMIP type simulations with prescribed SST and sea ice climatologies. Simulations with the above models were analyzed at various horizontal resolutions, although results are shown here only for the standard resolution (T42 for the spectral models ARPEGE and ECHAM3, (2.5° x 3.75°) for the HadAM gridpoint models), since the calculated radiative fluxes were shown to be insensitive to a change in horizontal resolution (Wild et al. 1995, Wild 1997). Thus the conclusions drawn in this study do not depend on a specific model resolution. All models include broad-band radiation schemes with two-stream approximation, as typically used in GCMs. The ECHAM3 and ARPEGE models further include simple aerosol climatologies based on WMO (1983), while HadAM2b does not include any aerosol effects. From the next generation model version HadAM3 simulations were available both with and without a (simple) aerosol climatology (Cusack et al. 1998), hereafter referred to as HadAM3 and HadAM3-NA, respectively.

For the comparison of model-calculated and observed fluxes, the model fluxes were interpolated to the observation sites using the four surrounding grid points weighted by their inverse spherical distances.

## 4. RESULTS

Global annual mean values of shortwave absorption at the surface, within the atmosphere and in the entire surface-atmosphere system are shown in Table 1 for the GCMs investigated in the present study, both for all-sky and clear-sky conditions. Note that "all-sky" includes all types of weather conditions, i.e. from totally overcast to completely cloudless. Additional estimates of the TOA fluxes from ERBE (Barkstrom 1990) as well as two estimates of surface and atmospheric absorption (Ohmura and Gilgen 1993, Wild et al. 1998), which make use of the GEBA observations, are displayed in Table 1. Wild et al. (1998) provide separate estimates of surface and atmospheric absorption under both clear-sky and all-sky conditions. They are based on a blend of direct observations and model calculations with well-established bias structures.

Compared to these estimates, the GCMs show significantly higher absorption of solar radiation at the surface while significantly less in the atmosphere. On the other hand, at the TOA, the GCMs agree quite well with the (satellite) estimates in their net fluxes. This is found for both all-sky and clear-sky conditions.

To get more insight into these discrepancies, the (all sky) radiation climatologies of the GCMs are assessed in the following using the 720 surface sites and their collocated satellite-measured TOA climatologies. The radiation budgets of the models ECHAM3, ARPEGE and HadAM2b are discussed in Section 4.1, while the HadAM3 models with and without aerosol are discussed in Section 4.2.

*Table 1.* Global annual mean absorption of solar radiation at the surface, within the atmosphere and in the total system (TOA) in the GCMs ECHAM3, ARPEGE, HadAM2b, HadAM3 with aerosol and withouth aerosol (HadAM3 NA). Units $Wm^{-2}$

| | ALL-SKY | | | CLEAR-SKY | | |
|---|---|---|---|---|---|---|
| | Surface | Atmosphere | TOA | Surface | Atmosphere | TOA |
| *GCMs:* | | | | | | |
| ECHAM3 | 164 | 71 | 235 | 222 | 63 | 285 |
| ARPEGE | 174 | 68 | 242 | 215 | 67 | 282 |
| HadAM2b | 175 | 67 | 242 | 234 | 61 | 295 |
| HadAM3 | 168 | 76 | 244 | 219 | 71 | 290 |
| HadAM3 NA | 174 | 71 | 245 | 228 | 66 | 294 |
| *Other estimates:* | | | | | | |
| ERBE | | | 240 | | | 287 |
| Ohmura+Gilgen (1993) | 142 | 98 | 240 | | | |
| Wild et al (1989) | 154 | 86 | 240 | 216 | 72 | 287 |

## 4.1 The models ECHAM3, ARPEGE, HadAM2

### 4.1.1 Assessment of surface radiation budgets

To obtain a reference dataset for the assessment of the GCM absorbed shortwave radiation at the surface, the observed values of the incoming shortwave radiation from GEBA had first to be weighted with their associated surface albedos. They were taken from the albedo climatology provided by the Surface Radiation Budget Project SRB (Darnell et al. 1992) as described in Section 2.

The differences between the absorbed surface solar radiation calculated in the ECHAM3, ARPEGE, and HadAM2b GCMs and the observed estimates at the 720 sites are shown in Figure 2c. The differences have been averaged over the sites in latitudinal belts of 5°. All three GCMs show very similar tendencies: Too much solar energy is absorbed at the surface, particularly in low latitudes. The global mean surface absorption in these models is also 10 - 30 $Wm^{-2}$ higher than the estimates of Ohmura and Gilgen (1993) and Wild et al. (1998) (Table 1). At high latitudes, the models tend to underestimate the absorption of solar energy. This implies an excessive meridional gradient of absorbed solar energy at the surface in all three models, a typical feature of current GCMs.

### 4.1.2 Assessment of Top-of-Atmosphere budgets (TOA)

To determine the origins of the above biases at the surface, the total absorbed solar energy in the surface-atmosphere column at the 720 GEBA locations is examined, which has been directly measured by satellites during ERBE.

Differences between the annual mean model-calculated net shortwave fluxes at the TOA and the ERBE fluxes are shown in Fig. 2a, again averaged over the sites in the 5° latitude belts. The agreement is much better than at the surface with biases smaller than 10 $Wm^{-2}$ at most latitudes. The global mean values of the GCM shortwave TOA radiation budgets in Table 1 are close to the ERBE value of 240 $Wm^{-2}$, partly due to the tuning of the planetary albedo in the GCMs to the ERBE estimate. This indicates that the total amount of solar energy absorbed in the climate system is well captured in the models, and is not the main cause for the biases detected at the surface. Thus, the biases at the surface have to be attributed to deficiencies in the absorption within the atmosphere.

### 4.1.3 Assessment of atmospheric absorption

The differences between model-calculated atmospheric shortwave absorption and the observational estimates have been determined as residuals of the net flux differences at the top of atmosphere and at the surface, respectively (Fig. 2b). A lack of shortwave absorption in the atmosphere can be noted in all three GCMs, particularly at mid- and low latitudes, which amounts to more than 30 $Wm^{-2}$ near the Equator. Also, their global mean values of atmospheric absorption are 10 - 30 $Wm^{-2}$ lower than the (all sky) estimates of Ohmura and Gilgen (1993) and Wild et al. (1998) in Table 1.

The above results strongly suggest that the biases in the model-calculated fluxes found at the surface are not caused by deficiencies in the net amount of solar energy absorbed in the climate system, but rather due to errors in the atmospheric absorption. This is particularly evident in terms of a lack of shortwave absorption in the atmosphere of the mid- and low latitudes. The difficulty common to all these models thus lies in the partitioning of the absorption of solar radiation between atmosphere and surface. This problem could only be detected due to the combined use of satellite and surface data. With satellite data alone, this problem would still be unknown.

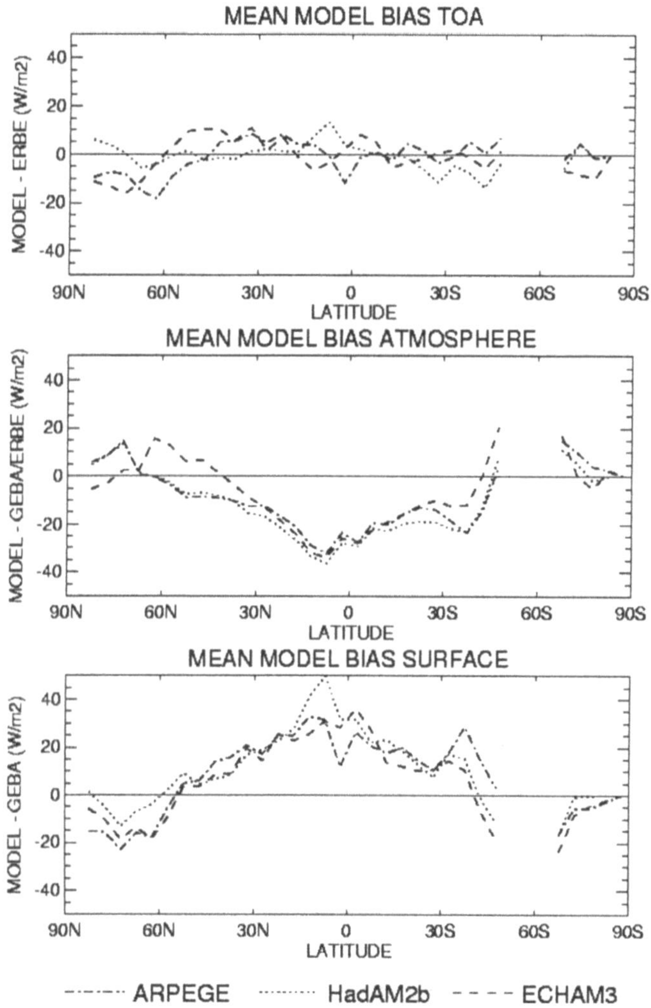


*Figure 2.* Mean model bias of shortwave absorption a) in the total surface/atmosphere system (TOA) b) within the atmosphere and c) at the surface, at 720 observational sites as function of latitude, for the 3 GCMs ARPEGE, HadAM2b and ECHAM3. Biases averaged over sites within 5° latitudinal bands. Units $Wm^{-2}$

## 4.2 The HadAM3 models

The same procedure as above is in the following applied to the HadAM3 model, which is the latest model version of the Hadley Centre. This model is particularly interesting, as it has been running both without aerosols and with a (simple) aerosol climatology. This allows the specific assessment of the effects of a simple aerosol climatology on the shortwave radiation budgets, as discussed in details in Cusack et al. (1998).

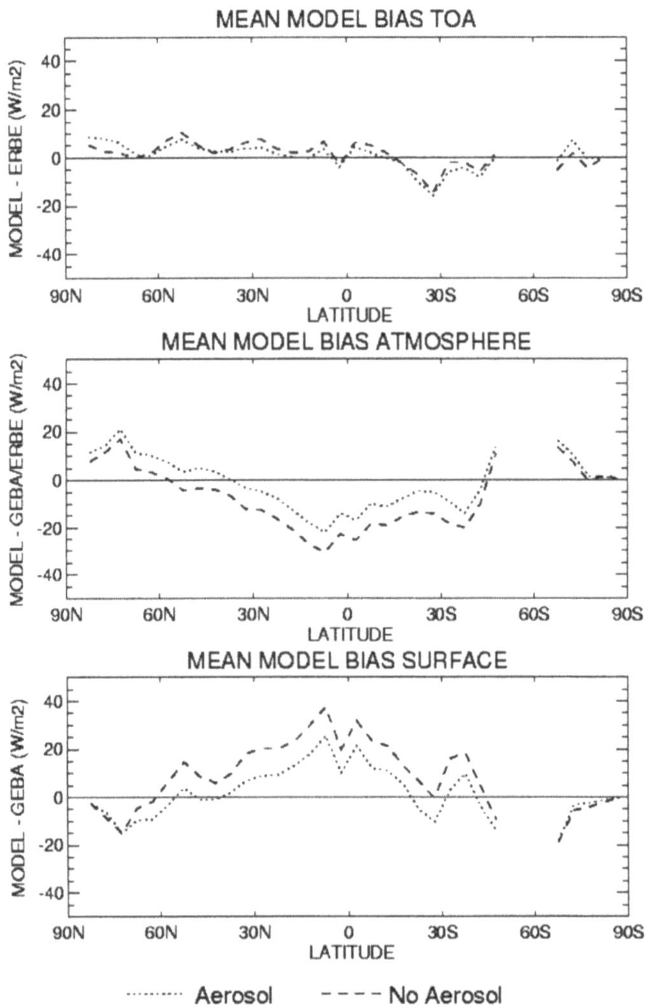


*Figure 3.* As Figure 2, but for HadAM3 with and without aerosol

At the TOA both model versions are in good agreement with the satellite data (Fig 3a), as the other GCMs before. The shortwave absorption within the atmosphere is enhanced in the model which uses an aerosol climatology. This helps to reduce the biases compared to the observational estimates in the atmosphere (Fig. 3b) and at the surface (Fig. 3c), at least at lower latitudes. This suggests that the high value of global mean atmospheric absorption in HadAM3 (76 Wm-2, cf. Table 1) is closer to reality than the lower values of the other GCMs. Still, a significant lack of atmospheric absorption remains at low latitudes also in the HadAM3 model with aerosol, similar to the other models in Fig. 2. The simple aerosol climatology is therefore not capable of entirely removing the absorption biases. A more detailed analysis (Wild and Slingo. in prep.) shows, that this lack of atmospheric absorption is largely restricted to areas and seasons with extreme loadings of aerosols, such as from biomass burning or desert storms (cf. Wild 1999). Such regional and seasonal aerosol peaks are not considered in the simple aerosol climatology used in HadAM3, which includes no seasonal and spatial resolution.

# 5. DISCUSSION

In general, the lack of absorption in the GCM atmospheres discovered above can either be due to a lack of absorption in the cloud-free atmosphere, or in the clouds. This aspect is further elaborated in the following.

## 5.1 Absorption in the cloud-free climate system

To estimate the contribution of the cloud-free atmosphere to the biases discussed above, we are currently constructing an observational dataset of clear-sky climatologies at selected sites worldwide (cf. Section 2). These climatologies are obtained from composites of cloud-free episodes at sites with measurements of high quality and high temporal resolution. So far, clear-sky climatologies have been established for 7 sites in Germany with long-term observational records of hourly data. The sites included are Norderney (53.72° N, 7.15° E), Hamburg (53.63° N, 10.00°E), Brauschweig (52.30° N, 10.45°E), Braunlage (51.72° N, 10.53°E), Trier (49.75° N, 6.67° E), Wuerzburg (49.89° N, 11.73°E) and Weihenstephan (48.40° N, 11.73°E) (Liepert et al. 1994). Combining thc surfacc clcar-sky climatologies at thesc sites with TOA clear-sky climatologies from ERBE allows an estimate of the shortwave absorption within the cloud-free atmospheric column above the sites.

The comparisons in the following are based on annual averages of model-calculated and observed fluxes. Since the biases at each individual site were found to be very similar, results will be shown in terms of averages over the seven German sites.

Annual mean shortwave absorption under cloud-free condition at the surface, averaged over the seven sites, are shown in Fig. 4a, for the five GCMs and the observations. The GCM values cover a range of 22 $Wm^{-2}$. There is a general tendency in the models to significantly overestimate the clear-sky surface absorption, with biases up to 25 $Wm^{-2}$, and an average overestimation of 13 $Wm^{-2}$.

The net fluxes at the TOA, i.e. the rate of absorption of the entire Earth system, cover a range of 13 $Wm^{-2}$ under clear-sky conditions, thus they are more consistent than the surface values (Fig. 4b). The observed estimate is only 1 $Wm^{-2}$ lower than the average over the model values, and the maximum bias is reduced to 7 $Wm^{-2}$, compared to 25 $Wm^{-2}$ at the surface. Hence, the problem in the models is not so much one of capturing the total amount of solar energy in the cloud-free surface-atmosphere column, but rather one of partitioning this energy between atmosphere and surface. This is visible in Fig. 4c, where a direct comparison of atmospheric absorption (the difference between the absorption at the TOA and at the surface) is shown. The models show a tendency to underestimate the absorption in the cloud-free atmospheric column, with biases up to 18 $Wm^{-2}$, and an average underestimation of 12 $Wm^{-2}$. Thus, the excessive surface absorption can be predominantly attributed to a lack of absorption of solar radiation in the cloud-free atmosphere and, only to a lesser extent, due to insufficient backscattering of solar radiation to space. This behavior under clear-sky conditions is thus very much the same as found previously under all-sky conditions on more global scales (Section 4). This further emphasizes the importance of the cloud-free atmosphere in the discussion of the "anomalous atmospheric absorption" phenomenon in the models.

The principal absorbers of solar radiation in the cloud-free atmosphere are water vapor and aerosols. Since water vapor is abundant in the GCMs at the German sites, as shown in Wild and Liepert (1998), the lack of atmospheric absorption in the GCMs must be due to the lack of aerosol absorption or deficiencies in the radiation codes themselves. Such deficiencies have been detected in stand-alone validations of the radiation scheme for the ECHAM3 GCM, which showed an excess insolation even with correctly prescribed atmospheric input profiles of humidity and temperature from radiosondes (Wild et al. 1995, 1998a). An overestimation of surface insolation under cloud-free conditions was also found in other radiation codes (e.g., Kato et al. 1997, Kinne et al. 1998). Similar deficiencies may therefore be present in many GCMs.

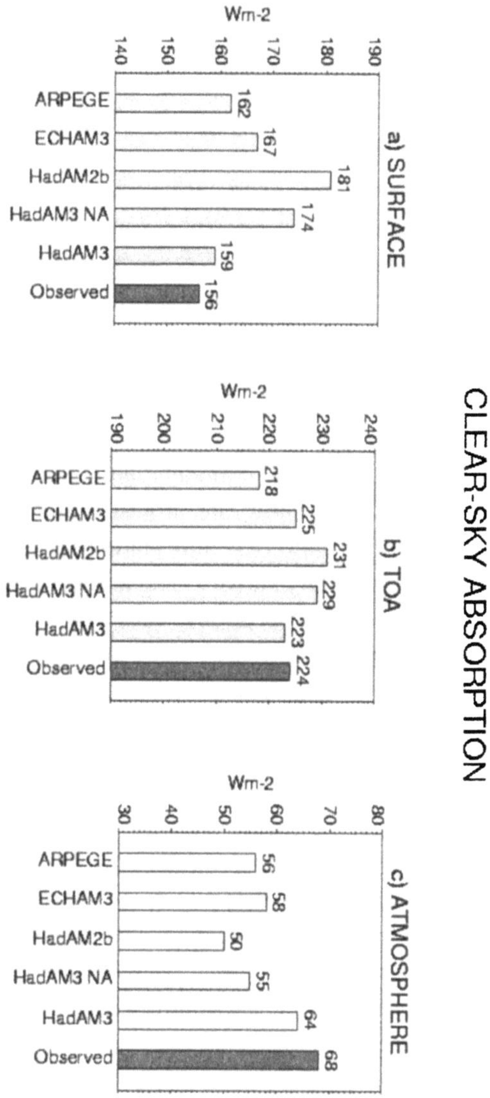


*Figure 4.* Annual mean shortwave absorption under clear-sky conditions averaged over 7 sites in Germany as calculated in different GCMs and derived from observations: a) at the surface, b) at the TOA, c) within the atmosphere. GCMs include ARPEGE (Déqué et al. 1994), ECHAM3 (Roeckner et al. 1992) HadAM2b (Stratton 1999) and HadAM3/ HadAM3 NA (Cusack et al. 1998) NA referes to HADAM3 without aerosols

The latest model version of the Hadley Centre, the HadAM3, shows an increased atmospheric clear-sky absorption when compared to the precursor version HadAM2b and is now in close agreement with the observational estimate (Fig. 4c). An increase in atmospheric absorption of 5 $Wm^{-2}$ is already noticeable in HadAM3 NA which, as HadAM2b, includes no aerosols. This is related to the improved radiation code of Edwards and Slingo (1996), which calculates higher water vapor absorption compared to the code used in HadAM2b (A. Slingo, S. Cusack personal communication, 1999). While in many GCMs the lack of water vapor absorption is a major cause for the lack of shortwave absorption in the atmosphere, this has no longer the same relevance for HadAM3. A higher water vapor absorption is also found in the radiation code used in the ECMWF model (Morcrette 1991), which was shown to be in excellent agreement with direct observations in stand-alone validation studies (Wild et al 1998b).

The above comparisons also suggest that, in addition to a proper treatment of water vapor absorption, the inclusion of aerosols is essential to bridge the gaps between model-calculated and observed estimates of atmospheric clear-sky absorption. The models which do not include aerosols (HadAM2b and HadAM3 NA) show the largest underestimates, while the introduction of an aerosol climatology into HadAM3 leads to an excellent agreement with the observational estimate. The aerosol effect in HadAM3 (9 $Wm^{-2}$ additional atmospheric absorption) is of comparable magnitude to other modeling studies (e.g., Garratt et al. 1998 and references therein). The inclusion of aerosols in HadAM3 leads also to closer agreement with observations at the TOA, due to the increased reflectance (Fig. 4b), in line with the findings in Cusack et al. (1998). However, it should be noted, that although this simple aerosol climatology may capture the annual mean aerosol effect at the German sites with moderate aerosol loading adequately, this no longer applies in areas with very high aerosol loading, particularly in the Tropics (Wild 1999, cf. Section 4.2).

In summary, the above indicates that GCMs should be equipped with both a state-of-the art radiation code and a sophisticated aerosol climatology in order to avoid the biases in clear-sky atmospheric absorption typically found in current GCMs.

## 5.2 Absorption in the cloudy atmosphere

Under all-sky conditions at the German sites, a similar comparison of the annual mean shortwave absorption at the surface, at the TOA and within the atmosphere is shown Fig. 5.

The presence of clouds tends to increase the inter-model differences, which is particularly evident at the TOA (Fig 5b). Again significant deviations from the observed estimates become apparent. The presence of clouds tends to increase somewhat the atmospheric absorption in the models (cf. Fig. 4c and 5c), whereas the observational estimates of atmospheric absorption derived under clear and cloudy conditions are very similar.

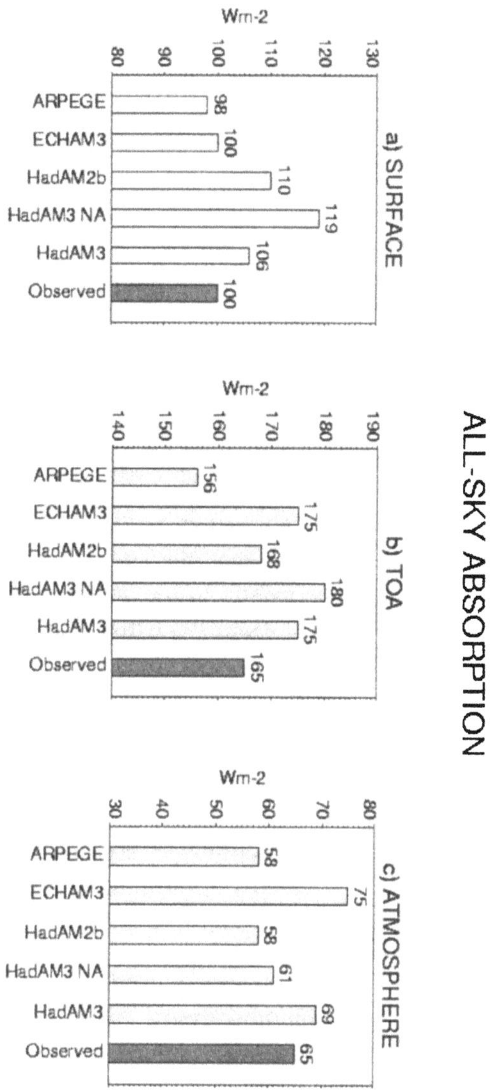


*Figure 5.* As Fig. 4 but for all-sky conditions

The increase in atmospheric absorption in the cloudy model atmospheres therefore partly compensate (for CNRM, HadAM2b, HadAM3 NA) or even overcompensate (ECHAM3) for the underestimated absorption in the cloud-free atmosphere.

A convenient measure for the overall effect of clouds on the shortwave atmospheric absorption is the ratio R of shortwave cloud radiative forcing at the surface to that at the TOA (Cess et al. 1995, for limitations of this concept see Li et al. 1995):

$$R = \frac{SW(surf.)_{all-sky} - SW(surf.)_{clear-sky}}{SW(TOA)_{all-sky} - SW(TOA)_{clear-sky}}$$

R equal one states that the presence of clouds does not alter the overall absorption in the atmospheric column. The magnitude of R is currently highly controversial, and previously published estimates range from R close to 1.0 (e.g., Li et al. 1995) to as high as 1.5 (e.g., Cess et al. 1995, Ramanathan et al. 1995), the latter suggesting a much higher absorption of solar radiation in the cloudy- than in the cloud-free column. Values of R calculated by the GCMs and derived from the observational estimates for the German sites are given in Fig. 6. The GCM values are in the range of 1.03 (ARPEGE) to 1.34 (ECHAM3), i.e. the inclusion of clouds leads to a certain increase in shortwave absorption in the GCM atmospheres which is, however, substantially below the 1.5 suggested in Cess et al. (1995). The observational estimates at the German sites, on the other hand, favor a value of R close to unity (Fig. 6), suggesting that the presence of clouds does not significantly alter the overall absorption in the atmospheric column, at least not for the region under consideration. Note that equal absorption in the cloudy and the cloud-free atmosphere does not exclude the possibility that the clouds themselves show an enhanced absorption. It rather states that a possible additional absorption by clouds is offset by the cloud shading which prevent the photons from entering deeper into the atmosphere, thereby lowering the chance of being absorbed (cf. e.g., Li et al. 1995). At the German sites, the latter effect even seems to dominate slightly, resulting in a somewhat lower absorption in the cloudy atmosphere than in the cloud-free atmosphere and R = 0.97. The present results are thus in line with the findings of Li et al. (1995) who also determined values for R lower than one particularly at high latitudes. They used a satellite-derived estimate as a reference for the clear-sky surface absorption, while the present study uses direct observations of the same quantity. For the German area specifically, Li et al. (1995) obtained R = 1.1, which is somewhat higher than the value of 0.97 found here. This difference may be explained by the neglect of aerosol in the algorithm used in Li et al. (1995) to derive surface clear-sky absorption from satellites, thereby slightly underestimating the shortwave absorption particularly in the cloud-free atmosphere (Z. Li., personal communication 1999).

On the other hand, there are no indications in the information provided by the direct measurements which would support R values as high as 1.5. A value of R equal 1.5 implies for the German sites a difference of 30 $Wm^{-2}$ between the cloudy and the cloud-free atmospheric absorption. It seems highly unlikely that the uncertainties inherent in the observational estimates could mask a signal of such magnitude. The observational estimates therefore provide no indications that cloud absorption should be higher than presently found in the GCMs.

This suggests that the underestimated shortwave absorption in the GCM atmosphere in Figs. 2 and 3 are predominantly caused by a lack of absorption in the cloud-free part of the atmosphere rather than in the cloudy part. This is related to an underestimation of water vapor absorption in GCMs which use radiation schemes based on older spectroscopic data, and additionally to an inadequate representation of aerosol absorption particularly in areas with high aerosol loading.

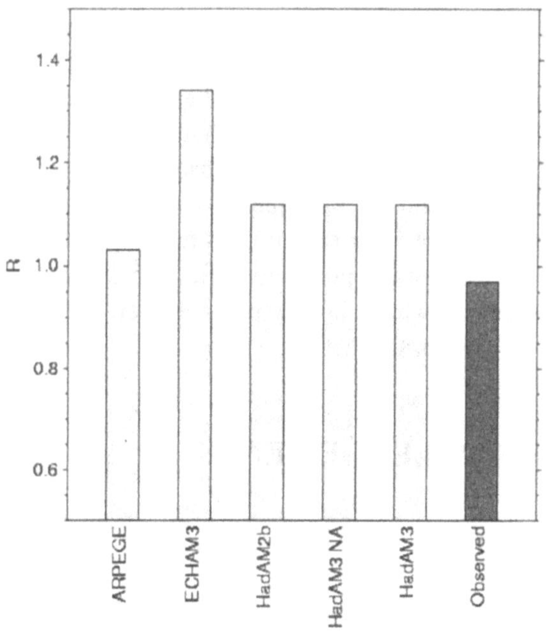


*Figure 6.* Ratio R of shortwave cloud radiative forcing at the surface to that at the TOA (annual mean). GCMs as referenced in Fig. 4

# 6. SUMMARY AND CONCLUSIONS

A comprehensive dataset of collocated surface and satellite observations has been used to assess the distribution of solar radiation in GCMs. Data from 720 sites present evidence that the GCMs have no problems in simulating the total amount of solar energy absorbed in the climate system correctly, but that the relative fraction of absorption within the atmosphere and at the surface is often substantially biased. The GCM atmospheres are generally too transparent for solar radiation: the global mean values of absorbed solar radiation within the atmosphere, typically around 70 $Wm^{-2}$ in GCMs, have been estimated to be too low by 10 - 20 $Wm^{-2}$. This puts the most likely value of global mean shortwave atmospheric absorption in the real world, a highly controversial number, to 80 - 90 $Wm^{-2}$, in line with the estimates derived in Wild et al. (1998).

The shortwave absorption was separately assessed for cloudy and cloud-free conditions in the atmospheric columns above selected observation sites in Germany. Based on collocated surface and satellite observations, estimates of shortwave atmospheric absorption for both clear-sky and all-sky conditions were obtained. The GCMs typically absorb too little solar radiation, not just in the cloudy, but also in the cloud-free atmosphere. Increased absorption in state-of-the-art radiation codes and the additional inclusion of absorbing aerosol help to narrow the gaps between simulated and observed estimates of clear-sky absorption. Under cloudy (all-sky) conditions, a significant increase of shortwave atmospheric absorption is not detectable in the observations. Therefore no direct evidence is found here that clouds in the GCMs absorb insufficient solar radiation as recently postulated in other studies. Rather, the present study points to deficiencies in the absorption in the cloud-free atmosphere which are responsible for the lack of shortwave absorption in the GCM atmospheres.

## 7. ACKNOWLEDGEMENTS

Dr. B. Liepert, Columbia University New York, kindly provided the clear-sky time series of the German sites. I am grateful to Prof. A. Ohmura for his support of this study. Dr. Hans Gilgen put enormous efforts into the build-up of the GEBA database. Thanks to Drs. M. Déqué and R. Stratton for making available the output of the Météo-France and UKMO GCMs within the framework of the EU project HIRETYCS. Special thanks to Drs. A. Slingo and S. Cusack, Hadley Centre for Climate Prediction and Research, for providing the results from their simulations with the HadAM3 model. The Swiss Scientific Computing Center (CSCS) generously provided the necessary computer resources for the ECHAM simulations. This study is supported by the ETH Schulleitung (Prof. A. Waldvogel) who financed the author's position. I would like to thank Prof. Martin Beniston and Dr. Michel Verstraete for the organization of the very stimulating workshop in Les Diablerets.

# 8. REFERENCES


Arking, A., 1996: Absorption of solar energy in the atmosphere: Discrepancy between model and observations. Science, **273**, 779-782.

Barkstrom, B.R., E.F. Harrison and R.B. Lee III, 1990: Earth Radiation Budget Experiment. EOS, **71**, 297-305.

Cess, R.D., M.H. Zhang, P. Minnis, L. Corsetti, E.G. Dutton, B.W. Forgan, D.P. Garber, W.L. Gates, J.J. Hack, E.F. Harrison, X. Jing, J. T. Kiehl, C.N. Long, J.-J. Mocrette, G.L. Potter, V. Ramanathan, B. Subasilar, C.H. Whitlock, D.F. Young and Y. Zhou, 1995: Absorption of solar radiation by clouds: observations versus models. *Science*, **267**, 496-499.

Cusack, S., A. Slingo, A., J.M. Edwards, M. Wild, 1998: The radiative impact of a simple aerosol climatology on the Hadley centre atmospheric GCM. *Quart. J. Roy. Met. Soc.*, **124**, 2517-2526.

Darnell, W.L., W.F. Staylor, S.K. Gupta, N.A. Ritchey, and A.C. Wilber, 1992: Seasonal variation of surface radiation budget derived from International Satellite Cloud Climatology Project C1 data. *J. Geophys. Res.*, **97**, 15741-15760.

Déqué, M., C. Dreveton, A. Braun and D. Cariolle, 1994: The ARPEGE/IFS atmosphere model: a contribution to the French community climate modelling. *Climate Dynamics*, **10**, 249-266.

Edwards, J.M., and A. Slingo, 1996: Studies with a flexible new radiation code: I: Choosing a configuration for a large scale model. *Quart. J. Roy. Meteor. Soc.*, **122**, 689-719.

Gates, W. L., 1992: AMIP: The atmospheric model intercomparison project. *Bull. Amer. Meteor. Soc.*, **73**, 1962-1970.

Garratt, J. R., 1994: Incoming shortwave fluxes at the surface - a comparison of GCM results with observations. *J. Climate*, **7**, 72-80.

Gilgen, H., M. Wild and A. Ohmura, 1998: Means and trends of shortwave irradiance at the surface estimated from Global Energy Balance Archive data. *J. Climate*, **11**, 2042-2061.

Gilgen, H., and A. Ohmura, 1999: The Global Energy Balance Archive *Bull. Amer. Meteor. Soc.*, **80**, 831-850.

Hense, A., M. Kerschgens, and E. Raschke, 1982: An economical method for computing radiative transfer in circulation models. *Quart. J. Roy. Meteor. Soc.*, **108**, 231-252.

Kato, S., T.P. Ackerman, E.E: Clothiaux, J.H. Mather, G.G.Mace, M.L. Wesley, F. Murcray, and J. Michalsky, 1997: Uncertainties in modeled and measured clear-sky surface short-wave irradiances. *J. Geophys. Res.*, **102** (D22), 25881-25898.

Kinne, S., R. Bergstrom, O.B. Toon, E. Dutton, and M. Shiobara, 1998: Clear-sky atmospheric solar transmission: an analysis based on FIRE 1991 field experiment data. *J. Geophys. Res.*, **103** (D16), 19709-19720.

Konzelmann, T., D.R. Cahoon, and C.H. Whitlock, 1996: Impact of biomass burning in Equatorial Africa on the downward surface shortwave irradiance: observations and calculations. *J. Geophys. Res.*, **101**(D1), 22833-22844.

Li, Z., H. Barker and L. Moreau, 1995: The variable effect of clouds on atmospheric absorption of solar radiation. *Nature*, **376**, 486-490.

Liepert, B., P. Fabian and H. Grassl (1994): Solar radiation in Germany- observed trends and an assessment of their causes. Part 1: regional approach. *Contrib. Atmos. Phys.*, **67**, 15-29.

Morcrette, J.J., 1991: Radiation and cloud radiative properties in the European centre for medium range weather forecasts forecasting system. *J. Geophys. Res .*, **96**, 9121-9132.

Ohmura, A, and H. Gilgen, 1993: Re-evaluation of the global energy balance. *Geophysical Monograph 75,* IUGG Volume **15**, 93-110.

Ohmura, A., H. Gilgen and M. Wild, 1989: Global Energy Balance Archive GEBA, World Climate Program - Water Project A7, Report 1: Introduction. Zuercher Geografische Schriften Nr. 34, Verlag der Fachvereine, Zuerich, 62pp.

Ramanathan, V., B. Subasilar, G. Zhang, W. Conant, R. Cess, J. Kiehl, H. Grassl and L. Shi, 1994: Warm pool heat budget and shortwave cloud forcing: a missing physics? *Science*, **267**, 499-503.

Roeckner, E., K. Arpe, L. Bengtsson, S. Brinkop, L. Dümenil, M. Esch, E. Kirk, F. Lunkeit, M. Ponater, B. Rockel, R. Sausen, U. Schlese, S. Schubert and M. Windelband, 1992: Simulation of the present day climate with the ECHAM3 model: impact of model physics and resolution. Max Planck Institute for Meteorology Report No. 93, 171 pp.

Rossow, W.B., and Y. C. Zhang, 1995: Calculation of surface and top of atmosphere radiative fluxes from physical quantities based on ISCCP data sets. Part II: Validation and first results. *J Geophys. Res.*, **100** (D1), 1167 - 1197.

Slingo, A., and R.C. Wilderspin, 1986: Development of a revised long-wave radiation scheme for an atmospheric general circulation model. *Quart. J. Roy. Meteor. Soc.*, **112**, 371-386.

Stephens, G., 1996: Correspondence - how much solar radiation do clouds absorb? *Science*, **271**, 1131.

Stratton, R.A., 1999: A high resolution AMIP integration using the Hadley Centre model HadAM2b. *Climate Dynamics* , **15**, 9-28.

Wild, M., A. Ohmura, H. Gilgen, and E. Roeckner, 1995: Validation of GCM simulated radiative fluxes using surface observations. *J. Climate*, **8**, 1309-1324.

Wild, M., L. Dümenil, and J.P. Schulz, 1996: Regional climate simulation with a high resolution GCM: surface hydrology. *Climate Dynamics*, **12**, 755-774.

Wild M., 1997: The heat balance of the Earth in GCM simulations of present and future climate. Zuercher Geografische Schriften Nr. 68, Verlag der Fachvereine, Zuerich, 188 pp.

Wild M., A. Ohmura, and U. Cubasch, 1997: GCM simulated surface energy fluxes in climate change experiments. *J. Climate*, **10**, 3093-3110.

Wild, M., A. Ohmura, H. Gilgen, E. Roeckner, M. Giorgetta, and J.J. Morcrette, 1998a: The disposition of radiative energy in the global climate system: GCM versus observational estimates.*Climate Dynamics* , **14,** 853-869.

Wild, M., A. Ohmura, H. Gilgen, and J.J. Morcrette, 1998b: The distribution of solar energy at the Earth's surface as calculated in the ECMWF Re-analysis. *Geophysical Research Letters*, **25**, 4373-4376.

Wild, M. and B. Liepert, 1998: Excessive transmission of solar radiation through the cloud-free atmosphere, *Geophysical Research Letters*, **25**, 2165-2168.

Wild, M., 1999: Discrepancies between model-calculated and observed shortwave atmospheric absorption in areas with high aerosol loadings. *J Geophys. Res.*, **104** (D22), 27361-27371.

World Meteorological Organisation (WMO), 1983: Report of the experts meeting on aerosols and their climatic effects, WCP-55, 107 pp.

# How well do aerosol retrievals from satellites and representation in global circulation models match ground-based AERONET aerosol statistics?

S. KINNE[1], B. HOLBEN[2], T. ECK[3], A. SMIRNOV[4], O. DUBOVIK[4], I. SLUTSKER[4], D. TANRE[5], G. ZIBOZDI[6], U. LOHMANN[7], S. GHAN[8], R. EASTER[8], M. CHIN[9], P. GINOUX[2], T. TAKEMURA[10], I. TEGEN[11], D. KOCH[12], R. KAHN[13], E. VERMOTE[14], L. STOWE[15], O. TORRES[1], M. MISHCHENKO[12], I. GEOGDZHAYEV[12] and A. HIRAGUSHI[16]

[1]*UMBC-JCET / NASA-Goddard*
[2]*NASA-Goddard*
[3]*Raytheon Corp. / NASA-Goddard*
[4]*SSAI / NASA-Goddard*
[5]*University of Lille, France*
[6]*JRC, Ispra, Italy*
[7]*Dalhousie University, Canada*
[8]*Batelle-PNNL*
[9]*GIT / NASA-Goddard*
[10]*University of Tokyo, Japan*
[11]*MPI Jena, Germany*
[12]*NASA-GISS*
[13]*NASA-JPL*
[14]*University of Maryland / NASA-Goddard*
[15]*NOAA / NESDIS*
[16]*NIES Tsukuba, Japan*

Abstract: Statistics from sky/sunphotometers at AERONET sites throughout the world provide the background for a comparison of monthly or seasonally averaged aerosol optical depths to retrievals by operational satellites and to representations in global models. Available data-sets, however, rarely relate to the same year(s). With strong year-to-year variations even for monthly averaged aerosol optical depths and open issues on sampling biases and regional representation by local measurements only larger discrepancies are investigated.

Aerosol optical depths retrievals of five different satellites and five different global models are compared. Quantitative accurate satellite retrievals over land remain a challenge and even their relative difference cannot provide clear answers on regional representation. Model predicted aerosol optical depth



*M. Beniston and M.M. Verstraete (eds.),*
*Remote Sensing and Climate Modeling: Synergies and Limitations,* 103–158.

averages are usually smaller than AERONET. The behavior of models is further explored on a component basis. For sulfate, dust, carbon and sea-salt optical depths, mass and assumed aerosol sizes are compared. For the conversion of the column (dry) component mass in optical depth in models, assumptions for component aerosol size and aerosol humidification are critical.
Statistical comparisons to ground-based monitoring will be more useful, if temporal differences are removed. This requires data from the same time-period and the use of sampling screens, to accommodate less frequent measurements. For the understanding of regional representation by local measurements, satellite data play a key role. Necessities to validate critical aerosol assumptions in models or satellite retrievals require field- experiments that focus on individual aerosol components plus continued and additional monitoring (e.g. AERONET) at sites, where a particular aerosol component dominates.

# 1. INTRODUCTION

Tropospheric aerosol imposes one of the least understood impacts on the Earth's climate. The limited lifetime (commonly on the order of days) of aerosol and uncertain strengths of aerosol sources and aerosol removal processes, including interactions with clouds and effects on heterogeneous chemistry, make it very difficult to define characteristic aerosol properties. Yet aerosol properties (and aerosol induced changes to cloud properties) are needed at high accuracy on global and seasonal scales to create confidence in model derived predictions on the aerosol climatic impact and the climate change attributed to anthropogenic aerosol.

The characterization of aerosol concentration and aerosol properties in these models is highly parameterized and frequently far from reality. Some of the aerosol information in these models due to their need for global coverage is based on satellite data. These data by themselves carry significant uncertainties. It is our goal to demonstrate uncertainties and to reveal misrepresentation as part of a collaborative intercomparison.

A central piece for this intercomparison is a statistics provided by the AErosol RObotic NETwork. AERONET is a network of automatic sun/sky radiometers distributed throughout the world, whose data are centrally monitored, maintained and archived at NASA-Goddard. Probably the most meaningful aerosol property (also from a visual [reduced visibility] point of view) is the mid-visible aerosol optical depth. Here, aerosol optical depths at selected AERONET sites are compared to representations in models and to satellite retrievals near those sites. A statistical approach was selected, because local (AERONET-) measurements only represent a sample within regions of either the footprint of a satellite pixel or of areas represented by a grid-point in global circulation models. Aside from spatial inconsistencies there are also usually differences in time. Most model simulations or satellite data-sets relate (back) to years for which AERONET data were not available. Monthly and seasonal time-averages were chosen, also in part as global aerosol satellite products and aerosol properties in global models are commonly presented for these averages.

First, AERONET data, which provide the basis for the aerosol optical depth intercomparison are introduced. Special attention is given to uncertainties regarding the statistics. Then, comparisons to and among currently available operational satellite data are discussed. Finally, comparisons to and among five global models are presented. Since all models distinguish among sulfate, carbon, dust and sea-salt, more details on the model-behavior could be deduced from comparisons on a component level.

## 2. AERONET

AERONET is a federated worldwide network of sun/sky-photometers that are monitored and maintained at the NASA-Goddard Space-Flight Center (Holben et al. 1998). Data have been collected since 1993.

### 2.1 Selected sites

Eight continental and eight near-ocean sites with better statistics have been chosen for the intercomparison. Figure 1 illustrates their position.

*Figure 1.* World-map with the location of the sixteen selected AERONET sites

Table 1 summaries site names, site locations and site altitudes (above sea-level). Table 1 also indicates the aerosol types that are expected to dominate at these sites. Furthermore given in Table 1 are seasons (by yearly quarters) and years, for which data are available. These data form the basis of the multi-year statistics.

*Table 1.* Selected AERONET sites

| LOCATION | lat(N) | lon(E) | z(m) | Aerosol-type | year-quarters | | | | years (19--) |
|---|---|---|---|---|---|---|---|---|---|
| Mongu | -15 | 23 | 1107 | Biomass | | | 3 | 4 | 95/96/97/98 |
| Seviletta | 34 | -107 | 1477 | Rural | 1 | 2 | 3 | 4 | 94/95/96/97/98 |
| Cuiaba | -15 | -56 | 250 | Biomass | | | 3 | 4 | 94 |
| Banizoumbou | 14 | 3 | 250 | Dust /Biomass | 1 | 2 | 3 | 4 | 96 |
| Waskesiu | 54 | -106 | 550 | Boreal Biomass | | 2 | 3 | | 94/95/96/97/98 |
| Goddard | 39 | -77 | 50 | Urban-Industrial | 1 | 2 | 3 | 4 | 94/95/96/97/98 |
| Ispra | 46 | 9 | 235 | Urban-Industrial | 1 | 2 | 3 | 4 | 97/98 |
| Cart-Site | 37 | -97 | 315 | Rural | 1 | 2 | 3 | 4 | 98/99 |
| Cape Verde | 17 | -23 | 60 | Dust | 1 | 2 | 3 | 4 | 95/96/97 |
| Bermuda | 32 | -23 | 10 | Mixed | 1 | 2 | 3 | 4 | 96/98/99 |
| Dry Tortugas | 25 | -83 | 0 | Mixed | 1 | 2 | 3 | 4 | 98 |
| Lanai | 21 | -157 | 80 | Maritime | 1 | 2 | 3 | 4 | 96/98/99 |
| Dakar | 14 | -17 | 0 | Dust | 1 | 2 | 3 | | 97 |
| Kaashidhoo | 5 | 73 | 0 | Martime /Urban | 1 | 2 | 3 | 4 | 98/99 |
| Bahrain | 26 | 51 | 0 | Dust /Urban | 1 | 2 | 3 | 4 | 98/99 |
| Barbados | 13 | -60 | 0 | Dust /Maritime | 1 | 2 | 3 | 4 | 97/98/99 |

## 2.2 Measurements

The AERONET statistics is based on up to (about) 50 daytime measurements with CIMEL sun/sky-photometers (Holben et al., 2000). Measurement samples at multiple solar spectral sub-bands (.34, .38, .44, .50, .67, .87, 1.02 μm) are always immediately repeated twice in order to help identify and eliminate poor or cloud-contaminated data. These triplet samples are taken with every 0.5 airmass change at lower sun-elevations (for the 8 to 30 degree range), whereas at higher sun-elevations (above 30 degree and air-mass factors smaller than 2) triplet samples are taken (less frequently) every 15 minutes (– weather conditions permitting). Based on sharp discontinuities among triplet values and adjacent triplet averages, inadequate data (due to instrument malfunction or due to μcloud contamination) are removed (Smirnov et al. 2000). The remaining triplet averages are the basis for the AERONET statistics.

CIMEL sun/sky-photometers have a 1.2degree field of view. There are two major measurement modes, a direct mode and a scanning mode. In the direct mode the instrument is turned toward the sun and measures the attenuation of direct sun-light. In the scanning mode the instrument conducts an upward polar scan and a complete azimuth scan. The additional information from the scanning mode enables radiative transfer inversion techniques (Dubovik et al., 2000 a,b) to derive (multi-modal) aerosol size distributions and to provide estimates for aerosol absorption. Future AERONET statistics will include these properties. Here however, AERONET statistics are solely based on two properties from the direct measurement mode:

$\tau$ the aerosol optical depth at a mid-visible wavelength of 0.55μm and
$\alpha$ the Angstrom parameter for the 0.44μm to 0.87μm spectral range.

The Angstrom parameter $\alpha$ (defined as the negative slope in log{optical depth}-log{wavelength}-space) captures the change in optical depth with wavelength. At visible wavelengths (0.3μm to 0.7μm) this spectral dependency is also indicative of particle size: Clouds and large dust aerosol display little to no spectral dependence ($\alpha$ is close to 0). Atmospheric particles smaller than cloud droplets, however, display a decrease in optical depth with increases to (visible) wavelengths. The magnitude of the spectral dependency depends on particles size and absorption ($\alpha$ between 0.5 and 2 are common to aerosol). Very small and non-absorbing particles have the largest spectral dependency ($\alpha = 4$). Angstrom parameters approaching such a large value are characteristic for scattering of sun-light on air-molecules, and the strong spectral dependency of these scattering processes is demonstrated by the blue color of the sky.

Data of the Angstrom parameter $\boldsymbol{\alpha}$ will be used to rescale measured or retrieved aerosol optical depths taken at wavelengths other than 0.55μm.

## 2.3 Monthly statistics

AERONET site statistics for average and standard deviation in this study are based on all measurements (not daily averages) during a month independent of year. However, the number of accepted measurements and the number of days were tracked on which triplet averages contributed to the monthly statistics - in an effort to eliminate poor monthly statistics. Monthly averages were accepted only, if at least 8 different days and a minimum number of measurements contributed. The minimum number of measurements for a month was set to 100. With restrictions to selected daytime periods fewer measurements were permitted. Based on these selection criteria examples are given in Figures 2a to 2d for a sulfate aerosol dominated urban-industrial site (GSFC or Goddard), for a carbon aerosol dominated (seasonal) biomass burning site (Mongu), a dust aerosol dominated site (Cape Verde) and a maritime site (Lanai) with significant sea-salt aerosol.

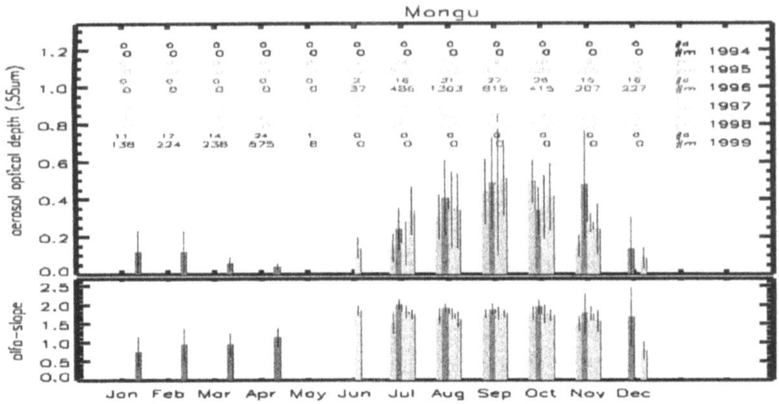
Mongu
aerosol optical depth (.55um)
alfa-slope
1994
1995
1996
1997
1998
1999
Jan Feb Mar Apr May Jun Jul Aug Sep Oct Nov Dec

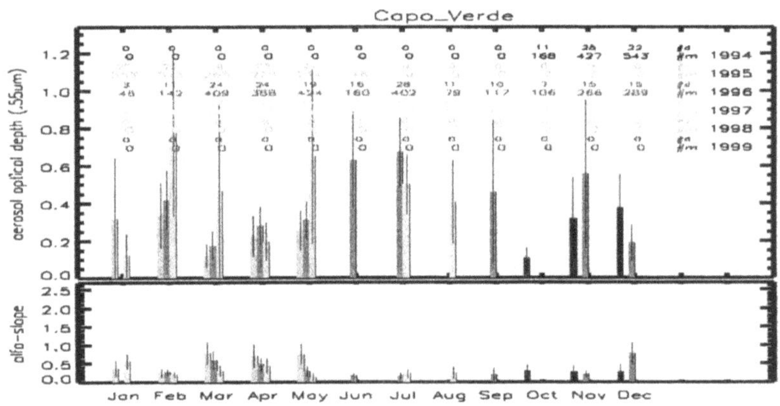
Capo_Verde
aerosol optical depth (.55um)
alfa-slope
1994
1995
1996
1997
1998
1999
Jan Feb Mar Apr May Jun Jul Aug Sep Oct Nov Dec

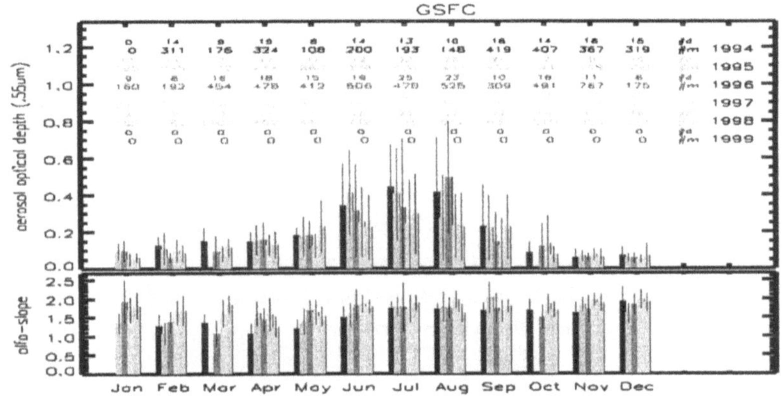
GSFC
aerosol optical depth (.55um)
alfa-slope
1994
1995
1996
1997
1998
1999
Jan Feb Mar Apr May Jun Jul Aug Sep Oct Nov Dec

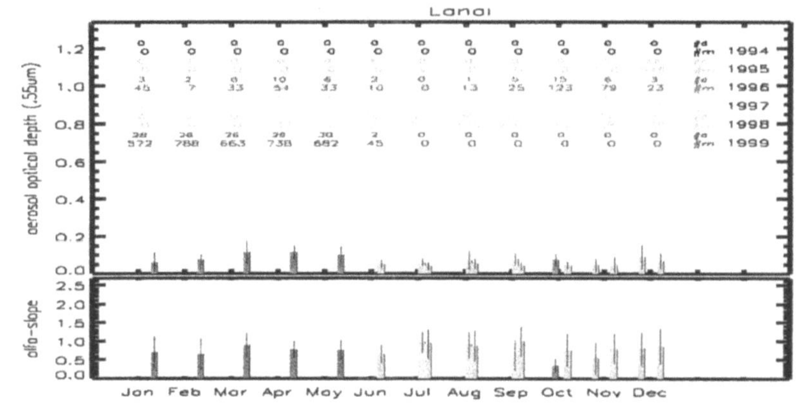


*Figure 2a-d.* Monthly averages (columns) and standard deviations (thin lines) for aerosol optical depth (upper panel) and Angstrom parameter (lower panel) from AERONET measurements between 1994 and 1999. Numbers in the upper panels indicate the contributing number of days (#d) and measurements (#m). Monthly averages, based on a minimum requirement of contributions by 100 measurements and 8 days, are displayed for four AERONET-sites: (a) a carbon aerosol dominated biomass burning site [**Mongu**], (b) a dust aerosol dominated site [**Cape Verde**], (c) a sulfate aerosol dominated urban-industrial site [**Goddard** or GSFC] and (d) a maritime site with significant sea-salt contributions [**Lanai**]

Figures 2 illustrate that the largest (monthly averages for) aerosol optical depths are found near major dust and biomass burning sources. Dust aerosol dominated sites are associated with small Angstrom parameters with values of less than 0.5 (often so small that dust at times is misinterpreted as clouds). These small values for dust across the visible spectrum indicate the presence of super-micrometer size aerosol particles. In contrast, biomass burning aerosol and urban-industrial aerosol are dominated by smaller particles with the sizes of a few tenth of a micrometer, posting typical Angstrom parameters between 1.5 ad 2.0 (Eck et al. 1999). Maritime sites and mixed sites, where no particular aerosol type dominates, display values between 0.5 and 1.5. Relatively stable values for the Angstrom parameter indicate only small changes to particle size (although a size-increase response due to a swelling under higher humidity may be partially lost by reduced absorption for sub-micrometer size particles).

Monthly averages for all sixteen AERONET sites are summarized later in Figures 3. Also shown is the range for monthly averages that was illustrated for four sites in Figures 2.

In contrast to the Angstrom parameter, optical depth variations are significant on a monthly, seasonally and on a year-to-year basis. The standard deviation for the monthly averaged optical depth is particular large (on the order of the average value) for dust and biomass sites. These strong variations on shorter time-scales reflect the event type character related to the limited lifetime of aerosol and changes atmospheric dynamics. Differences in dynamic patterns, however, also impact the variability on monthly and seasonal time-scales. This complicates efforts to establish characteristic monthly averages for aerosol optical depths as the baseline in comparisons to satellite data and model representations. Thus, tendencies of the AERONET statistics will be explored next. The non-representativeness of particular AERONET measurements will be termed ***bias*** (although a bias refers to a discrepancy between a measurement and a standard, which AERONET data [samples for a few years] are only in a relative sense).

## 2.4 Biases

AERONET statistics on aerosol optical depths may be biased due to its sampling nature. Thus, the following discussion focuses on trends with respect to a particular year, with respect to data from a particular daytime (e.g. satellite overpass) and with respect to regional representations (grid-points in global circulation models represent region on the order of 300km*300km). Understanding these trends or biases, separately near each site, is a major step in improving the significance of comparisons and validation efforts.

### 2.4.1 YEAR-to-YEAR-bias

Averages based on measurements of only one year may introduce a bias. For example higher averages for aerosol optical depths are expected after major volcanic eruption. Even without those years (e.g. 1992 and 1993 following the Mt.Pinatubo eruption) optical depth averages for the same month but different years can vary significantly, as illustrated in Figure 2. Primary explanations are differences in atmospheric dynamics (e.g. advection, winds, rainfall, temperature, humidity). Year-to-year variations are largest near dust and biomass sources, as differences in source-strengths also due to seasonal time-shifts contribute. Based on comparisons for months with AERONET optical depth averages from at least four different years, variations and averages appear correlated. Year-to-year variations are on the order of 20% of the monthly average with maximum differences reaching 50%. However, any identification of a yearly bias requires at least a decadal data record. Without it, AERONET averages will carry large uncertainties. Similar uncertainties apply to satellite data and model representations, if they are only based on one year. Moreover, data-sets for the same year are the exception. Thus, comparisons of AERONET statistics to satellite data and model representations will focus on major discrepancies.

### 2.4.2 DAYTIME-bias

For comparisons to the statistics of polar orbiting satellites daytime variations must be considered. These satellite data are linked to a particular local overpass time (e.g. TOMS, POLDER in the late morning, AVHRR during the afternoon). This requires a temporal filter for AERONET averages and the unfiltered use of all data can create a bias. Another daytime bias may be created since AERONET measurements are more frequent during sunrise and sunset. Based on a seasonal evaluation biases are summarized in Table 2.

*Table 2.* Daytime-biases for quarterly (1: J,F,M / 2: A,M,J / 3: J,A,S / 4:O,N,D) averaged AERONET aerosol optical depths. (σ: negative bias, π : positive bias, - : insufficient data or irregular deviations)

| | Sunrise | | | | Midday | | | | Sunset | | | | 10-11 am | | | | 2-3 pm | | | |
|---|---|---|---|---|---|---|---|---|---|---|---|---|---|---|---|---|---|---|---|---|
| Yearly quarter | 1 | 2 | 3 | 4 | 1 | 2 | 3 | 4 | 1 | 2 | 3 | 4 | 1 | 2 | 3 | 4 | 1 | 2 | 3 | 4 |
| Mongu | - | - | σ | - | - | - | π | - | - | - | - | π | - | - | π | - | - | - | π | - |
| Seviletta | - | - | - | σ | | - | π | π | - | - | σ | σ | - | - | π | π | - | - | - | - |
| Cuiaba | - | - | σ | - | - | - | π | - | - | - | σ | - | - | - | π | - | - | - | - | - |
| Banizoumbou | σ | - | - | σ | π | - | - | - | - | σ | - | - | - | - | - | - | - | - | - | - |
| Waskesiu | - | - | σ | π | - | - | π | - | - | - | σ | - | - | - | - | - | - | - | - | - |
| Goddard | - | - | π | - | - | - | σ | π | - | - | π | π | - | - | - | σ | - | - | - | - |
| Ispra | - | - | - | - | - | - | - | π | - | - | - | - | - | - | - | π | - | - | - | - |
| Cart-Site | - | - | - | - | - | - | - | - | - | - | - | - | - | - | - | - | - | - | - | - |
| Cape Verde | σ | σ | σ | σ | π | π | π | π | σ | - | - | - | π | - | - | - | - | - | - | - |
| Bermuda | - | - | - | - | - | - | - | - | - | - | π | - | - | - | - | - | - | - | σ | - |
| Dry Tortugas | - | - | - | - | - | - | - | - | - | - | - | - | - | - | π | - | - | - | σ | - |
| Lanai | - | - | σ | - | - | - | - | - | - | - | - | - | - | - | - | - | - | - | - | π |
| Dakar | - | - | - | - | - | - | - | - | - | - | - | - | - | - | - | - | - | - | - | - |
| Kaashidhoo | - | - | - | - | - | - | - | - | - | - | - | - | - | - | - | - | - | - | - | - |
| Bahrain | - | - | - | - | - | - | - | - | - | - | - | - | - | - | - | - | - | - | - | - |
| Barbados | - | - | - | - | - | - | - | π | - | - | - | σ | - | - | - | - | - | σ | - | - |

A bias for a particular time of day was identified if during two and more years deviations from daily averages (with no restrictions to the time of day) were of the same sign and exceeded 5%. The results show that daily variations are more common for land sites than for near-ocean sites. Biases of low aerosol optical depths during sunrise and largest aerosol optical depths around noon were determined for biomass and dust sites. Probable explanations are fewer fires (at biomass sites) and lighter winds (at dust sites) during the morning hours. For many sites, though, there are not sufficient data to extract a potential diurnal cycle. Nonetheless, our analysis revealed some interesting trends, such as a midday (solar elevations above 30 degree) summer minimum for an urban-industrial site (Goddard) or a midday summer maximum for a rural continental site (Seviletta).

For comparisons of AERONET statistics to model representations it must be considered that model averages include night-time data. It can be speculated that with reduced winds during the night and less biomass burning (at least towards the morning), AERONET data may be positive-biased near dust-sites and (unmanaged) biomass-sites. In contrast, higher relative humidities due to colder nighttime temperatures may cause AERONET data to be negative-biased, if water-uptake on aerosol is important (e.g. at urban-industrial sites). At a 10%-level the uncertainty introduced by ignoring day-time trends in monthly averages though is small in comparison to year-to-year variations.

#### 2.4.3 CLOUD-bias

The AERONET statistics does not include scenes, where effects from clouds interfered. In contrast, model results usually include aerosol optical depths in the vicinity of clouds. The removal of scenes with aerosol close to clouds in the AERONET data-set favors regions with lower relative humidities. This tends to bias AERONET statistics towards a lower aerosol optical depth and toward a larger Angstrom parameter, especially at urban-industrial regions, where aerosol humidification effects can be significant (Kothenruther et al., 1999).

#### 2.4.4 SPATIAL bias

Local aerosol properties are expected to differ from regional aerosol properties, even for time-averaged data. To investigate the regional representation of the local AERONET statistics, monthly and quarterly averages of satellite retrieved aerosol optical depths were compared at different regional resolutions near AERONET sites. Here, global satellite data-sets of TOMS and POLDER are applied (these are introduced and discussed later). The satellite data do not necessarily have to be quantitatively correct, because only relative changes are of interest. For each quarter impacts of two regional expansions on aerosol averages are summarized in Table 3. The left symbol categorizes the difference related to a regional expansion from about 40km*40km to 100km*100km (a commonly used resolution for global satellite data sets). The right symbol displays changes of a further expansion from 100km*100km to 300km*300km, the approximate grid size of global models.

Spatial variability is confirmed, but biases are difficult to detect. Scale related changes in aerosol optical depth averages often exceed 30%, yet they are often linked to only one of the two tested expansions. Satellite derived changes vary from quarter to quarter and sometimes even in sign. Variations in optical depth due to spatial changes appear larger for land retrievals than for ocean retrievals. The frequent disagreement among trends suggested by POLDER and by TOMS is puzzling, because data for yearly quarters 1 and 2 are even based on the same year. Also trends of the two different POLDER retrievals often differ, even though they relate to the same time-period. These initial results make it almost impossible to impose site related spatial biases. A more comprehensive study with decadal satellite data-sets are needed, to demonstrate to what degree differences between local and regional retrievals are related to variability and spatial averaging.

*Table 3.* Regional area related differences of quarterly (1: J,F,M / 2: A,M,J / 3: J,A,S / 4: O,N,D) averaged aerosol optical depths from satellite retrievals around AERONET sites. With data from POLDER and TOMS changes in averages related two different regional expansions were investigated, from 40*40km to 100*100km (left symbol) and from 100*100km to 300*300km (right symbol)

| Satellite | **POLDER** | | | **POLDER-ocean** | | | **TOMS** | | | |
|---|---|---|---|---|---|---|---|---|---|---|
| Yearly quarter | 1 | 2 | 4 | 1 | 2 | 4 | 1 | 2 | 3 | 4 |
| Mongu | ∃ = | = = | = = | ----- | ----- | ----- | ϑϑ | # = | ∃ ∃ | K ∃ |
| Seviletta | # = | ϑ = | = = | ----- | ----- | ----- | Kϑ | ∃ ϑ | = ϑ | ----- |
| Cuiaba | ∃ = | = = | = ϑ | ----- | ----- | ----- | ϑϑ | = = | = = | ϑϑ |
| Banizoumbou | = ∃ | = = | = ∃ | ----- | ----- | ----- | # ∃ | = = | ∃ = | = ∃ |
| Waskesiu | = # | = = | # ϑ | ----- | ----- | ----- | ----- | ϑϑ | = ∃ | ----- |
| Goddard | = ∃ | K ∃ | = ∃ | ----- | ----- | ----- | ϑ # | ∃ = | # ∃ | ϑ # |
| Ispra | KK | K ∃ | = ∃ | ----- | ----- | ----- | KK | # # | ∃ = | K ∃ |
| CART-site | = ∃ | = = | = = | ----- | ----- | ----- | ----- | ----- | ----- | ----- |
| Cape Verde | = = | = = | = = | = = | = = | = = | = = | = = | ∃ # | = # |
| Bermuda | = # | = = | Kϑ | = # | = = | K ∃ | ϑ = | = = | # # | ϑ = |
| Dry Tortugas | ∃ = | = = | K ∃ | = = | = # | = = | = = | # = | = ϑ | = = |
| Lanai | ----- | KK | ----- | = = | = = | = = | ϑ = | = = | # = | = = |
| Dakar | ∃ ∃ | K ∃ | # ∃ | = = | = = | = = | ϑ = | = = | K = | = = |
| Kaashidhoo | = = | K # | ∃ = | = = | = = | ∃ = | K = | # = | = = | ϑ # |
| Bahrain | = ∃ | = ∃ | = ∃ | ϑ = | # = | # = | ----- | ----- | ----- | ----- |
| Barbados | ϑϑ | ϑ = | = # | ϑ = | ϑ = | ϑ = | ∃ = | = = | ϑ ∃ | ∃ ∃ |

(**K**<*-30%*, *-30%*< ∃<*-10%*, *-10%*< = <*10%*, *10%*< #<*30%*, *30%*< **ϑ**, ----- no data)

**TOMS** data are based on a dual wavelength ultraviolet reflection retrieval for 1997.

**POLDER** data are based on a polarization retrieval for the period between Nov 1996 and Jun 1997.

**POLDER-ocean** data are based on a reflection retrieval (limited to over oceans) for the same time-period.

## 2.5 Reference data

Monthly averages for all sixteen AERONET sites are summarized in Figures 3. Also shown is the range of monthly averages, based on year-to-year comparisons, as illustrated for four sites in Figures 2.

In an effort to group the AERONET sites in sets of four by surface and dominant aerosol type, some mismatches could not be avoided. Based on the statistics for the Angstrom parameter, especially Banizoumbou and Bahrain appear misplaced. Banizoumbou is a dust-dominated site, but was added to biomass sites, because there are some carbon contributions from biomass burning between November and January (notice the increase in Angstrom parameters). Bahrain lies within a dust-dominated region, however, large Angstrom parameters indicate large contributions of sub-micrometer particles. This indicates that Bahrain data are strongly affected by local pollution, making statistics from this site less useful for regional comparisons.

Aerosol optical depth averages of Figures 3 will be the reference in comparisons to satellite data and model representations. In order to capture uncertainties from year-to-year variations, the presentations of deviations to satellite data and models in subsequent Figures will include the range of variations for monthly averages.

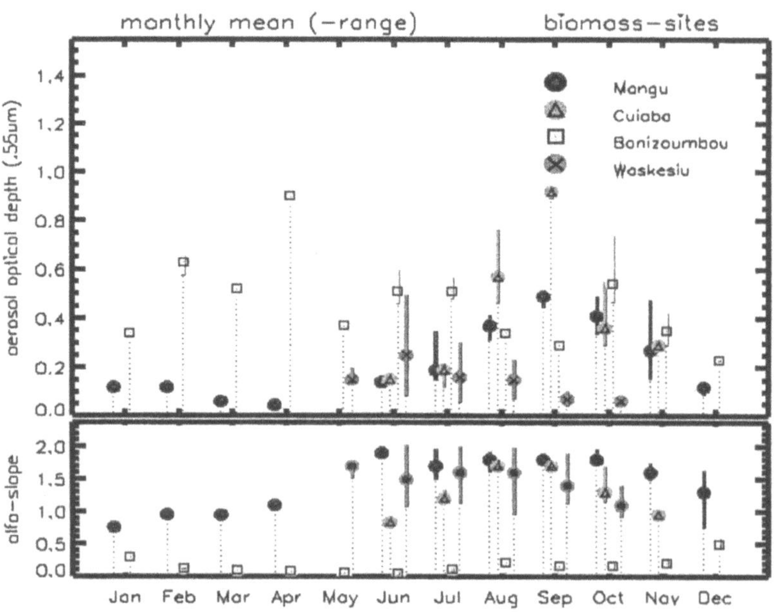

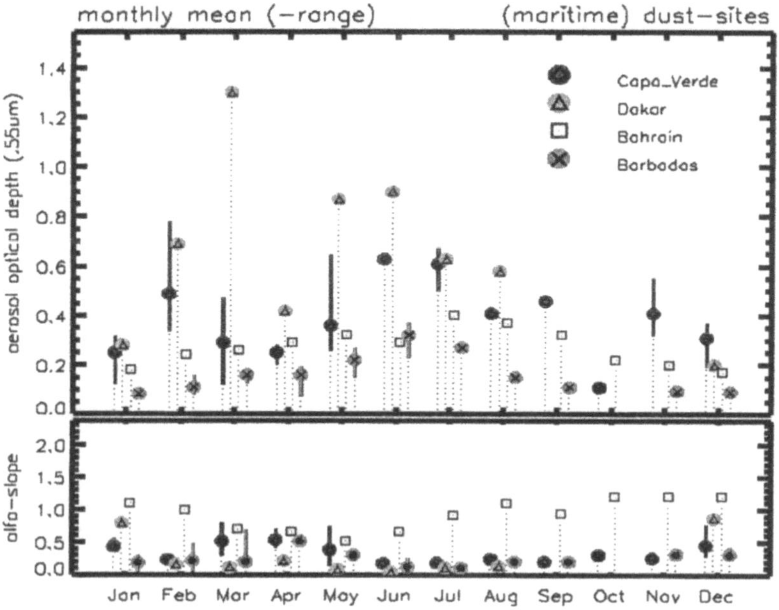
monthly mean (–range)
(maritime) dust–sites
aerosol optical depth (.55um)
alfa–slope
Capa_Verde
Dakar
Bahrain
Barbados
Jan Feb Mar Apr May Jun Jul Aug Sep Oct Nov Dec

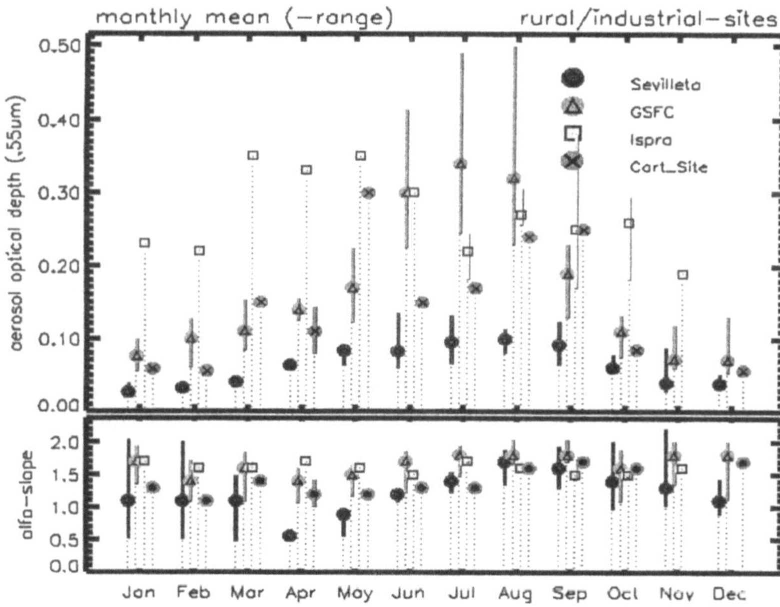
monthly mean (–range)
rural/industrial–sites
aerosol optical depth (.55um)
alfa–slope
Sevilleta
GSFC
Ispra
Cart_Site
Jan Feb Mar Apr May Jun Jul Aug Sep Oct Nov Dec

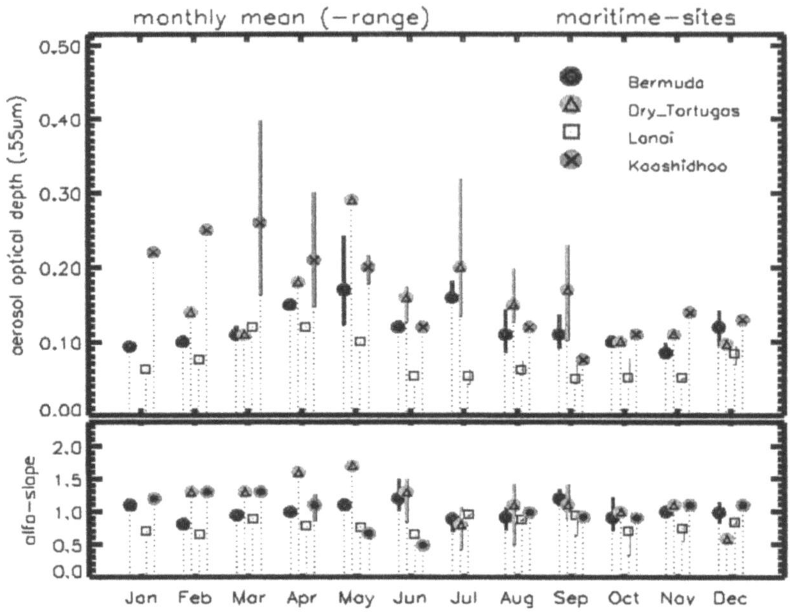


*Figure 3a-d.* Monthly averages (symbols) and range (vertical bars) for aerosol optical depths (upper panel) and Angstrom parameters (lower panel). Statistics of the sixteen AERONET-sites are presented in groups for four

# 3. SATELLITES

Remote sensing of tropospheric aerosol properties from space is difficult, as the aerosol associated backscatter signal is often at or below atmospheric (variability related) noise-levels. Considering the aerosol's (usually) submicrometer size, remote sensing is most promising at (atmospheric scattering dominated) regions of the solar spectrum. Multi-spectral radiometers and polarimeters (a few with multi-angular capabilities), as well as more recently CCD arrays, are used to probe changes in reflected sun-light in the ultraviolet, in the visible and/or in the near-infrared spectral region. However, before attributing changes in (solar) reflection to aerosol, impacts involving other contributors to solar reflection must be removed, most importantly reflections of clouds, molecular scattering and the earth's surface. Unfortunately, albedos from clouds and the earth's land surfaces are dominant modulators of solar radiation reflected to space. Thus, in aerosol retrievals from satellites these properties must be known at high accuracy. This is in sharp contrast to rather vague ideas on cloud cover and cloud microphysics or in contrast to a poor knowledge of surface conditions. To avoid these potential problems, aerosol retrievals from satellites are commonly limited to regions where albedo contributions below the aerosol can be neglected or can be expected to remain stable and small, such as over oceans.

## 3.1 Critical issues

Critical to the quality of satellite retrievals, aside from calibration issues of the instruments, are the identification of cloud-free scenes, the highly accurate representation (or elimination) of surface reflectance effects and the realism of a-priori assumptions.

### 3.1.1 CLOUD-FREE scenes

The identification and removal of cloudy scenes is usually based on combinations of spectral thresholds - including visible reflection and infrared blackbody temperatures. Nonetheless, detecting and removing all scenes with clouds can be challenge. Especially difficult is the detection of sub-pixel size clouds, whose reflection could be attributed to aerosol by mistake. This problem grows with the area of the satellite pixel and with the lack of simultaneous and co-located cloud-detecting spectral data (e.g. near-IR data for cirrus detection or far-IR data for a radiative background threshold). Techniques that import cloud-screen data from other sensors can introduce significant errors, especially if both data-sets are not co-located in time and space. Another problem arises from too stringent rules in the cloud-screening algorithm and the sub-sequent removal of aerosol scenes. A typical example is the misinterpretation of large dust particles off the African west-coast, which is commonly identified as low clouds (and which eventually leads to underestimates in aerosol optical depths for that region).

### 3.1.2 ALBEDO below aerosol

From a satellites perspective reflection from below the aerosol layer contaminates the aerosol signal. It is most desirable to eliminate these contributions. However, multi-spectral and dual-direction viewing methods have not yet become operational to provide monthly statistics (ATSR-2, MISR) and polarization methods are still in experimental stages (POLDER, see below). If impacts of surface reflectance have to be included, than it is desirable that the values for surface reflectance are small (higher sensitivity to a scattering aerosol signal) and accurate – and if possible invariant. As a rule, a 1% albedo error roughly corresponds to a visible optical depth retrieval error of 0.1. This is of the same order as the aerosol optical depth itself. Accurate surface reflectance values have to consider surface conditions (e.g.

soil, moisture, canopy, topography) and the dependence on viewing geometry. In addition, multi-spectral retrievals require information on spectral dependencies.

***Over land***, surface conditions (e.g. types of vegetation, snow, water, urban) and their associated reflectance are usually highly variable (spatially and temporally) within the footprint of a satellite pixel. Adding the dependence on viewing geometry, an accurate assessment of surface contributions to the detected signal is quite difficult. Somewhat successful are multi-spectral *dark pixel* methods that utilize for aerosol retrievals the low reflectance of green vegetation in the visible spectral region. Vegetation pixels are identified by retrievals at longer wavelengths, where contributions of (small) aerosol fade in the satellite signal. These methods are of limited success, because they assume fixed solar spectral reflectance relationships. In addition, spatial coverage of green vegetation for cloud-free scenes is usually very sparse. Other methods over land take advantage of a low surface reflectance in the ultraviolet, but requirements for data on aerosol altitude introduce added uncertainty. And, as already mentioned, polarization measurements are tested, as changes to polarization are less sensitive to surface contributions.

***Over water***, surface albedos are less variable and small, if sun-glint viewing geometries are avoided and if shallow water regions with uncertain sub-surface reflections are excluded. Then, retrieval algorithms (based on minimum reflection maps, as function of location, season, near-surface winds and viewing geometry) permit reliable estimates for surface contributions to the satellite signal.

Most current (thus, non-so) global satellite retrievals provide aerosol properties only over water. And it is usually left to models to fill the gaps over land, although over land the aerosol concentration and aerosol optical depths are largest.

### 3.1.3 Aerosol ASSUMPTIONS

The requirement for a-priori assumptions is caused by the lack of free parameters in satellite retrievals. Aerosol is defined by at least five different parameters (concentration, composition [defined by the real and the imaginary part of the refractive index], size [represented by size-distribution parameters] and shape). Characterizing aerosol is usually more complicated, because aerosol constitutes always a (frequently internally) mixture of many components. And each component has its own set of parameters. Added complexity comes from spectral dependencies (refractive index) and from dependencies on ambient relative humidity (refractive index, size) for some components. In addition, also nonsphericity must be considered for mineral aerosol (Mishchenko et al., 1997).

In contrast, a retrieval based on a single measurement contains only one piece of information, and this information is usually only valid for the spectral region of the measurement. Thus, assumptions are required for all remaining parameters that define the aerosol. Remote sensing from space at multiple spectral bands and/or multiple viewing angles and/or polarization may not provide unique solutions, but at least it can provide additional bounds for these a-priori assumptions (Mishchenko and Travis, 1997). Such advanced aerosol retrievals will possible with future satellites.

## 3.2 Satellite description

In the comparison to AERONET monthly statistics only satellite derived aerosol properties are included for which monthly averages were calculated on a routine basis in the past. Recent and future satellites and their capabilities (e.g. King et al. 1999) will not be discussed. Here, only satellites, whose data contributed with monthly averages to the intercomparison are introduced.

### 3.2.1 AVHRR

AVHRR is a 5 band (vis: 1, n-IR: 2, IR: 2) cross track scanning radiometer flown on many NOAA polar-orbiting satellites since 1978. Swath width and spatial resolution are about 2800km and 1km. Aerosol optical depths are derived from visible (0.63μm - as the nominal value for the .58-.68μm band) and near-infrared (0.83μm - as the nominal value for the 0.72-1.1μm band) reflection anomalies for cloud-free 1*4km GAC (Global Area Coverage) pixels over sun-glint free ocean scenes. Monthly averages are based on NOAA-9 (Feb.1985 - Oct.1988) data, which had fewer calibration drifts than AVHRR sensors on other NOAA satellites. Also during that time-period there was no major contamination by stratospheric aerosol from volcanic eruptions. Monthly averages include data from all four years.

***NOAA***: A 1-channel retrieval (NOAA) derives the aerosol optical depth at 0.63μm. The retrieval assumes a log-normal size distribution with a [concentration-] mode radius of 0.1μm and a standard deviation of 2.03. This size assumption translates into an effective radius $r_e$ of 0.35μm and an AERONET comparable .44μm/.87μm Angstrom parameter of 0.5 (see Figure 3 for comparisons). Other assumptions are no aerosol absorption (wo=1) and a spherical aerosol shape (Stowe et al., 1997). Cloud screening is done with the CLAVR-1 algorithm (Stowe et al., 1999) and the data were obtained from the Pathfinder ATMOSphere (PATMOS) electronic archive at NOAA. Clear-sky radiance statistics on a 110km*110km quasi equal area grid (1deg latitude by 1deg longitude) are used to derive aerosol optical thickness. Surface reflection is assumed to be Lambertian outside of a 40deg cone. Reflectivities over water within that cone are affected by sun-glint (specular reflection) and are not considered in monthly statistics. In addition to the NOAA-9 data monthly averages are provided based on a 8-year data-set from the Jul.1981-Jun.1991 period (NOAA-7, NOAA-9 and NOAA-11) minus the two years following the El Chichon volcanic eruption in Apr.1982.

***GISS***: A 2-channel retrieval (GISS) derives the aerosol optical depth at 0.55μm from an extrapolation with the Angstrom parameter. The Angstrom statistics is based on a power-law size-distribution, whose power exponent is inferred from the optical depth ratios of the two channels. Other retrieval assumptions are moderate absorption (.97<wo<.99) and a spherical aerosol shape (Mishchenko et al., 1999). Cloud screening is based on the ISCCP data-set (Rossow et al., 1993), with the additional constraints, as to include only the warmest pixels (only retaining pixels with IR temperatures warmer than the composite value).

### 3.2.2 OCTS

OCTS is a 12 band (vis: 6, n-IR: 3, IR: 3) cross track scanning radiometer flown on the ADEOS polar orbiting satellite. Swath width and spatial resolution are about 1400km and 0.7km. Although primarily designed to detect ocean color, two channels, a visible (0.67μm) and a near-infrared channel (0.87μm) are used to derive over sun-glint free ocean scenes from reflectances the aerosol optical depths at 0.5μm and the Angstrom parameter. The value for the Angstrom parameter is based on a bimodal log-normal size-distribution (effective radii of 0.14μm and 1.54μm), where weights of the two modes are chosen to match the ratio of retrieved aerosol optical depths in the two satellite channels. Other retrieval assumptions are moderate aerosol absorption (.97<wo<.99) and a spherical aerosol shape (Nakajima et al., 1998, Hiragushi et al., 1999). From eight months of available data (Nov.1996-Jun.1997) monthly averaged were only processed for April, May and June 1997.

### 3.2.3 POLDER

POLDER is 7 band (vis: 4, n-IR: 3) bi-dimensional CCD with filters and polarizers flown on the ADEOS polar orbiting satellite. Swath width and spatial resolution are about 2200km and 6km. Due to the limited lifetime of ADEOS, data were only available from Nov.1996 to Jun.1997. The optical thickness was derived using measuring changes to reflectance or using changes to polarization.

***Reflection 'ocean'***: A dual channel retrieval (0.67μm and 0.87μm) determines the near-IR optical depth at 0.87μm from reflection changes over sun-glint free 18*18km ocean scenes (Deuze et al., 1998). (The combination with the retrieval at 0.67μm provides values for the Angstrom parameter and estimates on aerosol size). A-priori assumptions on aerosol composition, size and shape are based on the 6S model (Vermote et al., 1996), which assumes regional dependent aerosol properties.

***Polarization***: A new (still experimental) 2-channel retrieval uses the same two channels, however, utilizing polarized radiances. Changes in clear-sky polarization at 0.67μm and 0.87μm are used to derive the aerosol optical depth at 0.87μm. The retrieval is based on the concept that the polarization signal detected by the satellite comes mainly from the atmosphere (Herman et al., 1997). Less important surface contributions to polarization are included based on a map with a minimum polarization. No assumption with respect to surfaces properties are required, thus retrievals of aerosol optical depth are not limited to ocean regions. The aerosol model assumes an Angstrom parameter of 1, creating a positive bias for urban-industrial aerosol and biomass burning aerosol. For (large) dust aerosol a negative bias is expected. This bias is further enhanced, because larger aerosol particles loose sensitivity to polarization. Aerosol a-priori assumptions are adopted from the 6S model.

### 3.2.4 SeaWifs

SeaWifs is a 8 band (vis: 6, n-IR: 2) cross track scanning radiometer flown on the SEASTAR polar orbiting satellite since 1997. Swath width and spatial resolution are (AVHRR-similar) about 2800km and 1km. Data are based on the Sep.1997-Nov.1998 time-period. SeaWiFS was designed to detect ocean color, thus it required the removal of atmospheric effects - including aerosol. Data for '1 deg latitude / 1 deg longitude' regions represent averages of sub-pixels values corresponding to minimum optical depths at 0.44μm - thus, a negative bias for retrieved aerosol optical depths can be expected. Different techniques were used to determine aerosol properties over water and land.

***Over water***, 1-channel retrievals derive aerosol optical depth from changes in reflectance at 0.67μm and 0.87μm. From both independently determined optical depths an Angstrom parameter was determined. The aerosol optical depth retrieval is based on the 6S model (Vermote et.al., 1996), which makes assumptions about aerosol composition, size and shape. Retrieved aerosol optical depths for a 1degree latitude / 1 degree longitude resolution are comprised of monthly sub-pixel 'minimum blue' data. Thus, these optical depths have a tendency to be too small.

***Over land***, the visible aerosol optical depth at 0.67μm was determined only over dark dense vegetation. These low reflecting regions were identified by 3.7μm AVHRR data from the same day. Based on the 'black pixel' approach, it is assumed that the aerosol optical depth can be neglected in the near-IR (except dust) and that spectral correlations for the surface reflectance apply (Kaufman et al., 1997). The optical depth retrieval, as for water, is based on the 6S model, which assumes regional dependent aerosol properties for size, composition and shape.

### 3.2.5 TOMS

TOMS is a 6-band (UV: 6) cross track scanning radiometer flown on many different polar-orbiting satellites (NIMBUS-7, Meteor-3, Earth Probe, ADEOS) since 1979. Swath width and spatial resolution are about 3000km and 50km. Although primarily designed to monitor ozone, measurements in channels with weak (0.34μm) or no ozone absorption (0.38μm) permit the retrieval of aerosol properties (Torres et al., 1998). The optical depth of scattering aerosol is derived from enhancements in molecular backscatter at 0.38μm. The optical depths of absorbing aerosol and also the aerosol single scattering albedo are derived from changes to the background spectral dependency of molecular scattering between 0.34μm and 0.38μm. These changes are a function of aerosol altitude so that altitude underestimates lead to overestimates in aerosol optical depth and vice versa. Also without sufficient background signal below the aerosol layer the detection of absorbing aerosol near the surface is difficult. The retrieval assumes surfaces at sea-level, which currently causes overestimates for absorbing aerosol optical depths at high altitude regions. For the properties of absorbing aerosol the current retrieval assumes an average altitude of 3km. Also a-priori assumptions for aerosol size (,shape) are necessary. The large 40km*40km pixel size of a TOMS image make it difficult to avoid contamination by clouds, which can have a strong impact on the retrieval of absorbing aerosol. Here a reflection threshold technique is assumed to remove clouds. Surface reflections, although small both over water and land in the ultraviolet, are based on minimum reflections for cases, where the spectral dependency of molecular scattering did not change. Data presented in the comparison are based on 1997 retrievals.

## 3.3 Comparisons to AERONET

For comparisons to AERONET data, all retrieved satellite aerosol optical depths are averaged spatially over regions of 1deg latitude by 1deg longitude (about 100km by 100km) and temporally over a month or season. All aerosol optical depths are scaled to a common mid-visible wavelength at 0.55μm, based on local statistics for the AERONET Angstrom parameter (see Figure 3). For a local comparison to AERONET-site statistics, the monthly averages of the 4 grid-points closest to each AERONET-site have been interpolated (with inverse distance weights). Aerosol optical depth differences of satellite data with respect to the AERONET statistics are summarized on a quarterly basis in Tables 4 and 5.

Table 4 illustrates, in reference to the AERONET statistics (Figures 3), differences in mid-visible optical depths for the two NOAA-9 based AVHRR retrievals, which are limited to near ocean sites, for two global retrievals with TOMS and POLDER and for a SeaWifs land-retrieval (employing an AVHRR dark pixel mask and cloud-screen). It should be stressed that the POLDER and the combined SeaWifs/AVHRR retrievals are highly experimental.

*Table 4.* Satellite based deviations for quarterly (1: J,F,M / 2: A,M,J / 3: J,A,S / 4: O,N,D) averaged aerosol optical depths with respect to AERONET statistics

| Satellite | **AVHRR**-noaa | | | | **AVHRR**-giss | | | | **TOMS** | | | | **POLDER** | | | | **SeaWifs**, l | | | |
|---|---|---|---|---|---|---|---|---|---|---|---|---|---|---|---|---|---|---|---|---|
| yearly quarter | 1 | 2 | 3 | 4 | 1 | 2 | 3 | 4 | 1 | 2 | 3 | 4 | 1 | 2 | 3 | 4 | 1 | 2 | 3 | 4 |
| Mongu | - | - | - | - | - | - | - | - | = | 6 | ∃ | ∃ | 5 | = | - | = | ϑ | ϑ | - | ϑ |
| Seviletta | - | - | - | - | - | - | - | - | # | # | # | - | 5 | = | - | = | ϑ | ϑ | - | ϑ |
| Cuiaba | - | - | - | - | - | - | - | - | - | 6 | K | ∃ | - | 6 | - | K | # | # | - | ϑ |
| Banizoumbou | - | - | - | - | - | - | - | - | = | K | 6 | ∃ | ∃ | K | - | 6 | - | - | - | - |
| Waskesiu | - | - | - | - | - | - | - | - | - | ∃ | 6 | - | - | 6 | - | = | - | ϑ | ϑ | - |
| Goddard | - | - | - | - | - | - | - | - | 6 | 6 | ∃ | 6 | # | 5 | - | # | ϑ | ϑ | - | ϑ |
| Ispra | - | - | - | - | ∃ | 6 | = | 6 | ∃ | K | ∃ | K | 5 | = | - | ∃ | - | - | - | - |
| Cart-site | - | - | - | - | - | - | - | - | - | - | - | - | 5 | 6 | - | 5 | - | - | - | - |
| Cape Verde | = | # | ∃ | 6 | 6 | 5 | 6 | = | 6 | 5 | ∃ | ∃ | K | K | - | K | - | - | - | - |
| Bermuda | 5 | # | = | = | # | 5 | 5 | 5 | = | = | = | 5 | 6 | 6 | - | 6 | - | - | - | - |
| Dry Tortugas | 5 | = | = | = | 5 | 6 | = | 5 | = | 6 | 6 | 5 | = | 6 | - | 6 | - | - | - | - |
| Lanai | = | = | = | = | 5 | 5 | # | 5 | 5 | # | # | 5 | 6 | 6 | - | - | - | - | - | - |
| Dakar | - | - | - | - | K | K | ∃ | # | K | K | ∃ | 6 | K | K | - | 5 | - | - | - | - |
| Kaashidhoo | 6 | 6 | 5 | - | 6 | = | # | 5 | ∃ | 6 | 5 | = | 6 | ∃ | - | 6 | - | - | - | - |
| Bahrain | 5 | 5 | = | 5 | 5 | 5 | 6 | 5 | - | - | - | - | = | 6 | - | = | - | - | - | - |
| Barbados | 5 | 5 | 5 | 5 | 5 | 5 | # | 5 | 5 | 5 | 5 | 5 | ∃ | K | - | 6 | - | - | - | - |

(K<-0.2, -0.2<∃<-0.1, -0.1<6<-0.02, -0.02< = <0.02, 0.02<5<0.1, 0.1<#<0.2, 0.2<ϑ / - : no data)

Table 5 illustrates, in reference to the AERONET statistics (Figure 3), difference of other satellites retrievals for near-ocean sites. It includes a NOAA AVHRR data-set for an extended time-period, a conventional (not polarization based) POLDER retrieval and a few samples from SeaWifs (0.67μm and 0.87μm) and OCTS single channel retrievals.

Deviations of Tables 4 and 5 are supported by a more detailed satellite comparison of monthly (rather than quarterly) averages, however only for the four sites of Figures 2, in Figures 5. Vertical bars over each symbol reflect the uncertainty of AERONET monthly averages based on the detected year-to-year variability (see Figures 2 and 3).

*Table 5.* Satellite based deviations for quarterly (1: J,F,M / 2: A,M,J / 3: J,A,S / 4: O,N,D) averaged aerosol optical depths with respect to AERONET statistics

| Satellite | **AVHRR**, 8yrs | | | | **SeaWifs**, o67 | | | | **SeaWifs**, o87 | | | | **POLDER**, o | | | | **OCTS** | | | |
|---|---|---|---|---|---|---|---|---|---|---|---|---|---|---|---|---|---|---|---|---|
| yearly quarter | 1 | 2 | 3 | 4 | 1 | 2 | 3 | 4 | 1 | 2 | 3 | 4 | 1 | 2 | 3 | 4 | 1 | 2 | 3 | 4 |
| Goddard | - | - | - | - | 5 | = | ∃ | 5 | = | 6 | ∃ | 6 | - | - | - | - | - | - | - | - |
| Cape Verde | # | 6 | ∃ | # | ∃ | 5 | K | ∃ | K | 5 | K | ∃ | = | # | - | 6 | - | ϑ | - | - |
| Bermuda | 5 | 6 | = | # | 5 | 5 | = | = | 5 | 5 | = | = | 5 | 5 | - | 5 | - | 5 | - | - |
| Dry Tortugas | 5 | 6 | 5 | 5 | - | - | - | - | - | - | - | - | 5 | # | - | 5 | - | = | - | - |
| Lanai | = | 6 | 5 | 5 | - | - | - | - | - | - | - | - | 5 | 5 | - | 5 | - | 5 | - | - |
| Dakar | - | - | - | - | - | - | - | - | - | - | - | - | ∃ | K | - | # | - | K | - | - |
| Kaashidhoo | 6 | 6 | 5 | 5 | - | - | - | - | - | - | - | - | # | = | - | 5 | - | = | - | - |
| Bahrain | # | 6 | 6 | ϑ | - | - | - | - | - | - | - | - | = | = | - | = | - | 6 | - | - |
| Barbados | # | ∃ | = | # | - | - | - | - | - | - | - | - | 5 | 5 | - | 5 | - | 5 | - | - |

(K<-0.2, -0.2<∃<-0.1, -0.1<6<-0.02, -0.02< = <0.02, 0.02<5<0.1, 0.1<#<0.2, 0.2<ϑ / - : no data)

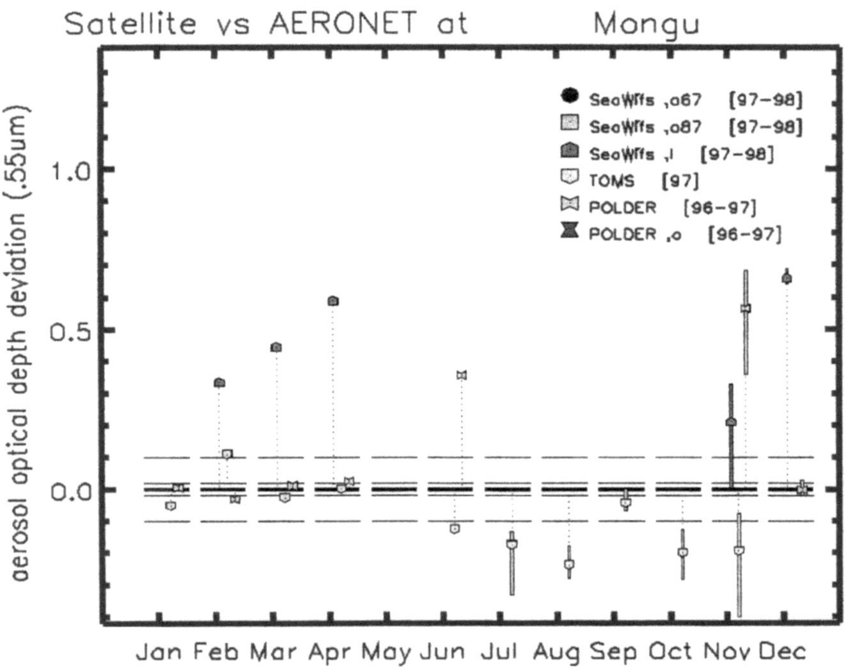

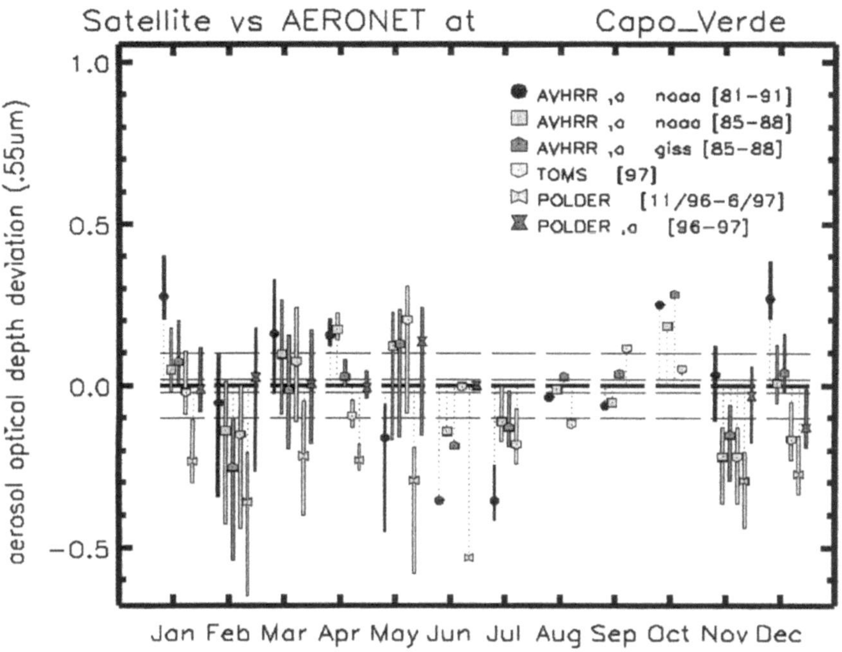
Satellite vs AERONET at Capo_Verde
AVHRR ,a noaa [81-91]
AVHRR ,a noaa [85-88]
AVHRR ,a giss [85-88]
TOMS [97]
POLDER [11/96-6/97]
POLDER ,a [96-97]
aerosol optical depth deviation (.55um)
1.0
0.5
0.0
-0.5
Jan Feb Mar Apr May Jun Jul Aug Sep Oct Nov Dec

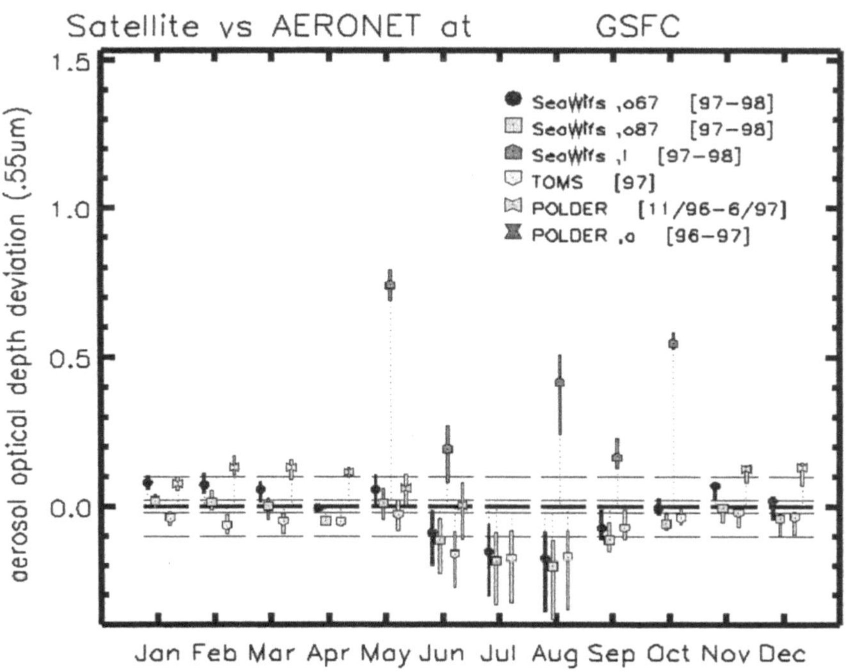
Satellite vs AERONET at GSFC
SeaWifs ,a67 [97-98]
SeaWifs ,a87 [97-98]
SeaWifs ,l [97-98]
TOMS [97]
POLDER [11/96-6/97]
POLDER ,a [96-97]
aerosol optical depth deviation (.55um)
1.5
1.0
0.5
0.0
Jan Feb Mar Apr May Jun Jul Aug Sep Oct Nov Dec

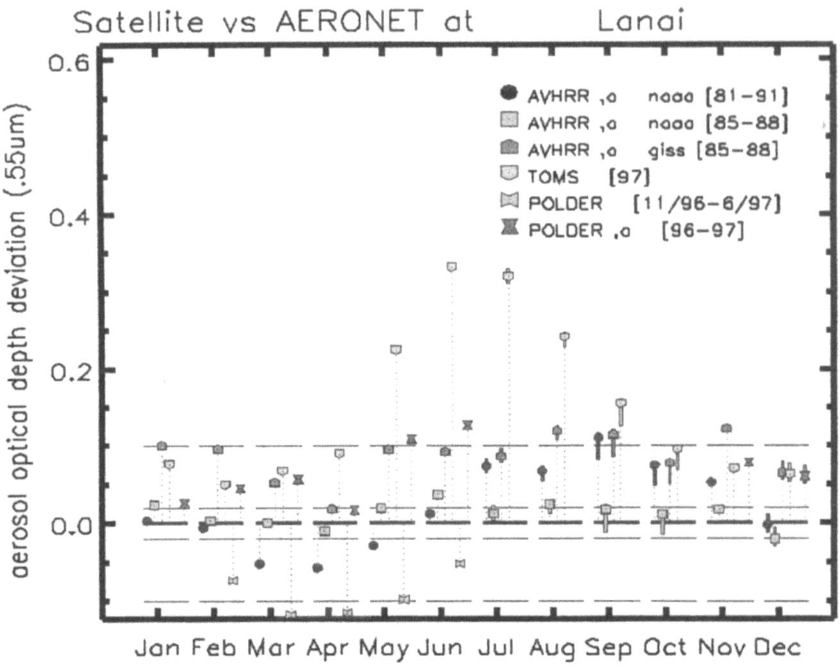


*Figure 4a-d.* Satellite based deviations in monthly statistics with respect to AERONET data for mid-visible aerosol optical depths. Spatially interpolated satellite data (from 1deg latitude by 1deg longitude data) were spectrally adjusted with local .44µm/.87µm Angstrom data from AERONET statistics (if aerosol optical depths were retrieved at wavelengths other than 0.55µm). Deviations to multi-year AERONET averages (large symbols) and deviations with respect to the year-to year range of AERONET monthly averages (vertical bars) are provided? - unless AERONET and/or satellite data were not available (small dark symbol). Data of global retrievals, of retrievals limited to ocean (,o) and of retrievals restricted to land-surfaces are given. Also outlined are deviation-distinguishing criteria [+/-0.02 and +/- 0.1] of Tables 4 and 5 (horizontal lines). Comparisons are presented for the four AERONET-sites of Figure 2

Quarterly averages for the mid-visible aerosol optical depth are rarely found to agree within +/- 0.02. This does not surprise, given spatial variability, sampling/screening biases and retrieval assumptions (even though a better retrieval performance may have benefited from partially offsetting biases or assumptions). For that reason the discussion will focus on cases, when the disagreement exceeded +/- 0.1 for the mid-visible aerosol optical depth (solid arrows in Tables 4 and 5).

An initial glance at these results indicates that over (deep water) ocean, even very simple aerosol retrievals can be accurate to within +/- 0.1 for mid-visible aerosol optical depths. If shallow water retrievals except for dust-sites, where, mainly due to problems with cloud-screening, retrievals under-predict aerosol optical depth. Over land accuracy for aerosol optical depth retrievals remain a problem.

### 3.3.1 AVHRR

AVHRR satellites, which provide data for now two decades, have become probably the most popular tool to derive aerosol optical depth. Although limited to regions over (deep water) oceans, AVHRR retrievals in this comparison are the only satellite data, whose monthly statistics is based on data from more then two years. In our comparison we present results from three different AVHRR statistics. Two monthly statistics, *AVHRR-noaa* and *AVHRR-giss* are based on the identical data of one satellite, NOAA-9, but employ different cloud-screening and retrieval assumptions. Two other monthly statistics *AVHRR-noaa* and *AVHRR, 8yrs* are based on the same retrieval but differ by considering once 4 years of data from one satellite (NOAA-9) and another time a total of 8 years of data from three satellites (NOAA-7, -9 and -11).

***NOAA***: The *AVHRR-noaa* statistics, despite its simple aerosol assumption, is usually in good agreement with the AERONET statistics. Underestimates or no data, however, occur at dust-dominated sites, primarily due to the inability to distinguish between thick dust and clouds and the subsequent removal of dust scenes during the cloud-screening process. It appears that when aerosol properties are close to assumptions of the retrieval (i.e. non-absorbing and a relatively large aerosol spheres) agreement is better (e.g. Lanai). A lack of absorption biases towards smaller optical depths, whereas overestimates in size tend to bias towards larger optical depths at viewing geometries with higher sun-elevations (Mishchenko et al. 1999). Stronger aerosol absorption is often associated with aerosol sizes that are smaller than the assumed size radius for the retrieval ($r_e$=0.35μm). Thus, the agreement to AERONET is often improved by partially compensating biases.

The *AVHRR, 8yrs* statistics includes in addition to the NOAA-9 data (2/85-10/88), data from NOAA-7 (7/81-3/82, 4/84-12/84) and data from NOAA-11 (11/88-6/91). Despite the additional data and using the same retrieval algorithm as *AVHRR-noaa*, more (not less) deviations to the AERONET statistics are detected. A comparison of monthly averages on a year-to-year basis showed that the variability increases as the additional years are added. It is unclear, to what degree these variations are real or at what degree calibration issues, the drift to later overpass times and discontinuities in data-set transitions contributed.

***GISS***: The *AVHRR-giss* statistics, although based on the same data-set as *AVHRR-noaa*, on average suggests slightly larger aerosol optical depths. The inclusion of (moderate) aerosol absorption certainly contributes, but differences with respect to cloud-screening may have contributed as well. (Recall that AERONET cloud-screening is conservative, thus, small overestimates by satellite retrievals could be expected). Large underestimates near a dust-site (Dakar) are again attributed to dust-removal during cloud-screening. As second parameter this retrieval determines aerosol size, which can be evaluated by comparisons of the derived Angstrom parameter to those from AERONET statistics. Sizing is well represented for sites dominated by urban-industrial outflow (re about 0.18μm). However, aerosol size is sharply underestimated at dust-dominated sites, as derived Angstrom parameters rarely fall below values of 0.7. Since the retreival's Angstrom parameter is used to convert the AVHRR optical depth to a smaller mid-visible wavelengths (0.55μm) *AVHRR-giss* underestimates of aerosol optical depths at these dust regions are even larger.

### 3.3.2 OCTS

OCTS data are high variable and data are too few to draw any conclusion.

### 3.3.3 POLDER

The *POLDER* polarization retrieval is highly experimental. This retrieval usually under-predicts aerosol optical depths over ocean sites. This is quite in contrast to common overestimates with the *POLDER-ocean* retrieval. More interesting are *POLDER* retrievals over land. Based on the limited amount of data, the polarization retrieval tends to exceed AERONET aerosol optical depth at urban-industrial sites. Aerosol optical depths are much smaller than AERONET averages at dust-aerosol dominated sites. This is at least in part related to a reduced sensitivity of polarization for larger aerosol sizes.

### 3.3.4 SeaWifs

Based on the few available samples for comparisons to the AERONET statistics, *SeaWifs* single channel ocean retrievals, *SeaWifs, o67* and *SeaWifs, o87*, underestimate optical depth. This was expected, as averages are composed from optical depths under minimum blue conditions. The *SeaWifs* land-retrieval, *SeaWifs, l*, was applied to cloud-free dark pixels (low reflecting) over land (e.g. green vegetation). Critical for this retrieval are a correct identification of dark pixels (assuming negligible aerosol contributions to satellite sensing at longer wavelengths) and the validity of solar spectral relationships for the surface albedo. Aerosol optical depths from this retrieval are clearly too large. Incorrect pixel identification, including a sub-pixel contamination by clouds and non-black surfaces, are believed to be largely responsible for optical depth overestimates, which at times are significant. Additional uncertainty is introduced by the screening. Dark-pixel information comes from another satellite (AVHRR) with a different local overpass time and cloud-scene removal is only based on visible thresholds.

### 3.3.5 TOMS

The *TOMS* aerosol retrieval is probably the most promising approach to global aerosol remote sensing, because surface contributions in the ultra-violet are small not only over oceans but also over land. However, due to unresolved discrepancies for derived surface (minimum) reflectance maps among different TOMS instruments, currently mid-visible aerosol optical depths are only accurate to within 0.1. The use of larger surface reflectance contributions in the retrieval for this comparison in part may explain the TOMS trend to smaller aerosol optical depths (than AERONET). Under-estimates are especially large, when optical depths are largest. This seems to indicate that the sub-pixel cloud-screening is extremely conservative and rejects the largest aerosol optical depths. Such rejection suggests that the largest aerosol optical depths are usually found in (TOMS-pixel) regions with clouds. Another important factor in TOMS retrievals is the strength of the background molecular scattering, requiring accurate data on aerosol altitude and site altitude. Thus, without orographic adjustments, aerosol optical depth overestimates are expected for high-altitude sites (e.g. Mongu).

## 3.4 Outlook

Better accuracies can be expected with multi-channel and multi-angle approaches, which will be an option with added instrumentation on newer satellites. Quantitative accurate aerosol retrievals over land remain a challenge. The dark pixel approach is not only rather limited spatially but may be limited to provide rough estimates only. The concept of polarization retrievals needs to be further explored, in particular with respect to regions where (large size) dust-aerosol dominates. The most promising approach for aerosol retrievals over land involves ultraviolet measurements, not by itself, but in combination with other spectral data ranging from the visible to the far-infrared. These measurements have to come from the same instrument or at least the same satellite, if improvements to retrievals on aerosol properties can be expected.

# 4. MODELS

The mid-visible (0.55μm wavelength) optical depth is commonly used to present model results from global aerosol simulations. Optical depths can be visualized, compared to ground measurements and are critical in the determination of aerosol climatic impacts. However, the determination of aerosol optical depths is quite complicated and depends on many assumption regarding emission, transport, removal, chemistry and aerosol properties. To account for the different nature of aerosol, with respect to absorption, size and humidification, it is desirable to process aerosol independently for type, with each aerosol type eventually contributing to the total aerosol optical depth. Thus, for models that distinguish by aerosol type, not only disagreement but also agreement to AERONET must be investigated due to the high probability for unnoticed cancellation of errors. First models are introduced, then, comparisons to AERONET statistics are presented, before model results are examined in view of their component contributions.

## 4.1 Model description

Five different global models participated in the comparison of monthly averaged aerosol optical depth:

- **ECHAM4**, which originated at the Max-Planck-Institute for Meteorology [Hamburg, GER], is a global circulation model (Lohmann et al., 1999). Wind-, temperature- and pressure-fields are generated without a link to a particular year (3 year simulation).
- **MIRAGE**, from PNNL [Richland, WA, USA], is a chemical transport model which is coupled on line with a global circulation model (R.Easter et al., 1999). To improve agreement on time-scales of days nudging can be (and in this case has been) applied. (Nudging towards analyzed winds, temperature and surface pressure is an elementary form of data assimilation, in order to reduce biases in simulated circulation and simulated winds). Results are based on ECMWF assimilated wind-, temperature- and pressure fields from June 1994 to May 1995.
- **GOCART**, from Georgia Institute of Technology and NASA-Goddard [Atlanta, GA / Greenbelt, MD, USA] is a chemical transport model (Chin et al. 2000) driven by assimilated meteorological fields from the GEOS DAS (Goddard Earth Observing System Data Assimilation System). Assimilated wind-, temperature- and pressure fields are based on 1990 data. In addition, simulations of 1996 data are presented, primarily to illustrate year-to-year variability from a modeling perspective.
- **CCSR**, from the Center for Climate System Research [Tokyo, JP], is a chemical transport model (Takemura et al., 2000) driven by NCEP / NCAR data (National Center for Environmental Predictions / National Center for Atmospheric Research). Wind-, temperature- and pressure fields are based on 1990 data.

- **GISS**, from NASA-GISS [New York, NY, USA] is a global circulation model (Tegen et al., 1997, Tegen et al., 2000; Koch et al. 1999). Wind-, temperature- and pressure-fields are generated without a link to a particular year (3 year simulation).

All models

- distinguish among five different aerosol components: (1) sulfate, (2) organic carbon, (3) black carbon, (4) soil dust and (5) sea-salt
- differentiate in aerosol emissions by type
- include sulfate chemistry (gas to particle conversions)
- simulate transport processes of advection, diffusion and convection and
- include aerosol processing.

The included aerosol processes are

- dry deposition (aerosol clustering from turbulent mixing)
- in-cloud scavenging (aerosol acting as cloud condensation nucleus or aerosol which diffuse in cloud drops with subsequent removal via precipitation)
- below cloud scavenging (aerosol capture and removal by rain)
- aerosol re-emission by evaporation (of rain or cloud drops) and
- gravitational settling.

While a detailed model comparison goes beyond the scope of this paper a comparison of model resolution and references for aerosol emission by type are given in Table 6.

*Table 6.* Comparison of model resolution and (major) reference for emission source for aerosol components

| Model | ECHAM4 MPI,Hamburg | MIRAGE PNNL, Richl. | GOCART NASA-godd. | CCSR NIES, Tokyo | GISS NASA-giss |
|---|---|---|---|---|---|
| Hori.grid (lon/lat) | 3.8*3.8 deg | 2.8*2.8 deg | 2.5* 2.0 deg | 5.6*5.6 deg | 4.0*5.0deg |
| Vertical layers | 28 | 23 | 20 | 11 | 9 |
| Simulation-period | 3 years | 6/94 - 5/95 | 1/90 – 12/90 | 1/90 – 12/90 | 3 years |
| Meteorology | generated | ECMWF* | GEOS/DAS | NCAR/NCEP | generated |
| Clouds | prognostic | prognostic | diagnostic | diagnostic | prognostic |
| Humidification | Koepke 97 | Koehler-th. | Kiehl 00 | Tang94,Ho97 | |
| Emis: dust | Ginoux 00 | Gilette 88 | Ginoux 00 | Gilette 78 | Tegen/Fung95 |
| Emis: org. carbon | Liousse 96 | Liousse 96 | Cooke 96, 99 | L96, Guent.95 | Liousse 96 |
| Emis: black carb. | Cooke 96 | Cooke 96 | Cooke 96, 99 | Cooke 96 | Cooke 96 |
| Emis: sea salt | Monohan86** | O'Dowd 97 | Monahan86** | Erickson 86 | Monahan86** |
| Emis: sulfur | Benkowitz 96 | Benkowitz 96 | Olivier 97 | Benkowitz 96 | Benkowitz 96 |
| Sulfur - oxidants | imported | CO-CH4 ch.. | imported | imported | imported |

* ***MIRAGE***: uses **E**-winds, calculates water-uptake from **E**- rel.humidity, applies nudging towards **E**-temp.

** ***GOCART*** uses **SSM/I** based winds, ***ECHAM4*** and ***GISS*** are based on CanadianGCM winds (Gong 97)

## 4.2 Comparisons to AERONET

For a local comparison to AERONET-site statistics, the monthly averages of models at the 4 grid-points closest to each AERONET-site have been interpolated (with inverse distance weights). Deviations of these interpolated aerosol optical depth averages on a quarterly basis with respect to the AERONET statistics are summarized from all five models in Table 7. Table 8 extends comparisons involving particular *ECHAM4* and *GOCART* versions. Deviations of Tables 7 and 8 are supported by more detailed model comparisons of monthly (rather than quarterly) averages, however only for the four sites of Figures 2, in Figures 5. Vertical bars over symbols in Figures 5 display the uncertainty of AERONET monthly averages based on year-to-year variability (see Figures 2 and 3).

The AERONET vs. model comparisons assumed (despite linear interpolation from the four closest grid-points) that local monthly averages are comparable to regional averages. The validity of this assumption was tested in comparing satellite retrieved optical depths for different spatial resolutions near each site (see Table 3). In that comparison, trends remained largely inconclusive. Thus, only deviations of mid-visible aerosol optical depths that exceeded +/-0.1 should be discussed (solid arrows in Tables 7 and 8).

*Table 7.* Model based deviations for quarterly (1: J,F,M / 2: A,M,J / 3: J,A,S / 4: O,N,D) averaged aerosol optical depths with respect to AERONET statistics. Results of five different models

| Model | ECHAM 4 | | | | MIRAGE | | | | GOCART | | | | GISS | | | | CCSR | | | |
|---|---|---|---|---|---|---|---|---|---|---|---|---|---|---|---|---|---|---|---|---|
| yearly quarter | 1 | 2 | 3 | 4 | 1 | 2 | 3 | 4 | 1 | 2 | 3 | 4 | 1 | 2 | 3 | 4 | 1 | 2 | 3 | 4 |
| Mongu | 6 | 6 | K | K | 6 | 6 | K | K | 6 | 6 | ∃ | ∃ | 6 | = | ∃ | K | 6 | = | = | 6 |
| Seviletta | = | = | 6 | = | 5 | = | = | = | 5 | 5 | 5 | 5 | 6 | 6 | 6 | 6 | 5 | 5 | 5 | 5 |
| Cuiaba | - | 6 | K | K | - | 5 | K | K | - | 6 | K | K | - | = | K | K | - | 6 | K | K |
| Banizoumbou | K | K | K | K | K | K | K | K | # | ∃ | 5 | ϑ | ∃ | K | K | K | ϑ | K | K | 6 |
| Waskesiu | - | ∃ | 6 | = | - | ∃ | 6 | 5 | - | 6 | = | 5 | - | ∃ | 6 | = | - | K | ∃ | 6 |
| Goddard | 5 | 6 | ∃ | 5 | # | # | # | 5 | 5 | = | 6 | 5 | = | 6 | ∃ | 5 | 6 | 6 | ∃ | = |
| Ispra | ∃ | ∃ | 6 | 6 | ∃ | ∃ | 6 | ∃ | ∃ | 6 | 5 | 6 | ∃ | 6 | = | ∃ | K | ∃ | ∃ | ∃ |
| Cart-site | = | 6 | ∃ | 5 | 5 | 5 | 6 | 5 | 5 | 6 | ∃ | 5 | = | = | 6 | 5 | 6 | ∃ | ∃ | 6 |
| Cape Verde | K | K | K | K | K | K | K | K | 5 | = | 6 | = | ∃ | ∃ | K | K | # | ∃ | K | ∃ |
| Bermuda | 5 | 6 | 6 | = | 5 | 5 | 6 | = | 6 | = | = | = | 6 | 6 | 6 | 6 | 6 | 6 | 6 | 6 |
| Dry Tortugas | 6 | ∃ | ∃ | = | = | 6 | 6 | = | = | 6 | ∃ | 6 | = | 6 | ∃ | 6 | 6 | ∃ | ∃ | 6 |
| Lanai | = | 6 | = | = | = | = | = | = | = | = | 5 | = | 6 | 6 | = | = | 6 | 6 | 6 | 6 |
| Dakar | K | K | K | 6 | K | K | K | 6 | ∃ | K | K | # | K | K | K | 6 | 5 | K | K | = |
| Kaashidhoo | ∃ | ∃ | 6 | 6 | ∃ | ∃ | 6 | 6 | K | 6 | 5 | 6 | ∃ | ∃ | 6 | 6 | ∃ | ∃ | = | 6 |
| Bahrain | ∃ | ∃ | 6 | 6 | ∃ | ∃ | ∃ | ∃ | # | # | # | # | 6 | 5 | 5 | 6 | 6 | 6 | ∃ | ∃ |
| Barbados | 6 | ∃ | ∃ | 6 | 6 | ∃ | ∃ | 6 | 5 | 6 | 5 | = | = | ∃ | ∃ | 6 | 6 | ∃ | ∃ | 6 |

(K<-0.2, -0.2<∃<-0.1, -0.1<6<-0.02, -0.02< = <0.02, 0.02<5<0.1, 0.1<#<0.2, 0.2<ϑ / - : no data)

*Table 8.* Model based deviations for quarterly (1: J,F,M / 2: A,M,J / 3: J,A,S / 4: O,N,D) averaged aerosol optical depths with respect to AERONET data. Deviations of modified ECHAM4 and GOCART versions. For *GOCART 5* the horizontal resolution of *GOCART* was relaxed from 2*2.5deg to 5*5deg lat/longitude. *ECHAM4-old* prescribes dust (Westphal 1988) and sea-salt (Monahan 1986) fields, rather than predicting the fields as in *ECHAM4*. *ECHAM4-clr,* a subset of *ECHAM4,* only considers grid-cells (ca. 300*300km) with low liquid water content (less than 33g/m2)

| Mod.-version | ECHAM4, old | | | | ECHAM4, 1yr | | | | ECHAM4, clr | | | | GOCART, old | | | | GOCART, 5 | | | |
|---|---|---|---|---|---|---|---|---|---|---|---|---|---|---|---|---|---|---|---|---|
| yearly quarter | 1 | 2 | 3 | 4 | 1 | 2 | 3 | 4 | 1 | 2 | 3 | 4 | 1 | 2 | 3 | 4 | 1 | 2 | 3 | 4 |
| Mongu | 6 | 6 | K | K | 6 | 6 | K | K | 6 | 6 | ∃ | ∃ | 6 | 5 | # | ∃ | 6 | 5 | # | ∃ |
| Seviletta | 5 | 6 | 6 | = | = | 6 | 6 | = | = | 6 | 6 | = | 5 | # | 5 | 5 | 5 | # | 5 | 5 |
| Cuiaba | - | 6 | K | K | - | ∃ | K | K | - | ∃ | K | K | - | 5 | ∃ | K | - | 5 | K | K |
| Banizoumbou | K | K | K | K | K | K | K | K | K | K | K | K | ϑ | 6 | # | ϑ | ϑ | ∃ | 5 | ϑ |
| Waskesiu | - | = | 6 | 5 | - | ∃ | 6 | = | - | ∃ | 6 | 6 | - | ∃ | 5 | # | - | 6 | 5 | 5 |
| Goddard | # | 5 | 6 | 5 | 5 | 6 | ∃ | 5 | = | ∃ | ∃ | = | 5 | 5 | 6 | # | 5 | = | 6 | # |
| Ispra | 6 | 6 | ∃ | 6 | ∃ | ∃ | 6 | 6 | ∃ | K | ∃ | ∃ | 6 | = | # | 5 | 6 | = | # | = |
| Cart-site | 5 | 6 | ∃ | = | = | ∃ | ∃ | 5 | 6 | ∃ | ∃ | = | # | 5 | = | 5 | 5 | 6 | ∃ | 6 |
| Cape Verde | K | K | K | K | K | K | K | K | K | K | K | K | # | = | 6 | 6 | 5 | 6 | ∃ | 6 |
| Bermuda | = | 6 | 6 | 6 | 5 | 6 | 6 | = | = | 6 | 6 | 6 | = | 5 | 5 | 5 | = | = | = | = |
| Dry Tortugas | 6 | ∃ | ∃ | 6 | 6 | ∃ | ∃ | = | 6 | ∃ | ∃ | 6 | = | 6 | 6 | = | = | 6 | 6 | = |
| Lanai | 6 | 6 | 6 | 6 | = | 6 | = | = | = | 6 | = | = | 5 | 5 | 5 | 5 | 5 | 5 | 5 | = |
| Dakar | K | K | K | ∃ | K | K | K | 6 | K | K | K | 6 | 5 | K | ∃ | # | 5 | K | K | # |
| Kaashidhoo | K | ∃ | 6 | 6 | K | ∃ | 6 | 6 | K | ∃ | 6 | 6 | ∃ | 6 | 5 | 6 | ∃ | 6 | 5 | 6 |
| Bahrain | ∃ | K | K | ∃ | ∃ | ∃ | 6 | 6 | ∃ | ∃ | 6 | 6 | # | ϑ | ϑ | # | 5 | # | ϑ | # |
| Barbados | 6 | ∃ | ∃ | 6 | 6 | ∃ | ∃ | 6 | 6 | ∃ | ∃ | 6 | 5 | 6 | = | # | 5 | 6 | = | = |

(K<-0.2, -0.2<∃<-0.1, -0.1<6<-0.02, -0.02< = <0.02, 0.02<5<0.1, 0.1<#<0.2, 0.2<ϑ / - : no data)

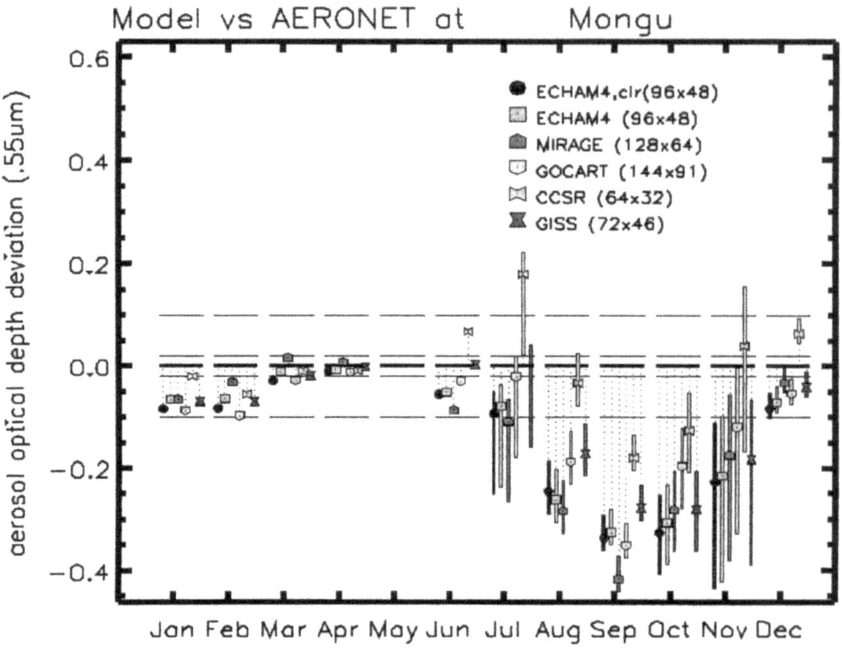
Model vs AERONET at Mongu
ECHAM4,clr(96x48)
ECHAM4 (96x48)
MIRAGE (128x64)
GOCART (144x91)
CCSR (64x32)
GISS (72x46)
aerosol optical depth deviation (.55um)
0.6
0.4
0.2
0.0
−0.2
−0.4
Jan Feb Mar Apr May Jun Jul Aug Sep Oct Nov Dec

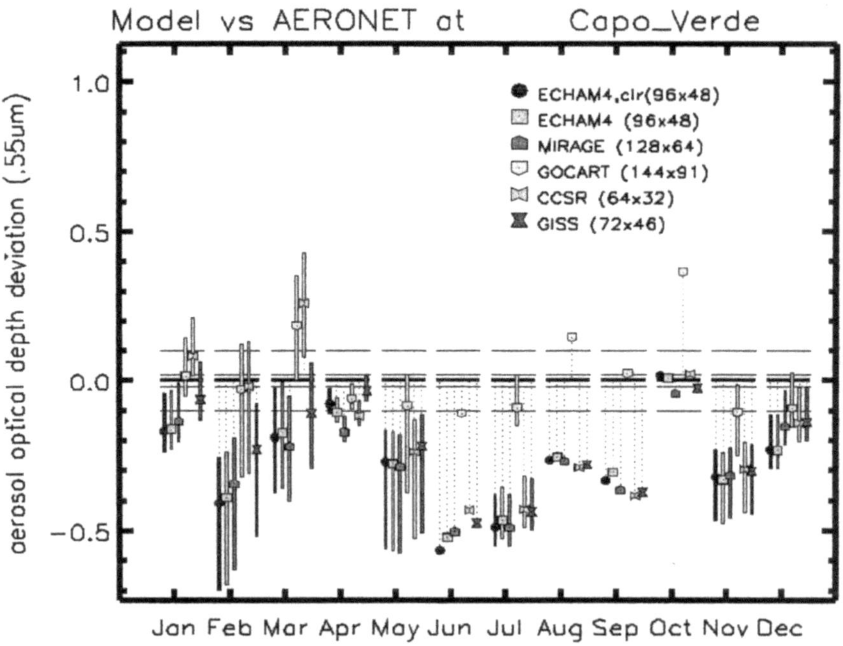
Model vs AERONET at Capo_Verde
ECHAM4,clr(96x48)
ECHAM4 (96x48)
MIRAGE (128x64)
GOCART (144x91)
CCSR (64x32)
GISS (72x46)
aerosol optical depth deviation (.55um)
1.0
0.5
0.0
−0.5
Jan Feb Mar Apr May Jun Jul Aug Sep Oct Nov Dec

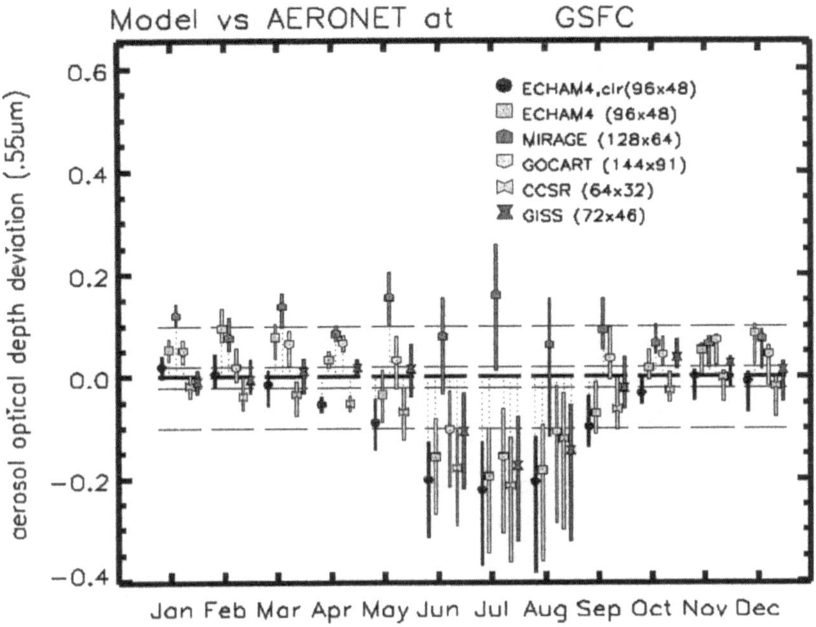


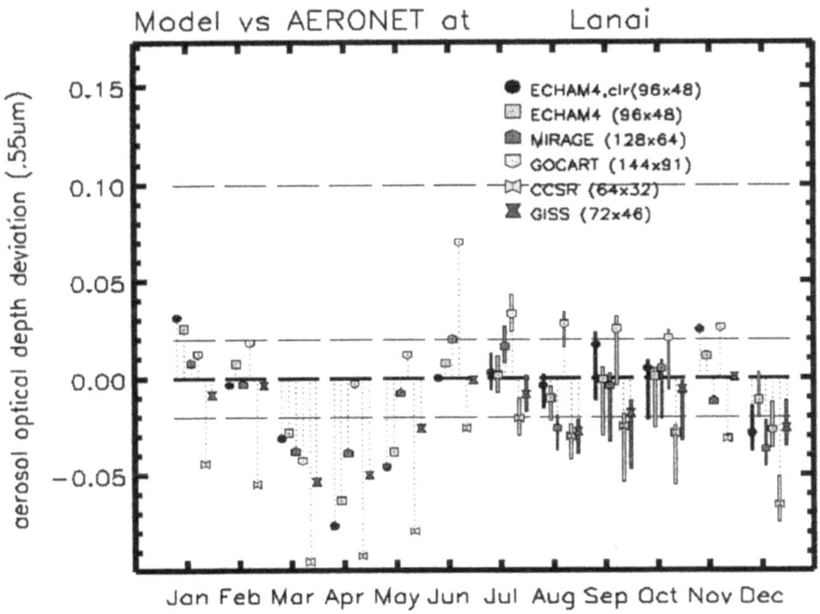


*Figure 5a-d.* Model deviations with respect to the statistics from AERONET for mid-visible aerosol optical depths monthly averages. Model data have been spatially interpolated (from the resolution of each models – see Table 6) Presented are model deviations to multi-year AERONET data (large symbols) and with respect to the range in AERONET averages (vertical bars) - unless AERONET monthly averages were not available (small dark symbol). For an improved comparison, deviation-distinguishing criteria [+/-0.02 and +/- 0.1] of Tables 7 and 8 are indicated (horizontal lines). Comparisons are presented for the four AERONET-sites of Figures 2 and 4

***FIRST IMPRESSIONS*** : The tested models tend to underestimate aerosol optical depths with respect to the AERONET statistics. The largest model deviations are underestimates and also underestimates are much more frequent than overestimates with respect to AERONET. All models have difficulties to reproduce the large aerosol optical depths at sites near biomass burning and dust sources, in particular the ECHAM4, MIRAGE and GISS models. The CCSR model has the least underestimates at biomass burning sites, the GOCART model performs best near dust sources, and the Mirage model usually suggests the largest optical depths at urban industrial sites.

Differences among the five models are large, often exceeding half of the value suggested by AERONET. Differences also change from month to month. In contrast, differences are less significant from changes in grid-spacing, based on a *GOCART* [2-by-2.5deg] vs. *GOCART5* [5-by-5 deg] comparison. Thus, differences in-grid spacing among of the five models seem less important. Also small in comparison to the model spread are differences between *ECHAM4* and its cloud-free subset, *ECHAM4-clr* . Larger reductions to aerosol optical depths of *ECHAM4* occur mainly at urban-industrial sites and mainly during the winter when ambient relative humidity is largest. Clear-sky data sub-sets, like *ECHAM4-clr*, were only available for the *ECHAM4* model, although such subsets seems a better match to AERONET statistics, with its conservative cloud-screening. Thus, all-sky data of models are expected to exceed AERONET averages, at least at urban-industrial sites. For most models this would increase the differences to AERONET data.

Prescribed fields for sea-salt and dust in *ECHAM4-old,* rather than predicted fields in *ECHAM4* provide on occasions a better match to AERONET data (see Table 8). This illustrates that an increased complexity will not necessarily reduce uncertainties, at least not initially, as physical processes and feedbacks need to be understood. An evaluation between *ECHAM4* and *ECHAM4-old* reveals that better agreement for a prescribed treatment at some sites is created by compensating errors (e.g. underestimates in dust- and carbon-aerosol are partially compensated by sea-salt overestimates). This illustrates, that an evaluation of the model performance has to be conducted on an aerosol sub-component basis.

## 4.3 Component treatment

A comparison of aerosol total optical depths between models and (AERONET) observations might be interesting from the point that these models are designed to provide estimates on aerosol forcing. Discrepancies in total optical depth, however, reveal little about models, if their total aerosol optical depth is derived from individual contributions of sulfate, black carbon, organic carbon, dust and sea-salt. Thus, in order to explain, understand and correct discrepancies to AERONET data and among models, evaluations on a sub-component basis become desirable, preferably at AERONET sites, where a particular component dominates. One of the major goals is the identification of off-setting errors. However, potential for off-setting errors even extends to the processing of each component. Most critical for the component treatment are model assumptions (of size and of humidification and ambient relative humidity) during the conversion of component mass into component optical depth.

After the presentation of model partitions of total aerosol optical depth into components, some background on the conversion of component mass to optical depth is provided. Then, each model's partition of aerosol mass is introduced, and from component data on optical depth and mass, effective mass-to-optical depth conversions are derived. Finally, in comparison of properties for large regions apparent trends of models will be summarized.

### 4.3.1 Component optical depth $\tau$

The partition of modeled total aerosol optical depth into components of dust, sea-salt, carbon (black carbon and organic carbon are combined) and sulfate are presented in Figures 6 and 7 (for the four sites that were introduced in Figure 2). In Figures 6 the component optical depths of all five models are presented, along with AERONET total optical depths for reference. In Figures 7 the component optical depths among different model versions are compared. Results from three pairs of model versions are presented to illustrate the (modeled) impact of nearby-clouds on aerosol optical depth (*ECHAM4* vs. *ECHAM4, clr*), to indicate the fraction of water in aerosol (*MIRAGE* vs. *MIRAGE-dry*) and to demonstrate year-to-year variations (*GOCART, 1990* vs. *GOCART 1996*).

Figures 6 and 7 demonstrate that the (modeled) total aerosol optical depths are always comprised of contributions from many aerosol components. Contributions of carbon, dust and sulfate are frequently comparable in magnitude. Near water, sea-salt contributions become comparable as well. Thus, an accurate treatment of each component is important. This also means that assuming aerosol component mixtures, rather than pure aerosol components, may be a much better choice for a-priori assumptions in satellite retrievals of aerosol (e.g. Kahn et al., 1999). Nonetheless, there are a few AERONET sites, where (at least for part of the year) the optical depth contribution of one aerosol component dominates. These sites are important for the evaluation of aerosol component modeling. For this reason comparisons at AERONET data are always given for four sites: Mongu has large **carbon** contributions in late summer and fall, Cape Verde has large **dust** contributions with two maxima in late winter and summer, at GSFC **sulfate** contributions are expected to dominate site and at Lanai **sea-salt** contributions should be significant.

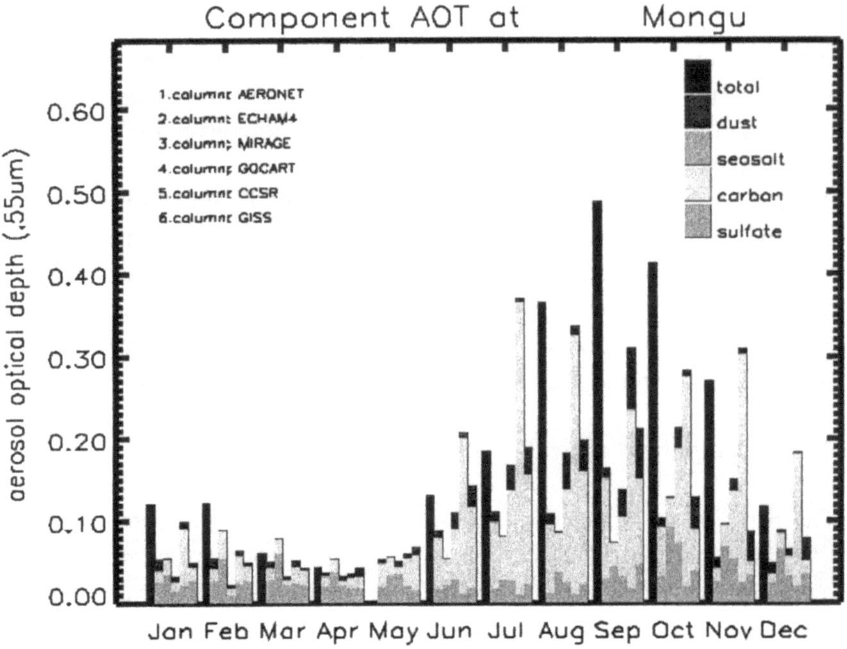


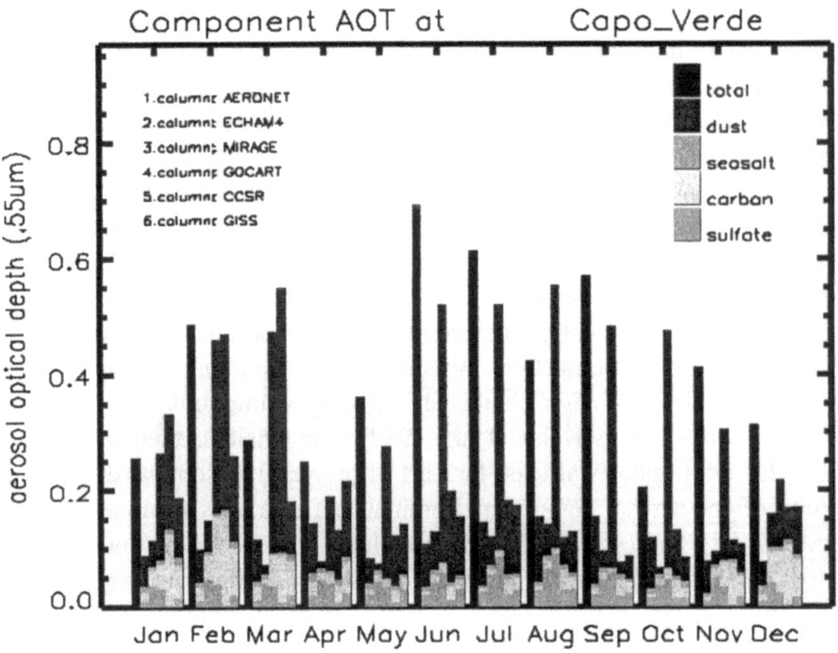

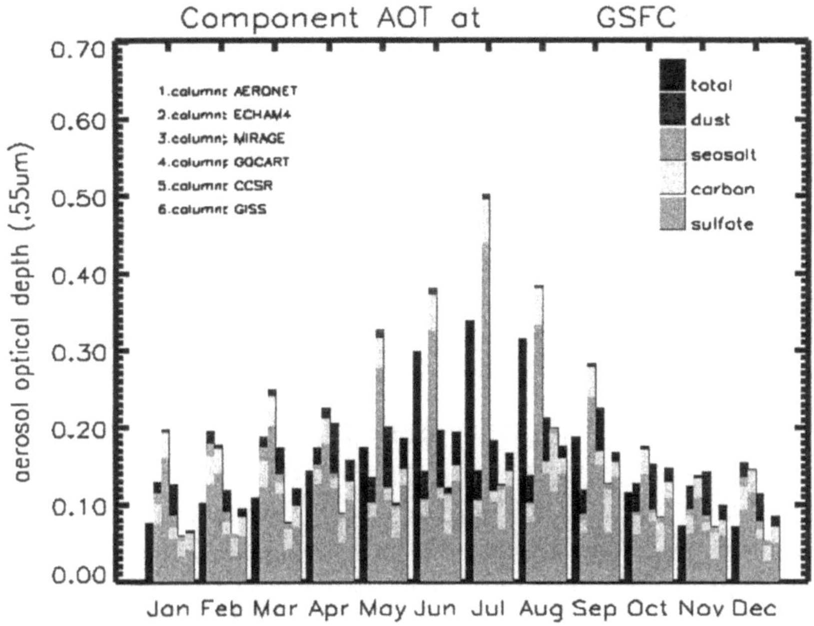


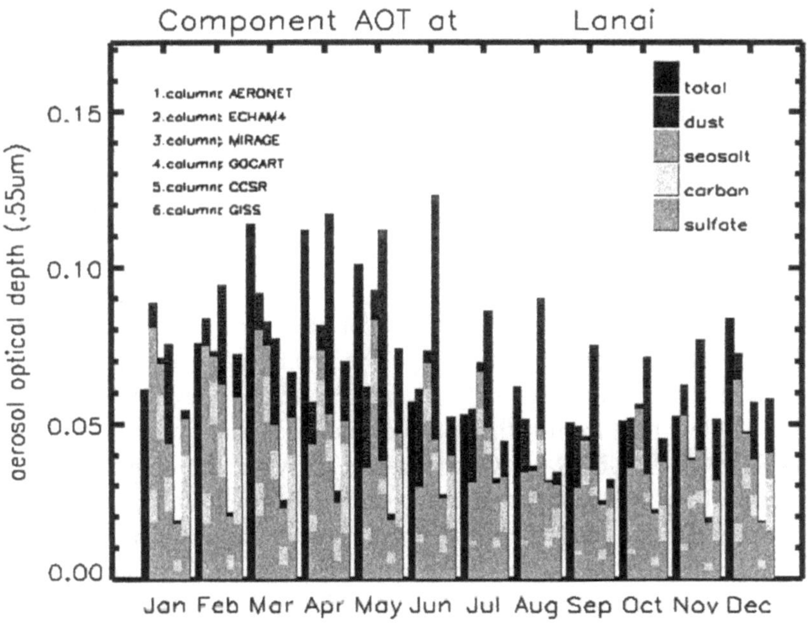


*Figure 6a-d.* Comparisons of modeled monthly averaged optical depth contributions for dust, sulfate, carbon and sea-salt among the five tested models (ECHAM4, MIRAGE, GOCART, CCSR and GISS). AERONET (total) optical depths are given for reference in the first (black) column. Comparisons are presented for the four AERONET sites of Figures 2, 4 and 5

Compositional differences among models with respect to locations are expected. Some of the differences in component optical depths are summarized next.

**CARBON-** $\tau$: Modeled carbon optical depths are usually too small. At sites near biomass burning only *CCSR* comes close to AERONET totals. The four other models, especially *MIRAGE*, miss in magnitude for these seasonal events. Away from regions of biomass burning, carbon contributions of *ECHAM4* and *GOCART* are smallest. Reductions in carbon optical depths with the cloud-removes sub-set remain small. Assimilations with meteorological data from different years are not too significant, although there is potential for significant differences with temporal shifts for dry and wet seasons.

**DUST-** $\tau$: Modeled dust optical depths are usually much too small. Near dust sources only *GOCART* and at times *CCSR* match the large optical depths of AERONET statistics. Suggestions of lower (often up to one order of magnitude lower) optical depths are quite common. *MIRAGE* dust optical depths are particularly low. Away from dust sources *GOCART* carbon optical depths are too large. Year-to-year variations can be significant due the dependence of dust-mass on near-surface winds and precipitation (soil wetness).

**SULFATE-** $\tau$: Sulfate optical depths differ strongly among models and with respect to AERONET averages at sulfate dominated sites. *MIRAGE* usually suggests the largest (and often too large) sulfate aerosol optical depths. In contrast, *CCSR* values are clearly too small. Sulfate optical depths in *ECHAM4* and *MIRAGE* display a high sensitivity to ambient relative humidity. This is also reflected in strong reductions of (too) large sulfate optical depths during northern hemispheric winters with the cloud-removed data subset.

**SEASALT-** $\tau$: Sea-salt optical depths over land are usually insignificant compared to dust carbon or sulfate. Even for near-ocean sites sea-salt optical depth contributions are not dominating. The strong seasonality for *ECHAM4* values is driven by its sensitivity to ambient relative humidity. However, there is no clear tend from with the cloud-removed data subset.

**AEROSOL-WATER-** $\tau$: Contributions of water are about half of the total aerosol optical depth at maritime and urban-industrial sites. Aerosol water contributions easily separate dry and wet seasons. For the northern hemisphere, optical depths related to aerosol water are often largest during winter, despite usually larger aerosol optical depths in summer.

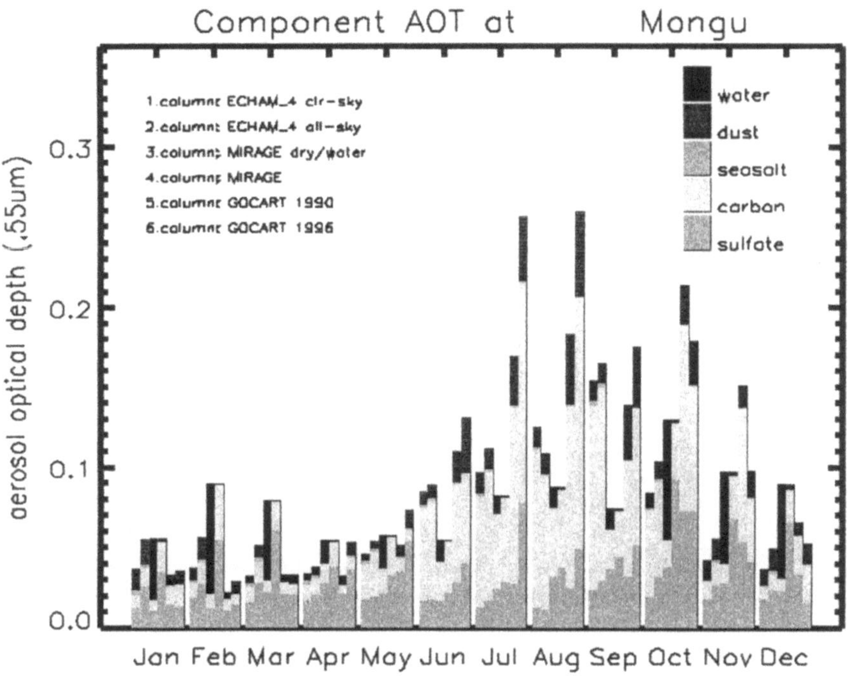
Component AOT at Mongu
1.column: ECHAM_4 clr-sky
2.column: ECHAM_4 all-sky
3.column: MIRAGE dry/water
4.column: MIRAGE
5.column: GOCART 1990
6.column: GOCART 1996
water
dust
seasalt
carbon
sulfate
aerosol optical depth (.55um)
0.0
0.1
0.2
0.3
Jan Feb Mar Apr May Jun Jul Aug Sep Oct Nov Dec

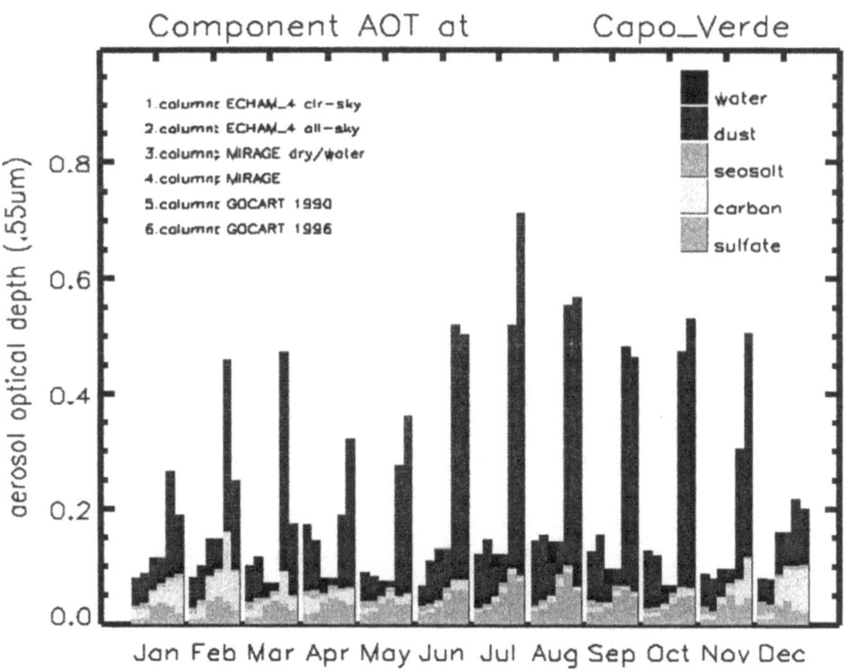
Component AOT at Capo_Verde
1.column: ECHAM_4 clr-sky
2.column: ECHAM_4 all-sky
3.column: MIRAGE dry/water
4.column: MIRAGE
5.column: GOCART 1990
6.column: GOCART 1996
water
dust
seasalt
carbon
sulfate
aerosol optical depth (.55um)
0.0
0.2
0.4
0.6
0.8
Jan Feb Mar Apr May Jun Jul Aug Sep Oct Nov Dec

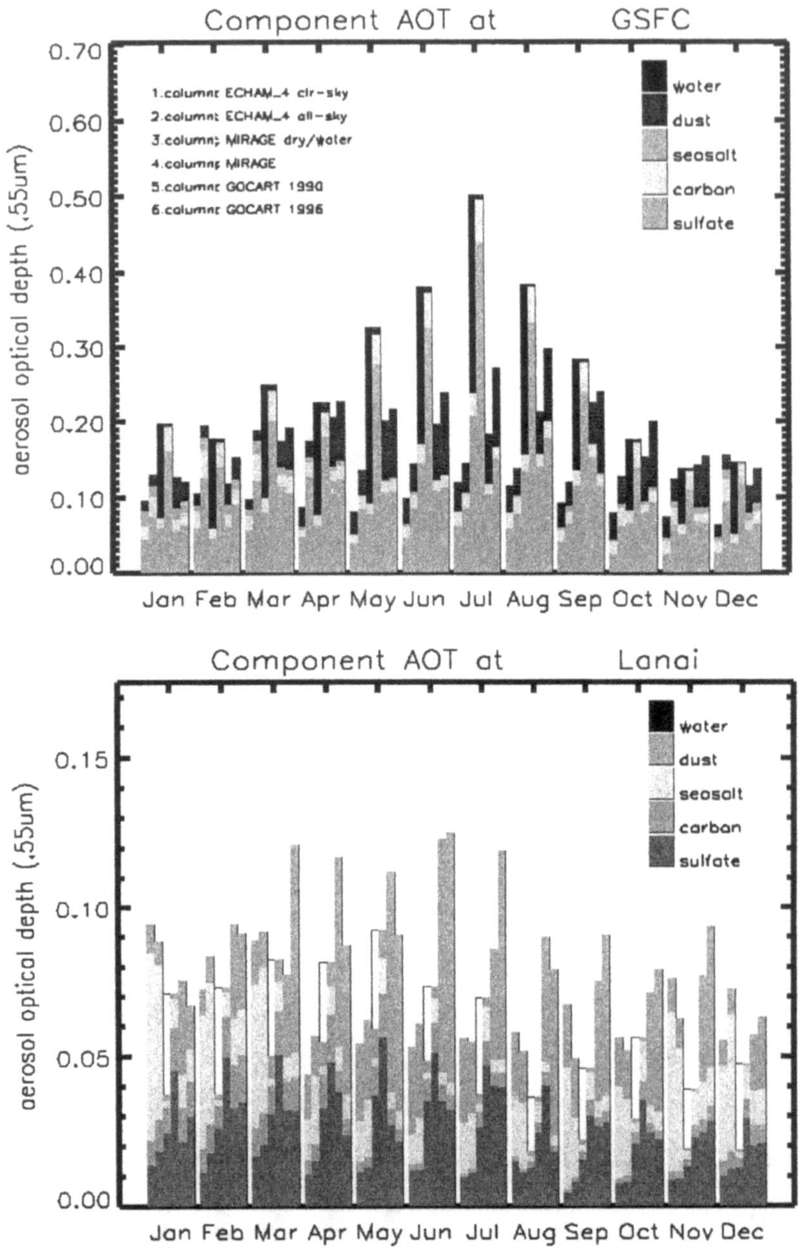


*Figure 7a-d.* Comparisons of modeled monthly averaged optical depth contributions for dust, sulfate, carbon and sea-salt among pairs of different model versions. Changes in component optical depths address (1) the impact of clouds on aerosol (by comparing ECHAM4 to a data-subset without grid-points with precipitating clouds, ECHAM4-clr), (2) the importance of water in aerosol (by separating in MIRAGE dry/water optical depths related to aerosol water and dry aerosol – vs. MIRAGE) and (3) the importance of assimilated meteorology (simulations for 1990 and 1996 with GOCART). Comparisons are presented for the four AERONET sites of Figures 2, 4, 5 and 6

Comparisons of Figures 6 and 7 addressed interpolated model values. To demonstrate that the sampled model tendencies apply on larger spatial scales, model comparisons were also conducted for large regions. Component aerosol column optical depths of all five models are compared for four different ocean regions in Figures 8: the **NW-Atlantic** is heavily influenced by urban-industrial pollution (sulfate and carbon), the **E-Atlantic** is dominated by dust, the **SE-Atlantic** has strong contributions from biomass burning during the first months in the years and the **W-Pacific** was added as a mixed site. For comparison purposes in Figures 8 *AVHRR-giss* optical depths averages based on NOAA-9 data (explained earlier) are presented.

AVHRR regional averages are usually larger than averages suggested by models. Note that E-Atlantic AVHRR averages are likely underestimates (due to removal of optically thicker dust cases during cloud-screening). This confirms the trend of models to underestimate aerosol optical depth from comparisons at AERONET sites. Total aerosol optical depths among models are often similar but quite different in their component partition. The optical depth component trends of models from local comparisons are largely confirmed: *GOCART* provides the largest dust optical depths. *MIRAGE* stands out with the largest sulfate optical depths, which in part offset deficiencies for carbon and dust. *CCSR* suggests one of the larger carbon contributions and compensates in part for small sulfate values. *ECHAM4* and *MIRAGE* display the strongest humidification effects. To understand component optical depth tendencies of models, the component aerosol column mass is compared and assumptions for their conversion into optical depth are examined.

### 4.3.2 Mass conversion theory

The conversion of a component aerosol mass $\boldsymbol{m}$ into a component aerosol optical depth $\tau$ by a model is expressed by the mass extinction efficiency $\boldsymbol{b}$, defined as

$$\boldsymbol{b} = \tau / \boldsymbol{m} = 0.75 * \boldsymbol{Q} / (\rho * \boldsymbol{r_e})$$

where $\boldsymbol{r_e}$ is the effective radius, $\rho$ is the aerosol density and $\boldsymbol{Q}$ is the extinction efficiency.

Given a component (dry) mass $\boldsymbol{m}$ (no aerosol water), these three aerosol properties, $\boldsymbol{r_e}$, $\rho$ and $\boldsymbol{Q}$ (in addition to ambient relative humidity and to permitted aerosol humidification) are critical for the derivation of component optical depths.

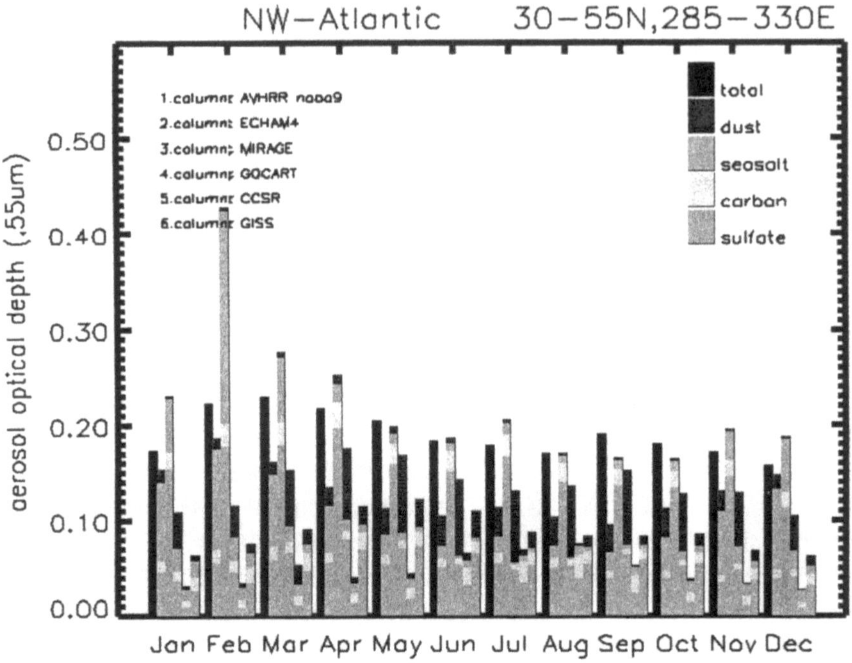
NW-Atlantic 30-55N,285-330E
1.column: AVHRR noaa9
2.column: ECHAM4
3.column: MIRAGE
4.column: GOCART
5.column: CCSR
6.column: GISS
total
dust
seasalt
carbon
sulfate
aerosol optical depth (.55um)
0.00
0.10
0.20
0.30
0.40
0.50
Jan Feb Mar Apr May Jun Jul Aug Sep Oct Nov Dec

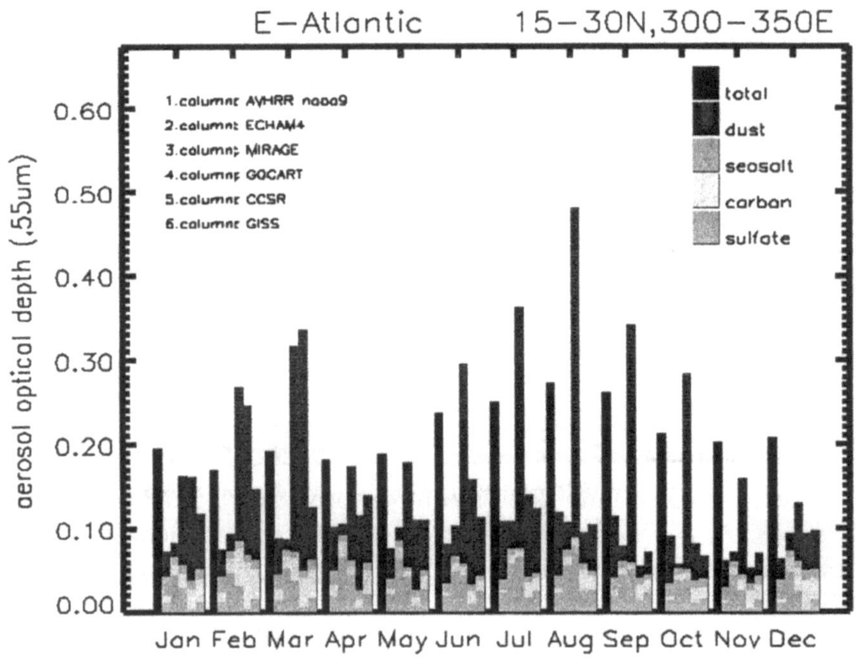
E-Atlantic 15-30N,300-350E
1.column: AVHRR noaa9
2.column: ECHAM4
3.column: MIRAGE
4.column: GOCART
5.column: CCSR
6.column: GISS
total
dust
seasalt
carbon
sulfate
aerosol optical depth (.55um)
0.00
0.10
0.20
0.30
0.40
0.50
0.60
Jan Feb Mar Apr May Jun Jul Aug Sep Oct Nov Dec

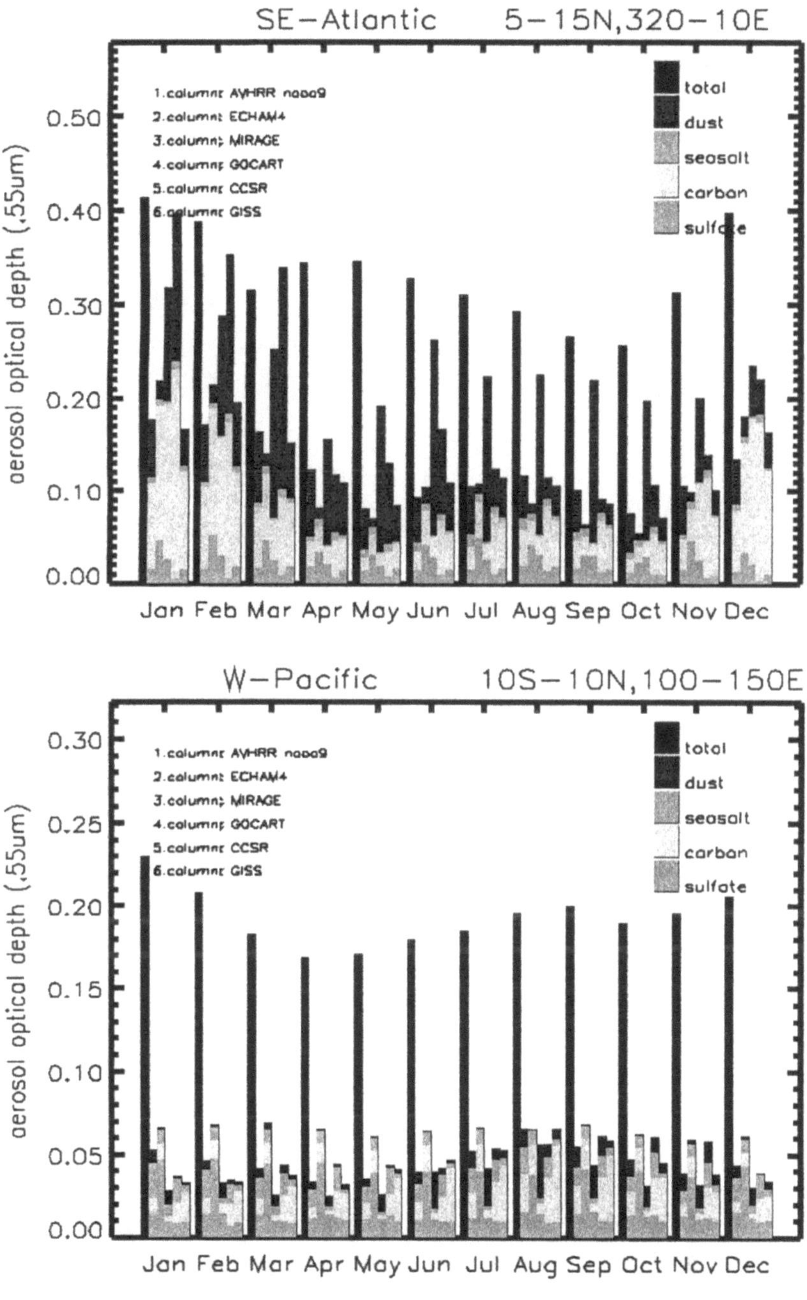


*Figure 8a-d.* Regional comparisons of modeled monthly averaged optical depth contributions from dust, sulfate, carbon and sea-salt for five models (ECHAM4, MIRAGE, GOCART, CCSR and GISS). Component comparisons are presented for regions over oceans: (1) for the north-western Atlantic (30-55N deg latitude and 285-330E deg longitude), (2) for the eastern Atlantic (15-30N, 300-350E), (3) for the south-eastern Atlantic (5-15N, 320-10E) and for the western Pacific (10S-10N, 100-150E). AVHRR-giss retrievals of (total) aerosol optical depths based on NOAA9 data are given for reference in the first (black) column

$r_e$ : The effective radius $r_e$ is the representing size for a distribution of sizes ($r_e$ is the ratio between third moment ($r*r*r$) and second moment ($r*r$ ) integrals of a size-distribution). Suggestions for $r_e$ based on inversion methods of AERONET sun/sky-photometer data are of limited use, because they relate to aerosol mixtures, rather than pure components. (Nonetheless, AERONET inversions can provide good estimates at sites and seasons, when a component clearly dominates). The component size assumptions of the tested models are summarized in Table 9. For assumed mono-modal size-distributions the associated effective radius $r_e$ is presented, for model inter-comparison. Comparisons of size-assumptions, however, are not possible for size-classes (many $r_e$) without knowing the associated weights.

*Table 9.* Comparison of size-assumptions for aerosol components in models. Either assumed aerosol size classes or effective radius $r_e$ of assumed size-distributions are listed for dry aerosol [$r_e$ at 100% relative humidity]

| Aerosol-type | ECHAM 4 | MIRAGE | GOCART | CCSR | GISS |
|---|---|---|---|---|---|
| Dust | *2 size-classes* | *2 size-classes* | *4 size-classes* | *10 size-classes* | *8 size-classes* |
| | .21μm | .13μm | .50μm | .13μm, .20μm, | .15μm, .25μm, |
| | 1.3μm | 1.2μm | 1.4μm | .33μm, .52μm, | .40μm, .80μm, |
| | | | 2.4μm | .82μm, 1.3μm, | 1.5μm, 2.5μm, |
| | | | 4.5μm | 2.0μm, 3.2μm, | 4.0μm, 8.0μm |
| | | | | 5.1μm, 8.0μm, | |
| Org. Carbon | .11μm[.27μm] | .13μm[.30μm] | .11μm[.27μm] | .29μm [.55μm] | *6 size-classes* |
| Black Carbon | .04μm | .023μm | .04μm | (in org.carb) | None |
| Sea Salt | *2 size-classes* | *2 size-classes* | *4 size-classes* | *10 size-classes* | None |
| | .73μm[3.5μm] | .13μm[.34μm] | .26μm,1.2μm, | 0.1 – 10μm | |
| | 5.9μm[17.μm] | 3.3μm[10.μm] | 2.4μm,7.6μm | [ca 3times larger] | |
| Sulfate | .24μm[.81μm] | .13μm[.34μm] | .24μm[.81μm] | .49μm[1.61μm] | .50μm |

**ECHAM4** and **GOCART** model aerosol sizes are based on the **GADS** data-set (Koepke et al., 1997)
**MIRAGE** predicts size for each mode within size-bounds – this table lists nominal sizes of the modes

Aerosol sizes for carbon and sulfate aerosol in *CCSR* are much larger than for *MIRAGE* or *ECHAM4*. Sulfate aerosol sizes of *MIRAGE* are smaller than those of *ECHAM4*.

- For hydrophilic aerosol components (sulfate, sea-salt and organic carbon) increases in ambient relative humidity cause a non-linear increase in aerosol size. Size increases are larger at higher relative humidity. In Table 9, expected effective radii $r_e$ at 100% relative are given in square brackets next to the assumed effective radii of the dry aerosol.

$\rho$: The densities of assumed dry aerosol components are compared in Table 10.

*Table 10.* Comparison of assumed density $\rho$ in models for components of dry aerosol (in g/cm3)

| Aerosol-type | ECHAM 4 | MIRAGE | GOCART | CCSR | GISS |
|---|---|---|---|---|---|
| Dust | 2.6 | 2.6 | 2.6 | 2.5 | 2.5 |
| Org. Carbon | 1.9 | 1.7 | 1.9 | 1.55 | None |
| Black Carbon | 1.0 | 1.0 | 1.0 | 1.25 | None |
| Sea Salt | 2.2 | 1.9 | 2.2 | 2.25 | None |
| Sulfate | 1.7 | 1.77 | 1.7 | 1.77 | None |

Densities of explicitly size-resolving aerosol components usually agree among models. Only different *CCSR* carbon densities and a smaller *MIRAGE* sea-salt density stand out.

- For hydrophilic aerosol components (sulfate, sea-salt and organic carbon) water uptake in aerosol decreases the aerosol density. Ultimately at high ambient relative humidities aerosol densities will be close to that of water (of 1g/cm3). Assuming aerosol sizes of the GADS data-set (Koepke et al., 1997), they are used in *ECHAM4* and *MIRAGE*, densities at 80% relative humidity reduce to 1.15g/cm3 for sulfate, 1.50g/cm3 for organic carbon and 1.18g/cm3 for sea-salt.

***Q***: The extinction efficiency ***Q*** is the ratio between extinction cross-section and the geometric cross-section. ***Q*** depends on aerosol size and composition (e.g. Lacis and Mishchenko, 1994, Tegen and Lacis, 1996). ***Q*** is largest, if particle radius and interacting wavelength have similar values. Maximum values for ***Q*** of near 3 are common for size-distributions with effective radii of about 0.5μm at mid-visible wavelengths. ***Q*** converges towards 2 for increasingly larger radius-to-wavelength ratios. For increasingly smaller radius-to-wavelength ratios, ***Q*** decreases sharply (inverse proportional to 4-th power) for scattering aerosol but only moderately (inverse proportional) for absorbing aerosol.

- For hydrophilic aerosol components (sulfate, sea-salt and organic carbon) water uptake impacts ***Q*** in two ways: Water uptake increases the aerosol size, thereby increasing the radius-to-wavelength ratio. And water uptake decreases the aerosol absorption, which is less important if the effective aerosol radius is larger than the wavelength.
- ***b*** : Mass extinction efficiency ***b*** (as illustrated by the formula above) is proportional to ***Q*** and inverse proportional to $\rho$ and $r_e$ . Moreover, all three properties ***Q***, $\rho$ and $r_e$ are function of relative humidity. Thus, for the evaluation of ***b*** assumptions on aerosol component humidification and data for ambient relative humidities are important as well.

Observed and calculated values for ***b*** are presented in Table 11. Calculations are based on assumptions for size, humidification and density by the GADS data-set and by the *CCSR* model (see Tables 9 and 10).

*Table 11.* Comparison of observed and calculated mass extinction efficiencies ***b*** (m2/g) at visible wavelengths

| Aerosol-type relative humidity | Observations 'ambient' conditions | **Calc.**, GADS **sizes** ** 0% / 80% / 100% | **Calc.**, CCSR **sizes** ** 0% / 80% / 100% |
|---|---|---|---|
| Dust < 1um | 1.-1.7 (Malm 94) | 0.6-2.4 * | |
| Dust > 1um | 0.25 (Carlson 77) | 0.1-0.4 * | |
| Org. Carbon | 4 (Liousse 96) | 2 / 13 / 82 | 5 / 11 / 25 |
| Black Carbon | 9 (Liousse 96) | 10 * | 5.7 * |
| Sea Salt < 1um | 0.4 (Andreae 95) | 1.3 / 8.6 / 18 | |
| Sea Salt > 1um | | .12 / .69 / 2.2 | |
| Sulfate | 5 (Liousse 96) | 3.6 / 21 / 90 | 2.4 / 11 / 40 |

* hydrophobic aerosol type,
** sizes are given in Table 9 [**G**lobal **A**erosol **D**ata **S**et (Koepke et al., 1997)]

Calculated values for ***b*** are not necessarily correct. However, the range and differences for ***b*** demonstrate the importance of assumptions for dry aerosol size selection and for ambient relative humidity in the determination of the component aerosol optical depth (for sulfate, see also Kiehl et al., 1999). Thus, a model with accurate data for aerosol mass is going to compromise its ability to predict aerosol optical depth (and climatic impacts) with a poor selection for ***b***. The assumption of a constant for ***b***, as for a few components in some models ('None' in Tables 9 or 10), will certainly introduce errors. For hydrophilic aerosol, in addition, the strong sensitivity to size at higher ambient relative humidities creates a big problem: Already small deviations in ambient relative humidity create large differences in component aerosol size and optical depth. Thus, rather than predicting ambient relative humidity (e.g. ECHAM4), less variable data from assimilations (e.g. GOCART) are often adopted or simply prescribed (e.g. *MIRAGE*).

The effective ***b*** can be deduced from the ratio between component optical depth and component mass. Thus, modeled component mass are introduced next.

### 4.3.3 Component MASS *m*

All models provided data on mass for sulfate, organic carbon, black carbon, sea-salt and dust. The mass does not include aerosol water, which adds to the aerosol size of sulfate, organic carbon and sea-salt as a function of ambient relative humidity. Mass data are also provided for the cloud-removed subset, *ECHAM4 clr*. Component aerosol mass of these models is compared in Figures 9 for the four AERONET sites of Figures 2.

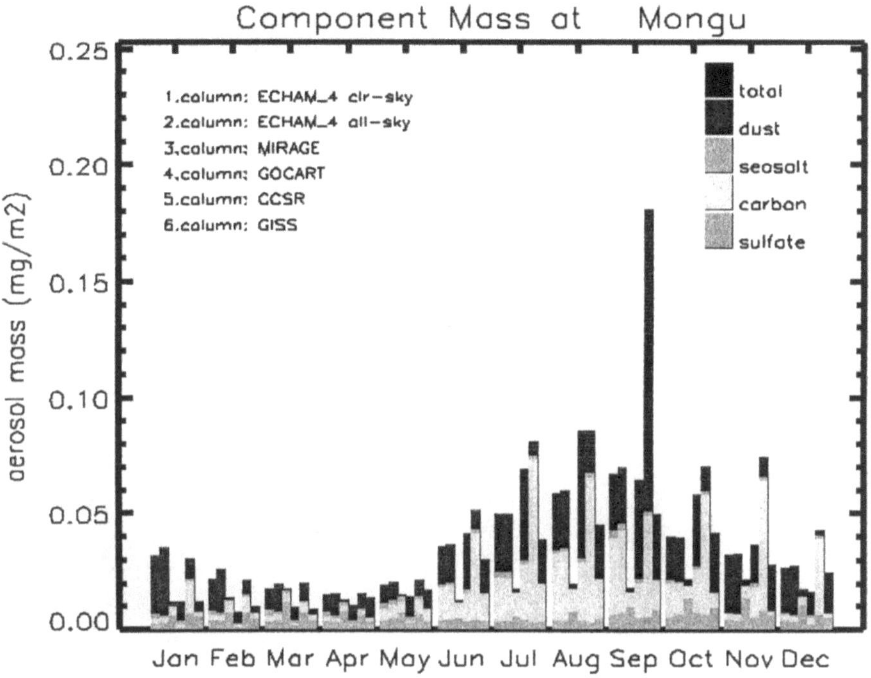
Component Mass at Mongu
0.25
0.20
0.15
0.10
0.05
0.00
aerosol mass (mg/m2)
1.column: ECHAM_4 clr-sky
2.column: ECHAM_4 all-sky
3.column: MIRAGE
4.column: GOCART
5.column: CCSR
6.column: GISS
total
dust
seasalt
carbon
sulfate
Jan Feb Mar Apr May Jun Jul Aug Sep Oct Nov Dec

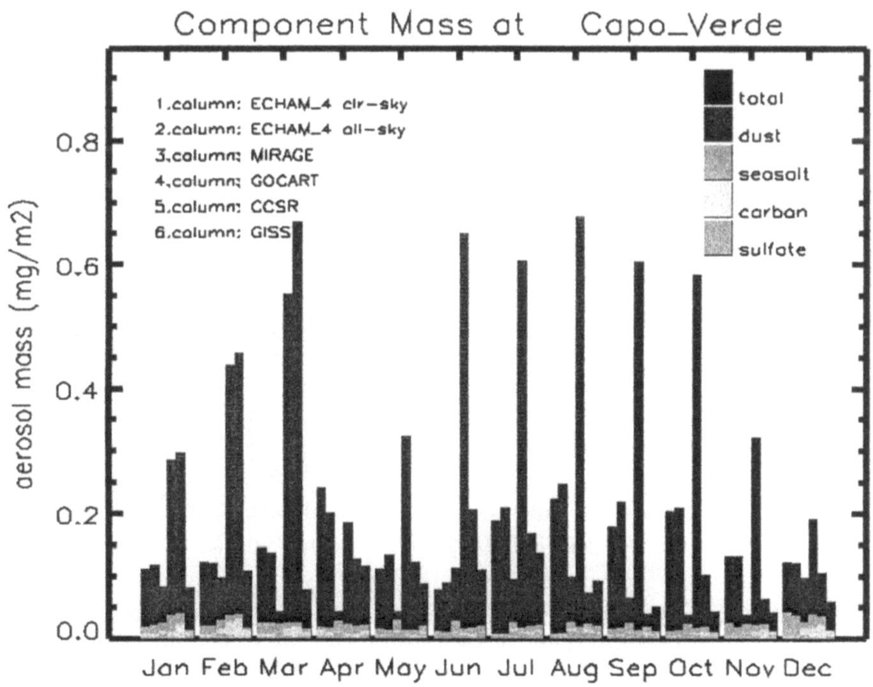
Component Mass at Capo_Verde
0.8
0.6
0.4
0.2
0.0
aerosol mass (mg/m2)
1.column: ECHAM_4 clr-sky
2.column: ECHAM_4 all-sky
3.column: MIRAGE
4.column: GOCART
5.column: CCSR
6.column: GISS
total
dust
seasalt
carbon
sulfate
Jan Feb Mar Apr May Jun Jul Aug Sep Oct Nov Dec

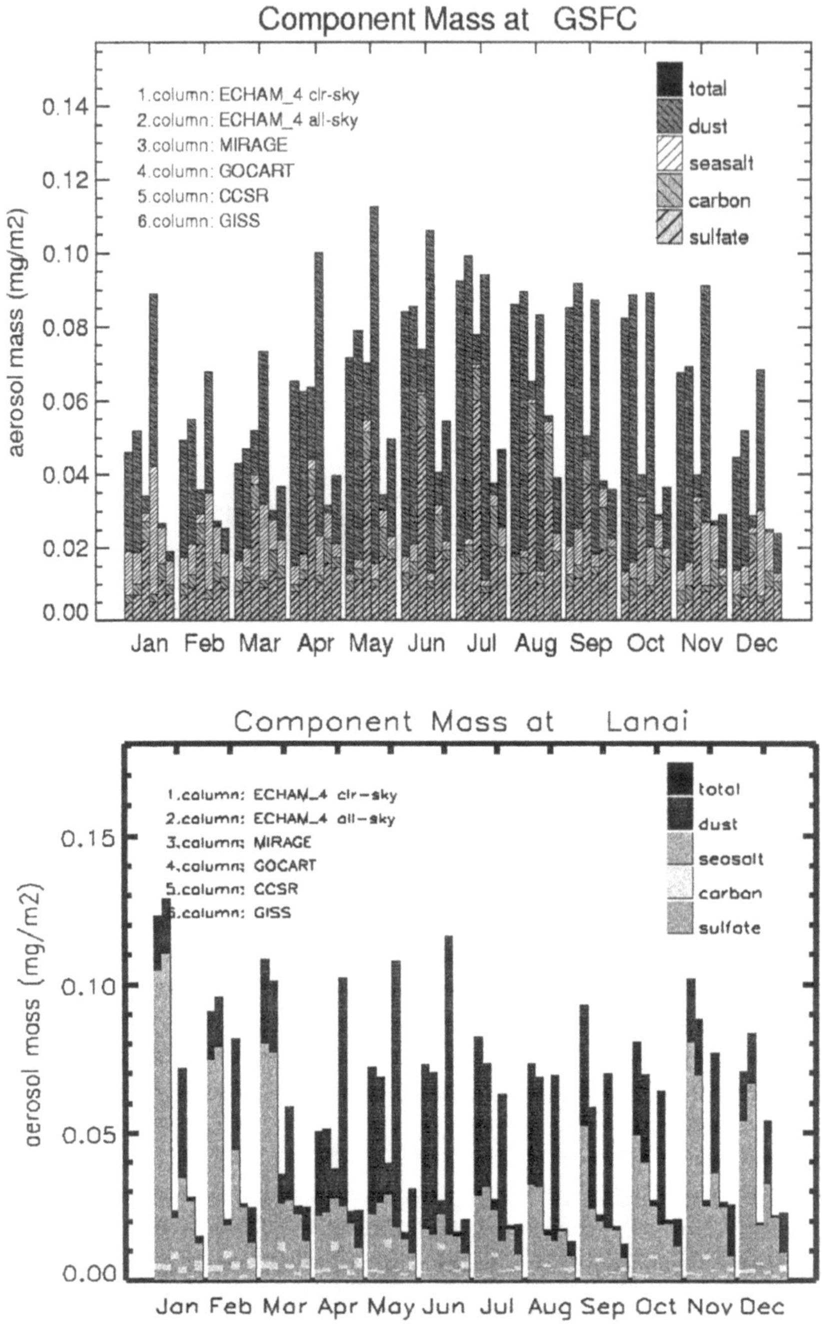


*Figure 9a-d.* Comparison of modeled monthly averages for aerosol mass. Individual contributions from dust, sulfate, carbon and sea-salt are presented for the five tested models (ECHAM4, MIRAGE, GOCART, CCSR and GISS). Comparisons are given for the four AERONET sites of Figures 2, 4, 5, 6 and 7

**O.CARBON- *m***: Despite the common reference for carbon emissions (Table 6), there are large differences in organic carbon mass. All models reproduce the seasonality of tropical biomass burning emissions, but quantitatively the range of monthly averages covers up to one order of magnitude. *CCSR* suggests by far the largest carbon mass. *MIRAGE* displays the lowest mass values in the tropics. At higher latitudes, though, *MIRAGE* suggests one of the larger carbon masses and *GOCART* values are now among the lowest.

**B.CARBON- *m***: The seasonal behavior for black carbon almost mirrors that of organic carbon. Black carbon mass is smaller than organic carbon mass, but there are large model differences for the ratio between organic and black carbon. Near biomass burning, ratios of all models are close to 8:1, however, ratios range from 2:1 to 12:1 at higher latitudes. Larger ratios are common for *GISS* and smaller ratios are common for *GOCART,* with the consequence that at high latitudes, *GISS* black carbon mass is lowest among models.

**DUST- *m***: Dust modeling in *GOCART* (based on Ginoux, 2000) produces the largest dust mass. The neglect of larger dust sizes in *ECHAM4* reduces mass to about half its value. This is still sufficient to exceed the dust mass of the other three models, near sources. There, *MIRAGE* dust mass is usually lowest among models. Away from dust sources, dust mass of *GOCART* and *ECHAM4* stand out even more, dominating by up to one order of magnitude. Away from dust sources *CSSR* dust mass at times is very small.

**SULFATE- *m***: By far the largest sulfate mass is usually suggested by *MIRAGE*. At higher latitudes in the northern hemisphere *MIRAGE* sulfate mass is about 2 to 4 times larger than for other models. Seasonal sulfate mass trends among models are similar.

**SEASALT- *m***: Differences in sea-salt mass among models are large, as the range of monthly averages often exceeds one order of magnitude. Sea-salt mass in especially low in the *GISS*-model. The seasonal cycle between winter maximum and summer minimum for the northern hemispheric is most pronounced in *GOCART* and *ECHAM4*.

### 4.3.4 Effective Mass Extinction Efficiency *b*

The mass extinction efficiency ***b*** for each aerosol component represents the conversion factor, which multiplied with the aerosol component mass, determines the component aerosol optical depth. Inversely, data for aerosol mass and for the associated aerosol optical depth permit a derivation of the 'effective' mass extinction efficiency. These 'effective' ***b*** of models are compared in Table 12.

*Table 12.* Comparison of mass extinctions efficiencies ***b*** (m/g2) in models ([..-..] indicate typical range)

| Aerosol-type | **ECHAM4** | | **MIRAGE** | | **GOCART** | **CCSR** | **GISS** | |
|---|---|---|---|---|---|---|---|---|
| Version: | all-sky | cl-free | | no H2O | | | Koch | Tegen |
| Dust | .5 * | .5 * | [.7-.8] | [.7-.8] | [.8-.9] | [.1-10] | [1-2] | [1-2] |
| Org. Carbon | [3-10] | [3-6] | [6-9] | [2-4] | [5-8] | [4-5] | 8 * | 8 * |
| Black Carbon | 7.7 * | 7.7 * | [3-5] | [3-5] | [9-11] | [4-5] | 9 * | 9 * |
| Sea Salt | [.6-2] | [.6-1.5] | [.7-2] | [.2-.4] | 0.4 * | [.3-5] | 2 * | 0.3 * |
| Sulfate | [5-11] | [4-5] | [5-10] | [2-4] | [8-12] | [3-4] | [3-7] | 5 * |

* assumed constant

**O.CARBON-** ***b***: *GISS* is the only model that prescribes the organic carbon conversion at 8m2/g. This is a large value in comparison to that of size-resolved conversions, which usually range between 3 and 6m2/g, The *GISS* value is only reached and slightly exceeded in *GOCART* or *ECHAM4*, when ambient relative humidity in models is higher.

**B.CARBON-** ***b***: *GISS* and *ECHAM4* prescribe the black carbon conversion with values of 9m2/g and 7.7m2/g. The size-resolved conversion of *GOCART* is slightly larger between 7.5 and 9m2/g. The surprising low values of *MIRAGE* (between 2 and 4m2/g) are an artifact in the derivation, as *MIRAGE* considers aerosol internally mixed.

**DUST-** ***b***: *ECHAM4* prescribes a dust conversion value of 0.5m2/g. Values near 0.8m2/g are typical for the *GOCART* and *MIRAGE*. The lack of seasonal variability suggest that changes to dust aerosol size are small. This is surprising and quite in contrast to *GISS*. Strong seasonal variations but also the largest values, ranging from 1.0 to 1.8m2/g, in *GISS* seem to indicate a frequent dominance of sub-micrometer dust sizes.

**SULFATE-** ***b***: All tested models use size-resolving schemes for the sulfate conversion. The largest conversions at about 10m2/g are suggested by *GOCART* and on occasion by *ECHAM4*. Both models are based on the same assumption for size and humidification, thus, differences largely reflect the ambient relative humidity of the model. Values for GISS at 4m2/g and for CCSR at 3m2/g reflect assumptions of a larger sulfate aerosol size

**SEASALT-** ***b***: Model values for sea-salt mass conversion are near 0.5m2/g. *MIRAGE* and *ECHAM4* values are slightly larger and GOCART values are slightly lower. *GISS* assumes a relatively large constant conversion of 2m2/g.

## 4.4 Model - Performance

Based on comparisons to the local aerosol optical depth statistics of sixteen AERONET-sites, all tested models have a tendency to underestimate aerosol optical depths. Although there are a few exceptions, underestimates are much more frequent (and larger) than overestimates. Based on comparisons of component mass and component optical depth characteristic tendencies of each model are summarized.

**ECHAM4** underestimates dust, carbon and sulfate optical depth. The model is very sensitive to the (predicted) ambient relative humidity. This is demonstrated by larger optical depths sulfur and sea-salt during northern hemispheric winters. Larger sulfur mass conversions of *GOCART*, which are based on the same assumptions for sulfur size and humidification, suggest that *ECHAM4* is often too dry. This is at least a contributing factor in underestimates to optical depths of sulfate, sea-salt and organic carbon aerosol. The removal of cloud scenes (with column water exceeding 33g/m3) in the *ECHAM4-clr* data-subset usually reduces the aerosol optical depths. Smaller optical depths are mainly the result of a reduced water uptake under smaller ambient relative humidities. However, often a lower aerosol dry mass in *ECHAM4-clr* appears to be a contributing factor.

**MIRAGE** stands out with the largest sulfate optical depths, while all other component optical depths are usually low. Small dust optical depths are most likely due to weak emissions. A lack in aerosol mass is also the major reason for underestimates in carbon optical depths near tropical biomass burning. Carbon mass at high latitude, however, is one of the largest among models. The large sulfate optical depths are attributed to sulfate mass overestimates, because the mass conversion is one of the smallest among models.

**GOCART** is the only model whose dust aerosol optical depths seem sufficient to match the AERONET statistics near dust sources. Away from dust sources, dust optical depths are usually too large, most likely due to the lack of removal processes. This offsets in part the lack in carbon and sulfate optical depth, however, it also delays summer-time maxima at northern hemispheric urban-industrial sites. Low sulfate optical depths are related to sulfate mass, because mass conversions for sulfate are the largest among models.

**CCSR** carbon optical depths are largest among models, however AERONET averages near biomass burning are rarely reached. Aerosol transport appears too weak (or removal processes too strong) as optical depths for dust and especially for carbon decline more rapidly away from sources than in other models. Sulfate aerosol optical depths are low primarily due to a relatively large assumption for sulfate aerosol size.

**GISS** dust and carbon optical depths are too low, despite mass conversions that are large in comparisons to other models. These large values in part compensate for the lack in mass. Overestimates for sulfur aerosol size cause a smaller sulfur mass conversion, which in part is responsible for underestimates in sulfate optical depth. One of the smallest sea-salt masses among models is compensated by the largest sea-salt mass conversion.

# 5. CONCLUSIONS

Statistics from sky/sunphotometers at AERONET sites throughout the world provided the background for a comparison of monthly or seasonally averaged aerosol optical depths. The comparisons involved aerosol retrievals by operational satellites and representations of aerosol in global models.

Aerosol optical depths retrievals from five different satellites were compared. Critical issues are instrument calibration, the ability to remove scenes with clouds, assumptions on aerosol properties and contributions from background reflectance. Most retrievals are confined to ocean scenes, due to difficulties in representing background reflectance at high accuracy over land. Quantitatively accurate satellite retrievals over land remain a challenge and current efforts are often at an experimental stage. Still, these attempts provide data on spatial distribution, although initial applications to identify biases of (AERONET-) sites data in comparisons to regions (of models) remained inconclusive.

Aerosol optical depths representations of five global models were compared. Monthly averages among models vary and are usually smaller than AERONET averages. All models distinguish in optical depth contributions by carbon, sulfate, sea-salt and dust. Such component treatment promises better simulations for aerosol concentration and aerosol absorption, which are critical for accurate simulations of aerosol climatic impacts. A component treatment, however, complicates modeling and introduces new sources for errors. To identify model deficiencies and to circumvent offsetting errors, comparisons were conducted on a component basis for mass and optical depth. Critical in a derivation of a component optical depth is the conversion from mass into optical depth. Necessary assumptions for aerosol size and humidification were compared. Many poor assumptions were identified and several models are already represented by improved versions. The overall agreement among models has improved, but there are still many discrepancies that are better explored in more focused comparisons.

Co-location in time and space, are vital for more useful comparisons. Co-location in time for modeling means the use of identical data-sets for emission and assimilated meteorology. Co-location in time for comparisons to AERONET or satellite data means the application of data-screens that accommodate the less frequently sampled data-set. The regional representation of local measurements remains a problem. Sub-grid modeling is essential for the representation of non-linear processes in modeling (e.g. ambient relative humidity and its impact on aerosol size). Satellite data hold the key in connecting local (AERONET) data to regional averages (of models). New satellites with multi-angle viewing capabilities, finer resolution spatial and spectrally, and better calibration should provide more accurate retrievals.

Nonetheless, the need for in-situ data and data from field experiments remains in order to address assumptions for aerosol size and aerosol composition (including humidification). A better understanding of aerosol properties should also lead to a re-evaluation and possible combination of twenty year long data-sets from AVHRR and TOMS, which provide a global and seasonal framework in which model-output can be tested. As future remote sensing from space will be providing a more detailed characterization of the earth and its atmospheres, as simultaneous monitoring of the atmosphere from the ground is spreading, and as global communication and data-processing is accelerating, modelers are challenged to test their models, with simulations in a 'nowcast' mode.

## 6. REFERENCES

Andreae. M., Climate change of changing atmospheric aerosol levels, in World Survey of Climatology, 16 *Future Climate of the World,* Elsevier, New York, 1995.

Benkowitz, C., M.Scholtz, J.Pacyna, L.Tarrason, J.Dignon, E.Voldner, P.Spiro, J.Logan and T.Graedel, Global gridded inventories of anthropogenic emissions of sulfur and nitrogen, *J.Geophys.Res. 101*, 29239-29253, 1996.

Carlson, T. and R.Caverly, Radiative characteristics of saharan dust at solar wavlengths, *J.Geophys.Res, 82*, 3141-3152, 1977.

Chin, M., R. Rood, S.-J. Lin, J.-F. Muller, and A. Thompson, Atmospheric sulfur cycle simulated in the global model GOCART: Model description and global properties, *submitted to J.Geophys.Res*, 2000.

Chin, M., D.Jacob, G. Gardner, P.Spiro, M.Foreman-Fowler and D.Savoi, A global three dimensional model of tropospheric sulfate, *J.Geophys.Res. 101*, 18667-18690, 1996.

Cooke,W. and J.Wilson, A global black carbon aerosol model, *J.Geophys.Res. 101*, 19395-19409, 1996.

Cooke,W., C.Liousse, H.Cachier and J.Feichter, Construction of 1deg*1deg fossil fuel emission data set for carbonaceous aerosol and implementation and radiative impact in the ECHAM4 model, *J.Geophys.Res. 104*, 22137-22162, 1999.

Deuze, J., M.Herman, P.Goloub, D.Tanre and A. Marchand, Characterization of aerosols over ocean from POLDER, *Geophys.Res.Lett., 26*, 1421-1424, 1998.

Eck, T., B.Holben, J.Reid, O.Dubovik, A.Smirnov, N.O'Neill, I.Slutsker and S.Kinne, Wavelength dependence of the optical depth of biomass burning, urban and desert dust aerosols, *J.Geophys.Res. 104*, 31333-31349, 1999.

Dubovik,O, B.Holben,Y.Kaufman, M.Yamasoe, A.Smirnov, D.Tanre and I.Slutsker. Single scattering albedo of smoke retrieved from the sky-radiance and solar transmission measured from the ground, *J.Geophys.Res. 103*, 31903-31924, 1998.

Dubovik, O. and M.King, A flexible inversion algorithm for the retrieval of aerosol optical properties from sun and sky radiance measurements, *submitted to J.Geophys.Res.* 2000.

Dubovik, O., A.Smirnov, B.Holben, M.King, Y.Kaufman, T.Eck and I.Slutsker, Accuracy assessments of aerosol optical properties retrieved from AERONET sun and sky radiance measurements, , *J.Geophys.Res.* 105, 9791-9806, 2000.

Easter,R., S.Ghan, R.Saylor, E.Chapman, H.Abdul-Razzak, R.Leung, Y.Zhang, X.Bian, and A.Zaveri, MIRAGE: A Model for Integrated Research on Atmospheric Global Exchanges, *submitted to J.Geophys.Res.* 2000.

Gilette, D. A wind tunnel simulation of erosion of soil: Effect of soil texture, sandblasting, wind-speed, and soil consolidation, *Atmos.Environ. 12*, 1735-1743, 1978.

Gillette D. and R.Passi, Modeling dust emission caused by wind erosion, *J.Geophys.Res. 93*, 14233-14242, 1988.

Ginoux, P., M. Chin, I. Tegen, D. Savoie, J. Prospero, B. Holben, and S.-J. Lin, Global simulation of tropospheric dust: 1. Model description and evaluation, *submitted to J.Geophys.Res.* 2000.

Gong, S., L.Barrie, J.Prospero, D.Savoie, G.Ayers, J-P.Blanchet, L.Spacek, Modeling sea-salt in the atmosphere, atmospheric concentration and fluxes, *J.Geophys.Res. 102*, 3819-3830, 1997.

Guenther, A., C.Hewitt, D.Erikson, R.Fall, C.Geron, T.Graedel, P.Harley, L.Klinger, M.Lerdau, W.McKay, T.Pierce, B.Scholes, R.Steinbrecher, R.Tallamraju, J.Taylor and P.Zimmermann, A global model for natural volatile organic compound emissions, *J.Geophys.Res. 100*, 8873-8872, 1995.

Herman, M., J. Deuze, C.Devaux, P.Goloub, F. Breon and D.Tanre, Remote sensing of aerosols over land surfaces including polarization measurements and applications to POLDER measurements, *J.Geophys.Res. 102*, 17039-17049, 1997.

Hiragushi, A. and T.Nakajima, Development of a two channel aerosol retrieval algorithm on global scale using NOAA AVHRR. *J.Atmos.Sci. 56*, 924-941, 1999.

Hobbs P., J.Reid, R.Kothenruther, R.Ferek and R.Weiss, Direct radiative forcing by smoke from biomass burning, *Science* 275 1776-1778, 1997.

Holben, B., D.Tanre, A.Smirnov, T.Eck, I.Slutsker, N.Abuhassan, W.Newcomb, J.Schafer, B.Chatenet, F.Lavenue, Y.Kaufman, J.Vande Castle, A.Setzer, B.Markham, D.Clark, R.Frouin, R.Halthore, A.Karnieli, N.O'Neill, C.Pietras, R.Pinker, K.Voss, G.Zibordi, An emerging ground-based aerosol climatology: Aerosol Optical Depth from AERONET, *submitted to J. Geophys. Res.* 2000.

Holben, B., T.Eck, I.Slutsker, D.Tanre, J.Buis, E.Vermote, J.Reagan, Y.Kaufman, T.Nakajima, F.Lavevau, I.Jankowiak and A.Smirnov, AERONET, a federated instrument network and data-archive for aerosol characterization, *Rem.Sens.Environ. 66*, 1-66, 1998.

Kahn, R., P.Banerjee and D.McDonald, The sensitivity of multi-angle imaging to natural mixtures of aerosols over ocean, *J.Geophys.Res.* in print, 2000.

Kaufman, Y, D.Tanre, L.Remer, E.Vermote, A.Chu, B.Holben, Operational remote sensing of tropospheric aerosol over land from EOS-Moderate Resolution Imaging Spectrometer. *J.Geophys.Res. 102*, 17051-17067, 1997.

Kiehl, J., and B.Brieglab, The relative role of sulfate aerosol and greenhouse gases in climate forcing, *Science 260*, 311-314, 1993.

Kiehl, J., T.Schneider, P.Rasch and M.Barth, Radiative forcing due to sulfate aerosols from simulations with the NCAR CCM3, *J.Geophys.Res.* 105, 1441-1457, 2000.

King, M., Y.Kaufman, D.Tanre and T.Nakajima, Remote sensing of tropospheric aerosol from space: past, present and future. *Bull.Am.Met.Soc.* 80, 2229-2259, 1999.

Koepke, P., M.Hess, I.Schult and E.P.Shettle, Global aerosol data set, *Tech.Rep. 243, Max-Planck-Institut fuer Meteorolgie*, Hamburg, 1997.

Kothenruther, R. and P.Hobbs, Humidification factors for aerosol from biomass burning in Brazil, *J.Geophys.Res. 103*, 32081-32089, 1998.

Koch, D., D.Jacob , I.Tegen, D.Rind and M.Chin, Tropospheric sulfur simulation and sulfate direct forcing in the Goddard Institute for Space Studies (GISS) general circulation model. *J.Geophys.Res. 104* 23799-23822, 1999.

Kothenruther, R., P.Hobbs and D.Hegg, Humidification factorsfor atmospheric aerosols off the midatlantic coast of the United States, *J.Geophys.Res. 104*, 2239-2251, 1999.

Lacis, A. and M.Mishchenko, Climate Forcing, Climate Sensitivity, and Climate Response: A radiative modeling perspective on atmospheric aerosols. *Aerosol Forcing and Climate. Eds. Charlson and Heintzenberg, J.Wiley & Sons Ltd.*, 1994.

Liousse C., J.Penner, C.Chuang, J.Walton, H.Eddleman and H.Cachier, A global three-dimensional model study of carbonaceous aerosol. *J.Geophs.Res. 101*, 19411-19432, 1996.

Lohmann,U., J.Feichter, C.Chuang and J.Penner, Predicting the number of cloud droplets in the ECHAM GCM, *J.Geophys.Res. 104*, 24557-24563, 1999.

Malm, W., J.F.Sisler, D.Hoffmann, R.Eldred and T.Cahill, Spatial and seasonal trends in particle concentration and optical extinction in the United States, *J.Geophys.Res. 99*, 1347-1370, 1994.

Mishchenko, M., L.Travis, R.Kahn and R.West, Modeling phasefunctions for dust-like tropospheric aerosol using a shape mixture of randomly oriented polydisperse speriods, *J.Geophys.Res. 102*, 16831-16847, 1997.

Mishchenko, M. and L.Travis, Satellite retrieval of aerosol properties over ocean using polarization as well as intensity of reflected sunlight, *J.Geopys.Res. 102,* 16898-17013, 1997.

Mishchenko, M., I.Geogdzhayev, B.Cairns, W.Russow and A.Lacis, Aerosol retrievals over ocean using channel 1 and 2 AVHRR data: a sensitivity study and preliminary results, *Appl.Opt. 36*, 7325-7341, 1999.

Monahan, E., D.Spiel and K.Davidson, A model of marine aerosol generation via whiecaps and wave-disruptions, *Oceanic Whitecaps, Reidel Publish.Co.*, 1986,

Nakajima,T and A.Hiragushi, A use of two-channel radiances for an aerosol characterization from space. *Geophys.Res.Lett. 25*, 3815-3818, 1998.

O'Dowd, C., M.Smith, I.Consterdine and J.Lowe, Marine aerosol, sea-salt and the marine sulfur cycle, a short review, *Atmos.Environm. 21*, 73-80, 1997.

Olivier, J., A.Bouwan, C.Van der Mass, J.Berdowski, C. Veldt, J.Bloss, A.Vesschedijk, P.Znaweld and J.Haverlag, Description of EDGAR version 2.0, a set of global emission inventories of greenhouse gases and ozone depleting substances for all anthropogenic and most natural sources on a per country basis on a 1x1 grid, *RIVM/TNO report 771060 002, RIVM*, Bilthoven, Dec. 1996.

Smirnov, A., B.Holben, O.Dubovik, D.Westphal, A.Goroch, C.McClain, T.Eck and I.Slutsker, Atmospheric aerosol optical properties in the Persian Gulf region, *submitted to J.Geophys.Res.* 2000.

Smirnov, A., B.Holben, T.Eck, O.Dubovik and I.Slutsker, Cloud screening and quality control algorithms for the AERONET database. *accepted to Rem.Sensing of Environm.*, 2000.

Stowe, L., A.Ignatov and R.Singh, Development, validation and potential enhancements to the second generation operational aerosol products at the National Environmental Satellite, Data, Information Service of the National Oceanic and Admospgeric Administration (NESDIS/NOAA). *J.Geophys.Res. 102*, 16923-16934, 1997.

Stowe, L,. P.Davis, E.McClain, Scientific basis and initial evaluation of the CLAVR-1 global clear/cloud classification algorithm for thc advanced very high resolution radiometer, *J.Geophys.Res. 16*, 656-681, 1999.

Takemura,T., H.Okamoto, Y.Maruyama, A.Numaguti, A.Hiragushi and T.Nakajima, Global three dimensional simulation of aerosol optical thickness distribution of various origins, *accepted to J.Geophys.Res* 2000.

Tang I. and H.Munkelwitz, Water activities, densities, and refractive indices of aqueous sulfate and sodium nitrate droplets of atmospheric importance, *J.Geophys.Res. 99*, 18801-18808, 1994.

Tegen I., and A.Lacis, Modeling of particle size-distribution and its influence on the radiative properties of mineral dust. *J.Geophys.Res. 101*, 19237-19244, 1996.

Tegen, I., P.Hollrig, M.Chin, I.Fung, D.Jacob and J.Penner, Contribution of different aerosol species to the global aerosol extinction optical thickness: Estimates from model results, *J.Geophys.Res. 102*, 23895-23915, 1997.

Tegen, I., D.Koch, A.Lacis and M.Sato, Towards a global climatology: preliminary trends in tropospheric aerosol amounts and corresponding impact on aerosol radiative forcing between 1950 and 1990, *submitted to J.Geophys.Res.* 2000.

Torres, O., P.Barthia, J.Herman, Z.Ahmad and J.Gleason, Derivation of optical properties from satellite measurementsof backscattered ultraviolet radiation: Theoretical basis. *J.Geophys.Res. 103*, 17099-17110, 1998.

Vermote. E., D.Tanre, J.Deuze, M.Herman, J.Morcrette, Second simulation of the satellite signal in the solar spectrum(6S), *6S user's guide version 1*, Lab. d'Optique Atmospherique, Lille, France, 214pp.

Westphal, D., O.Toon, and T.Carlson, A study of mobilization and transport of saharan dust, *J.Atmos.Sci. 45*, 2145-2175, 1988.

# Remote Sensing of Snow and Characterization of Snow Albedo for Climate Simulations

ANNE W. NOLIN and ALLAN FREI
*National Snow and Ice Data Center Cooperative Institute for Research in Environmental Sciences. The University of Colorado, Boulder, USA*

Abstract: Accurate estimates of the spatial distribution and albedo of snow cover are needed for climate models, that use surface albedo as a lower boundary condition. We perform a sensitivity study that shows how model parameterizations of snow albedo affect computed snow-atmosphere fluxes. When albedo is calculated as a function of snow surface grain size, the variable albedo is significantly more realistic and representative than constant albedo values. We then describe new and planned satellite-derived products that will monitor seasonal changes in snow extent and albedo and have particular relevance to the climate modeling community.

## 1. INTRODUCTION

Over the past two decades, innovations in satellite technology and processing algorithms have generated improved regional and global snow mapping products. With a range of satellite-derived snow products available, we aim to clarify which products might best serve the climate modeling community. One way to examine this question is to first investigate how climate models portray snow, specifically looking at how modeled snow-atmosphere fluxes are affected by the type of snow albedo parameterization. The objective of this paper is to provide a better understanding of the sensitivity of surface-atmosphere fluxes to albedo parameterization and, given these results, recommend and describe appropriate remote sensing snow products for climate simulations.

From the perspective of traditional climate simulations, seasonal snow cover is considered primarily a passive responder to climate, where the



*M. Beniston and M.M. Verstraete (eds.),*
*Remote Sensing and Climate Modeling: Synergies and Limitations*, 159–180.

annual pattern of snow accumulation and ablation are controlled by atmospheric processes. However, there exists ample evidence that snow cover affects atmospheric temperatures and circulation at local, regional and even hemispheric scales (Dickson and Namias, 1976; Dewey, 1977; Walsh and Ross, 1988; Barnett et al., 1989; Cohen and Rind, 1991; Walsh, 1993; Groisman et al., 1994; Walland and Simmonds, 1997; Gutzler and Preston, 1997; Ellis and Leathers, 1998; Clark, 1998; Cohen and Entekhabi, 1999; Frei and Robinson, 1999). Although this feedback involving snow albedo and air temperature has long been considered important, the magnitude of its effect on surface-atmosphere fluxes is only just now becoming known.

Snow albedo ($\alpha$) in the optical region (0.4 - 2.5 μm) ranges from 0.98 in the visible part of the spectrum to near zero in the shortwave infrared region. Spectral albedo also varies strongly as a function of surface layer grain size with near-infrared values decreasing markedly as snow grains become larger. Figure 1 shows both the spectral and grain size dependencies of snow albedo. For clean, deep snow, grain size is the primary physical property governing snow albedo (Wiscombe and Warren, 1980; Warren and Wiscombe, 1980). The rate of grain growth depends strongly on snowpack temperature and temperature gradient (Colbeck, 1979; 1982) and multi-temporal observations of grain growth have been related to changes in snowpack thermodynamics and snow albedo (Nolin and Stroeve, 1997).

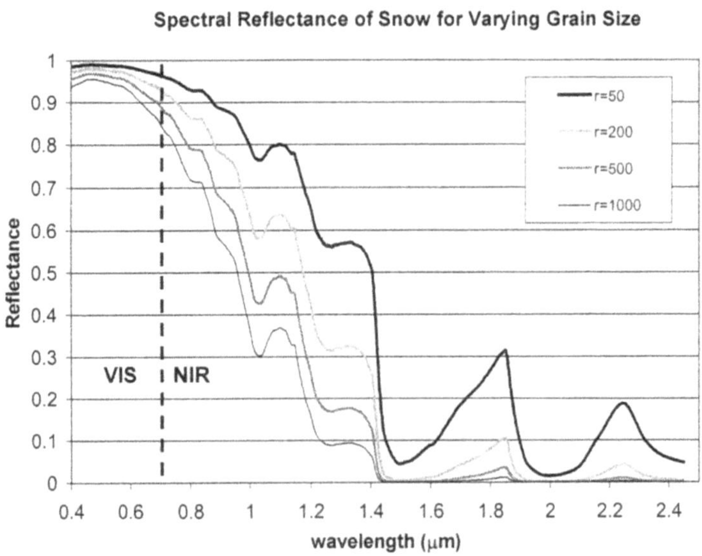


*Figure 1.* Spectral variation of modeled snow albedo for a range of snow grain sizes. Grain radius ranges from 50 μm (fine-grained new snow) to 1000 μm (coarse-grained melting or refrozen snow). These values were computed using the DISORT (DIScrete Ordinates Radiative Transfer) model of Stamnes et al. (1988). Illumination conditions in the model were for direct beam illumination only with a solar zenith angle = 60°

In Section 2 we demonstrate the sensitivity of snow surface-atmosphere fluxes to selected parameterizations of snow albedo. We describe how snow grain size affects albedo, thereby governing heat fluxes. In Section 3 we describe the complexities of mapping snow cover and albedo and discuss remote sensing limitations affecting the use of satellite-derived snow products in climate simulations. We identify new and planned remote sensing products that will be most useful for snow extent and albedo mapping.

# 2. SNOW ALBEDO IN CLIMATE MODELS

## 2.1 Albedo parameterizations

The albedo of snow in a climate model grid cell is portrayed in a variety of ways by different models. For example, some models use a simplistic approach, assuming that snow covers an entire grid cell while others have snow fraction as a function of the water equivalent depth of snow and vegetation type. Here, we categorize and evaluate snow albedo parameterizations that are commonly used in climate simulations. The Atmospheric Model Intercomparison Project (AMIP) is an ideal venue to evaluate and compare climate models (Gates, 1992). Over thirty model runs were submitted to the first phase of AMIP (AMIP-1) by an international array of climate modeling groups. In the AMIP models, snow albedo parameterizations range from constant, prescribed values (different values are used by various modeling groups) to more physically-based spectrally dependent values that vary non-linearly as a function of temperature (as a proxy for grain growth), snow age, depth, and direct/diffuse solar irradiance. For clarity, we group the albedo parameterizations into two categories: constant and varying. The varying parameterization is that of Marshall (1989), which is currently used in the NCAR Land Surface Model (LSM) Version 1.0 (Bonan, 1996).

This investigation is not meant to evaluate all snow albedo parameterizations used in climate models. Parameterizations exist that are more sophisticated than the constant albedo, but less physically-based than the Marshall parameterization. For example, Verseghy (1991) and Loth et al. (1993) parameterize snow albedo is a function of time but not grain size. These parameterizations are not considered in this analysis.

## 2.2 Sensitivity Analyses

We use the SNTHERM snowpack energy and mass balance model of Jordan (1991) to examine the sensitivity of shortwave radiation, latent and sensible heat fluxes and total surface energy balance to four different constant values of $\alpha$ and to the variable $\alpha$ parameterization. SNTHERM is a one-dimensional finite difference model with over 20 vertical computational snow layers that is driven by meteorologic data. Changes in albedo can be implemented in the code in a straightforward way. SNTHERM uses a thermodynamic growth function to calculate changes in grain size within each snow layer. For dry snow, the grain growth rate is a function of the mass vapor flux that is determined in large part by snow temperature and temperature gradient (Colbeck, 1979). For wet snow, grain growth is a function of the liquid water content (Colbeck, 1982).

The total energy flux for a snowcover, $\Delta Q$ ($Wm^{-2}$), is expressed as:

$$\Delta \mathbf{Q} = \mathbf{R} + \mathbf{LE} + \mathbf{S} + \mathbf{G} + \mathbf{M}$$

where,
**R** = Radiation balance (shortwave and longwave)
**LE** = Latent heat
**S** = Sensible heat
**G** = Ground heat
**M** = Heat flux by advection of mass (e.g. rain)

Positive energy fluxes are defined as going into the snowpack.. When $\Delta Q$ is negative the snowpack is cooling; when $\Delta Q$ positive the snowpack is either warming (when the temperature is below 0°C) or melting (when snow temperature is 0°C). Under typical conditions and for this sensitivity analysis, the ground heat flux is negligible and, because there is no precipitation, the advected heat component is zero.

We perform a series of model runs with a range of albedo treatments, seasonal boundary conditions and atmospheric conditions. The energy fluxes and melt rates for the different conditions and model formulations are then compared to evaluate the impact of different albedo parameterizations.

As input data to drive the model, we use a 12-day hourly meteorological data set collected in New Hampshire during the winter of 1987 (Jordan, 1991). Boundary conditions include air temperature, incoming solar radiation, incoming longwave radiation, relative humidity and wind speed. The model computes the reflected solar radiation based on the albedo parameterization. To simulate spring conditions, we modify the winter values by increasing the incoming shortwave radiation by 15%, daylight duration by 2 hours, and air temperatures (in Kelvin) by 6%. Figure 2 shows the solar irradiance and temperature trends (presented as daily mean values). Days 1-3 in the input data are cloudy and relatively warm, followed by one day of clearing skies and cooler temperatures. Day 5 is cloudy and relatively cold, followed by several days of clear skies and the coldest temperatures.

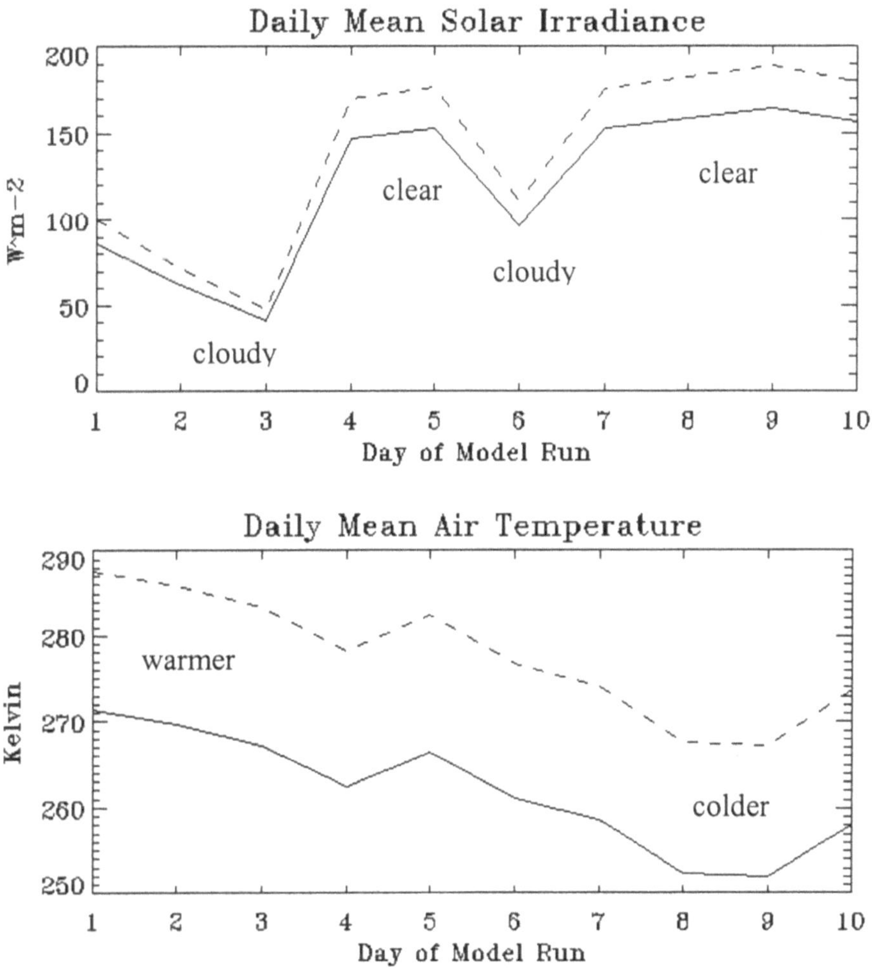


*Figure 2.* Daily mean air temperature and solar irradiance used for the winter (solid) and spring (dashed) "mixed" cases of the SNTHERM model runs. These values were computed from the hourly input values that were used in the model run

Table 1 shows the different albedo treatments and seasonal and atmospheric conditions used in this sensitivity exercise. The fixed albedo values all represent broadband albedo. The lowest albedo value is commonly used for melting snow while 0.70, 0.75, and 0.80 are commonly used as values for dry snow. The Marshall (1989) parameterization contains visible (0.4-0.7 μm) and near-infrared (0.7-2.5 μm) components and computes $\alpha$ as a function of surface grain size, solar zenith angle and diffuse/direct irradiance. Neither soot concentrations nor thin snow cases are considered. Positive feedbacks between snow albedo and air temperature are not included in these experiments. These feedbacks would tend to further enhance the flux effects so the differences described in Section 2.3 should be viewed as something of a lower bound. Initial conditions are identical for the winter and spring scenarios: snowpack temperature –2°C; snow bulk density 250 kg $m^{-2}$; and snow grain radius 160 μm.

This sensitivity study includes six unique 12-day sets of meteorological conditions: winter/clear, winter/cloudy, winter/mixed, spring/clear, spring/cloudy, and spring/mixed. To represent clear (cloudy) conditions, observations from Day 9 (Day 3) were replicated for 12 days. "Mixed" scenarios include the unaltered set of meteorological conditions. The cloudy case (Day 3) is representative of a 0.7 fractional mid-level altostratus cloud cover. For each 12-day run, the first two days are used for model spin up. Results shown in Section 2.3 are for the last 10-days of each run.

*Table 1.* SNTHERM model run treatments

| | |
|---|---|
| Albedo Values: | 0.60, 0.70, 0.75, 0.80, Variable |
| Seasons: | Winter, Spring |
| Atmospheric conditions: | Clear, Cloudy, Mixed |

## 2.3 Results and Discussion

Fluxes for the winter/clear sky scenario are shown in Figure 3. Maximum shortwave fluxes correspond to the lowest albedo value (i.e. maximum solar absorption into the snowpack). The difference in shortwave absorption between albedo=0.6 and albedo=0.8 is ≈38 $Wm^{-2}$. For the variable albedo case, albedo decreases from 0.83 to 0.73 over the course of the model run as a result of grain growth from a diameter of 360 μm to 900 μm. This alters the daily mean shortwave flux by as much as 12 $Wm^{-2}$, the daily mean latent heat flux by 3 $Wm^{-2}$, and the daily mean sensible heat flux by as much as 5 $Wm^{-2}$. Positive and negative values in the different fluxes tend to compensate each other resulting in very small values for $\Delta Q$ using any of the albedo parameterizations. However the individual flux terms are estimated differently depending on the albedo parameterization.

Under the winter/cloudy scenario (Figure 4), incoming shortwave radiation is lower than under clear skies. The sensible heat flux under cloudy conditions is also lower because the temperature gradient between snow and atmosphere is smaller. Also, under cloudy conditions, increased longwave fluxes (not shown) from the atmosphere decrease the cooling rate of the snowpack. As a result, $\Delta Q$ is negative and of even smaller magnitude ($-5 < \Delta Q < 0$ $Wm^{-2}$) than during clear conditions. The effects of the albedo parameterizations on energy fluxes are similar under cloudy and clear conditions.

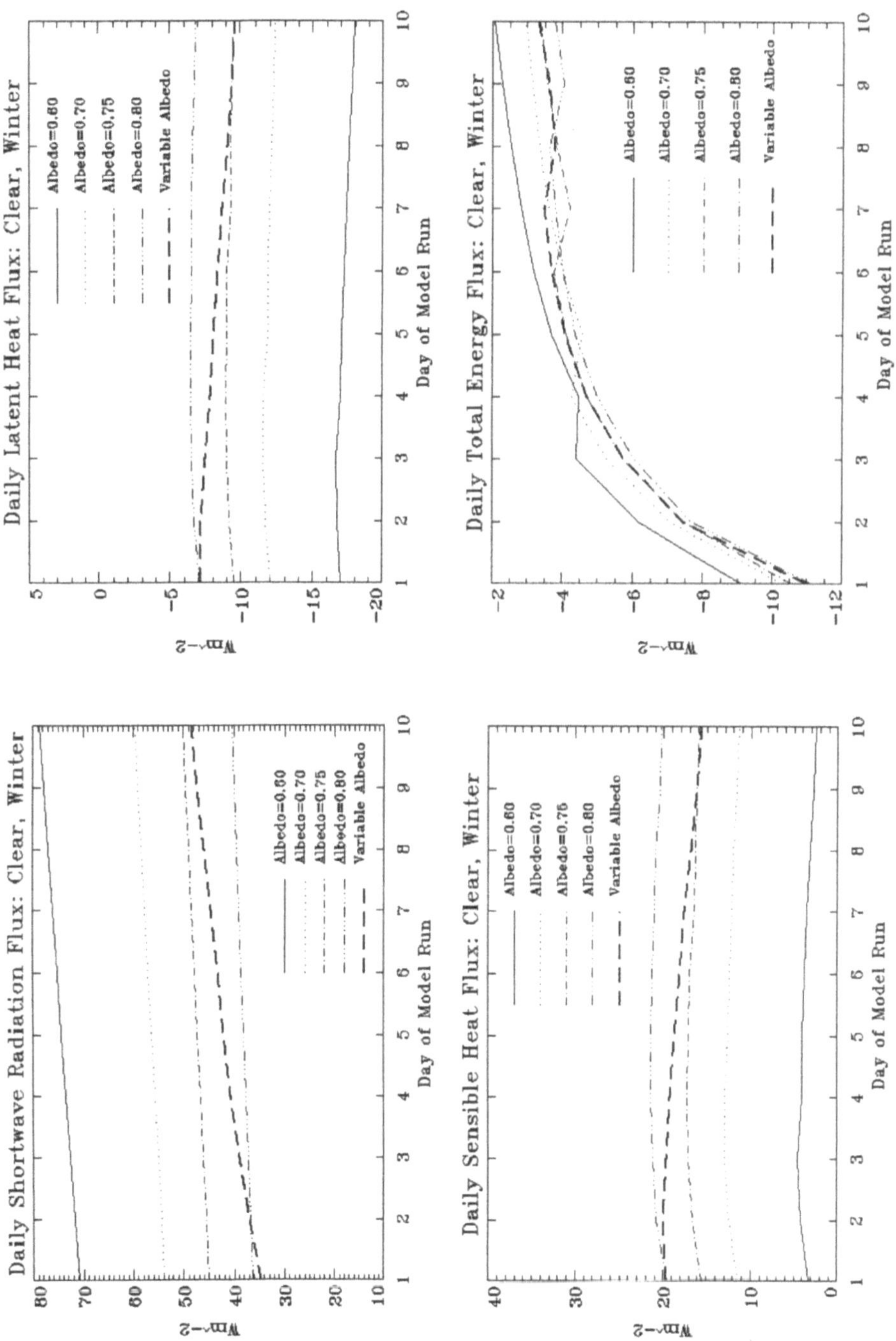


*Figure 3.* Shortwave, latent, sensible and total energy fluxes for clear sky, winter conditions

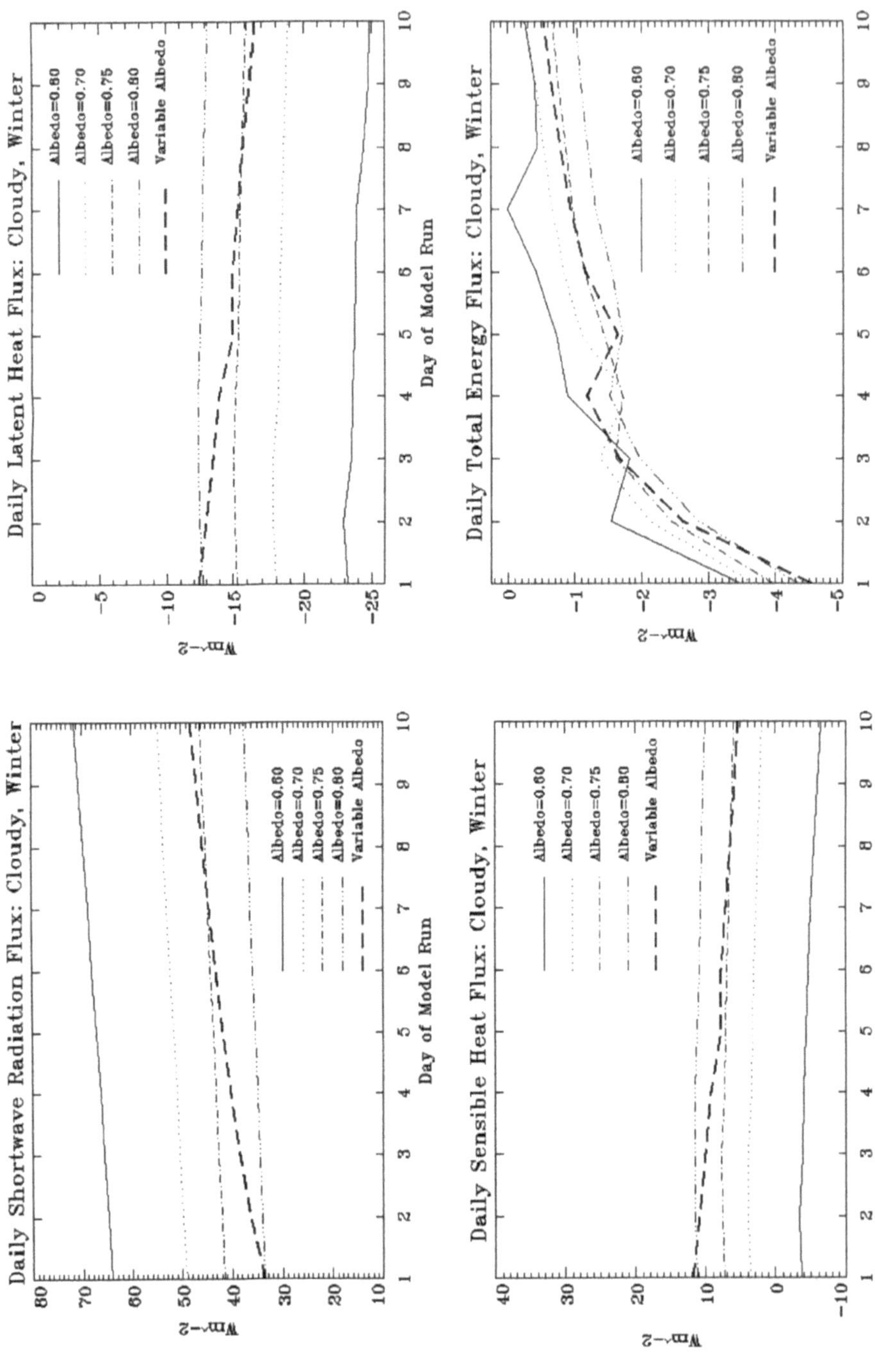


*Figure 4.* Shortwave, latent, sensible and total energy fluxes for cloudy sky, winter conditions

In the winter/mixed case (Figure 5) shortwave radiative effects are much larger for clear sky periods than for cloudy periods. On Day 5 when skies are clear and air temperatures slightly increase, the sensible heat flux increases strongly because of a larger temperature gradient between snow and the overlying atmosphere. Increased sublimation also occurs on Day 5 due to warmer air temperatures coupled with increased solar radiation. Under these winter conditions, the shortwave radiation and the turbulent fluxes are sensitive to the albedo parameterization. However, because solar insolation is relatively small during winter and turbulent fluxes tend to cancel each other out, in this case the total energy flux is relatively insensitive to albedo parameterization.

During spring, warmer air temperatures and higher insolation lead to more significant effects in surface-atmospheric fluxes (Figure 6). The temperature gradient between the atmosphere and the snowpack is larger than during winter, thereby increasing the sensible heat flux. The latent heat flux is positive for Days 1-3 indicating that deposition is occurring on the snow surface. However, when clear skies return on Day 4, the latent heat flux again becomes negative implying that sublimation is dominating over condensation. While the latent and sensible fluxes tend to be higher than in the winter scenarios, they are relatively insensitive to the choice of albedo parameterization. Shortwave radiation, which is most sensitive to albedo parameterization during clear periods with high insolation, controls the patterns seen in the mixed/spring plots. These relationships are similar to those observed and modeled by Marks (1989) in his investigation quantifying snow-atmosphere fluxes at sites in the Sierra Nevada, California.

Meltwater generation is another reflection of changes in $\Delta Q$. Spring meltwater production is plotted for the five different $\alpha$ parameterizations in Figure 7. In all scenarios, meltwater production for $\alpha$ =0.6 is far greater than for $\alpha$ =0.8 (in some cases is nearly double). Under both clear and cloudy conditions, the variable albedo case first tracks $\alpha$ =0.8 and then follows $\alpha$ =0.75. In the mixed/spring case, the largest differences between different parameterizations are evident during clear sky conditions. For a single day (Day 5), melt varies by as much as 15 mm, ranging from a minimum of 5 mm ($\alpha$ = 0.8) to a maximum of 20 mm ($\alpha$ = 0.6).

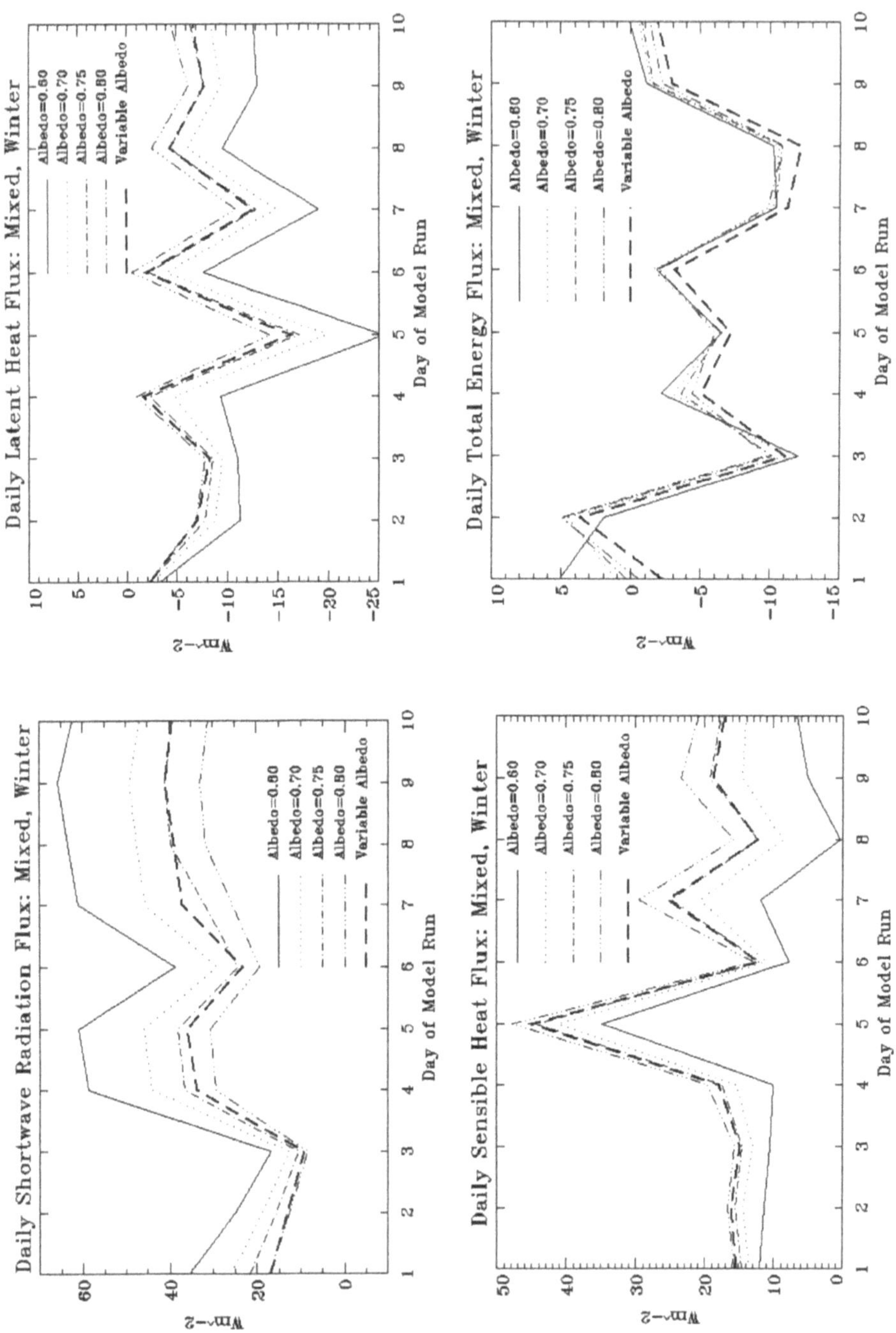


*Figure 5.* Shortwave, latent, sensible and total energy fluxes for mixed clear and cloudy skies, winter conditions

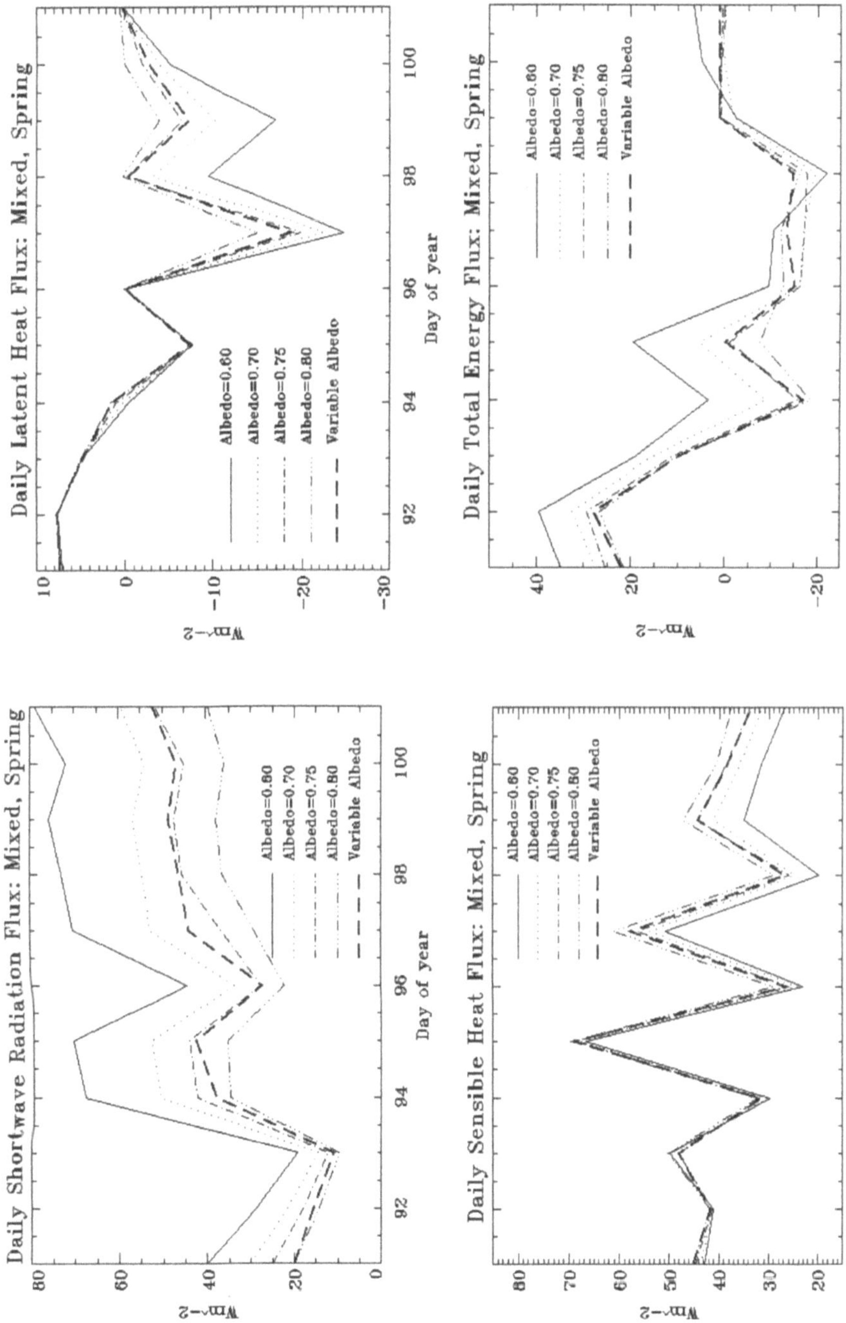


*Figure 6.* Shortwave, latent, sensible and total energy fluxes for mixed clear and cloudy skies, spring conditions

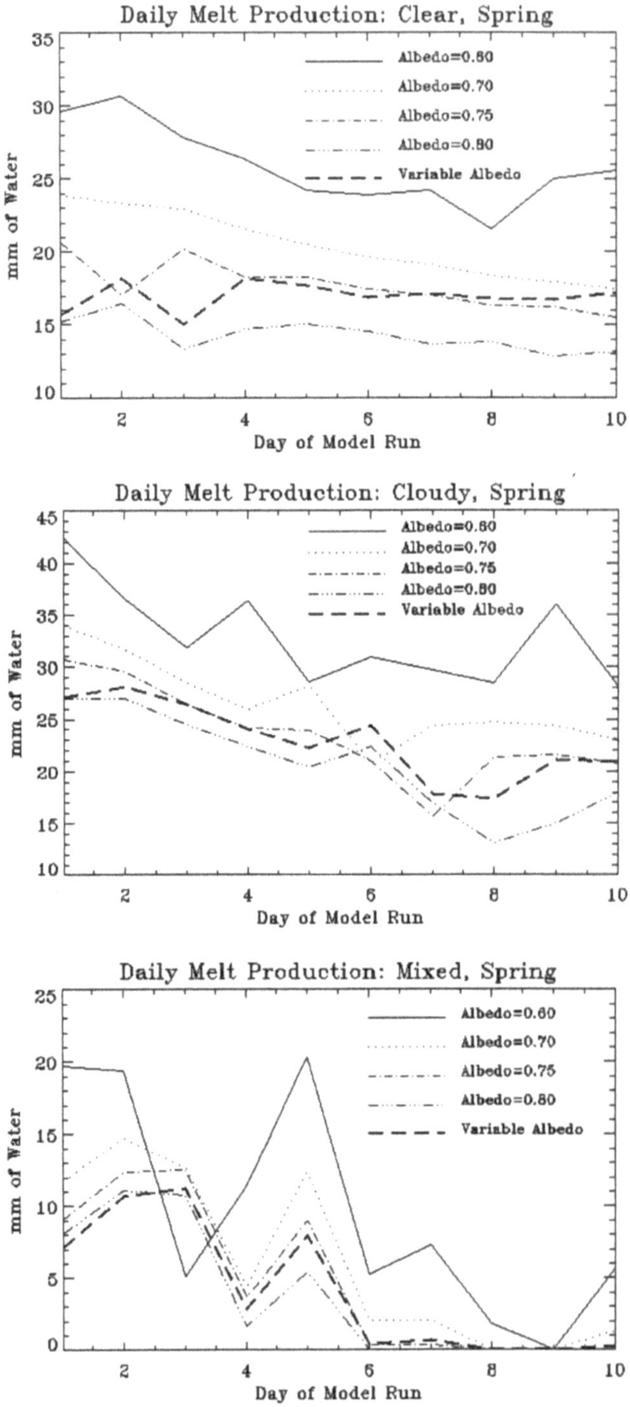


*Figure 7.* Daily total snowmelt (mm of liquid water) for clear skies, cloudy skies and mixed (clear and cloudy skies) under spring conditions

# 3. MAPPING SNOW COVER AND SNOW ALBEDO FROM SATELLITE

In this section, we describe some of the difficulties in providing remotely sensed snow covered area and albedo data, and offer information on some new and planned products that are suitable for climate modeling. The geophysical parameters (snow covered area, snow water equivalent, snow albedo, wet snow vs. dry snow) and the types of sensors used to map them (optical and passive microwave) are discussed below.

## 3.1 Snow Mapping Complexities

Snowstorms can blanket a region and significant ablation can occur as quickly as one day. Thus, a monthly or bi-weekly product may be unsatisfactory for many modeling purposes, particularly during the fall and spring transition seasons. Satellite mapping of snow cover offers regular, repeat coverage at spatial and temporal scales useful for climate simulations. However, while snow cover extent is a relatively straightforward parameter to retrieve using space-borne measurements, the detection of melt and estimation of snow albedo are more difficult. An optimal product would be one that concurrently maps snow-covered area, snow albedo, and snow water equivalent at 1-5 day intervals, all at the same spatial resolution. Current operational snow mapping is performed using optical sensors such as the Advanced Very High Resolution Radiometer (AVHRR) and passive microwave sensors such as the Defense Mapping Satellite Program (DMSP) Special Sensor Microwave/Imager (SSM/I). AVHRR records data at two spatial resolutions (1.1 km and 4 km) and is restricted to measurement during sunlit times. Therefore, winter observations at high latitudes are limited. Also, clouds interfere with AVHRR observations, a common problem during the snow season. Although albedo data from this sensor can be used for multi-temporal relative comparisons, the accuracy is insufficient for absolute snow albedo measurements (Stroeve et al., 1997).

Passive microwave estimates are based on the scattering effect of ice particles on microwave radiation emitted from the ground below. As snow accumulates, its scattering of microwave energy reduces the brightness temperature values recorded by satellite radiometers. Empirical algorithms have been developed to estimate snow extent and snow water equivalent from passive microwave sensors (Goodison, 1989; Chang et al., 1990; Chang and Tsang, 1992; Basist et al., 1997). Unlike optical region measurements, passive microwave observations are uninfluenced by darkness and non-precipitating clouds. Ice particles in clouds are transparent to the passive microwave frequencies being used to map snow. However, liquid water droplets in clouds will affect the algorithm results and warm storms may be mapped as snow. SSM/I is useful for mapping snow extent and is also effective at mapping snow water equivalent in regions without dense forests. However, in melting snow the presence of liquid water changes snow emissivity and masks the snow signal (Grody, 1991; Grody and Basist, 1996). A second disadvantage of SSM/I is its coarse spatial resolution (25 x 25 km) which makes it difficult to combine with albedo estimates from optical data. In addition, complex systems such as the northern boreal forest, where snow remains on trees and beneath them, present further snow detection complications. Basist et al. (1996) found that an AVHRR-derived operational snow cover product performed better than an experimental SSM/I-derived product when mapping snow extent under dense vegetation, whereas the SSM/I algorithm worked better over mountainous regions, under clouds, and during times when the snowpack is rapidly changing. Tait and Armstrong (1996) found that underestimation of snow depth from passive microwave data occurred in regions of boreal forest. Dense vegetation increases the brightness temperature, giving the false impression of less snow (Schweiger et al., 1987). To date, accurate mapping of snow cover in boreal forest regions remains problematic for passive microwave systems.

A new optical remote sensing instrument, the Moderate Resolution Imaging Spectroradiometer (MODIS), was launched in December 1999 and is mapping global snow covered area as one of its standard products (Hall et al., 1998). The MODIS Snowmap product provides daily and 8-day composite snow cover maps at 500-m spatial resolution. Snowmap exploits the reflectance contrast for snow between the visible and shortwave infrared wavelengths. The algorithm uses the Normalized Difference Snow Index (NDSI) defined as:

$$NDSI = \frac{VIS - NIR}{VIS + NIR}$$

where, *VIS* and *NIR* are the pixel reflectances in selected visible and near-infrared channels. The algorithm assumes a threshold, above which a pixel is determined to contain more than 50% snow cover. Currently, the NDSI threshold is set to 04. The MODIS Snowmap product represents an improvement over existing snow products because the MODIS-derived snow maps are produced at higher spatial and temporal resolution than the NOAA weekly snow cover charts (Matson, 1986) and the 1-km resolution regional snow products over North America from the National Operational Hydrologic Remote Sensing Center (Carroll, 1990). The Snowmap product offers improved snow detection accuracy by incorporating vegetation cover information over forested areas (Klein et al., 1998). Expected maximum error for the Northern Hemisphere is ≈7.5% with largest errors in the boreal forest regions, ≈9-10% (Hall et al., 1998). In addition, the MODIS 8-day snow product will contain statistics for snow cover duration and persistence for each 500-m grid cell.

# 4. MAPPING GLOBAL SNOW ALBEDO

Currently, no remotely sensed product provides accurate maps of snow albedo. However, a MODIS global snow albedo product is under development (D. Hall, personal communication). This albedo product is based on a conversion scheme developed by Nolin and Stroeve (1997), where a discrete-ordinates radiative transfer model is used to compute the coefficients for converting measured satellite-derived reflectances to surface albedo. MODIS has excellent radiometric calibration and will provide snow albedo estimates to within ±0.02. Prior to the snow albedo product release, another standard MODIS product will be available that will provide estimates of land surface albedo, including snow covered regions (Wanner et al., 1997). This land surface albedo product includes vegetation and soil albedos and, though the snow albedo part of this product is still simplistic, such a product could be useful for input to climate simulations.

## 4.1 Mapping Dry and Wet Snow

Although several snow extent products are currently available (Grody and Basist, 1996; Carrol et al., 1999; Ramsay, 1998; Robinson et al., 1999) there is only one product that maps both dry and wet snow extent at near real time on a global, daily basis. The Near real-time Ice and Snow Extent (NISE) product, produced by the University of Colorado's National Snow and Ice Data Center (NSIDC), is a daily, gridded map of sea ice concentrations and snow extent for both the Northern and Southern Hemispheres and is available from November 1997. The product currently uses passive microwave data from the SSM/I instrument on DMSP's F13 satellite. Vertically and horizontally polarized brightness temperatures from the F13 early morning (descending) orbits are used as input (early morning orbits are used to minimize the number of melting pixels). In addition to mapping sea ice concentration (0% - 100%), snow extent is mapped using a classification scheme that designates each land pixel as non-snow, dry snow or wet snow. Discrimination between dry and wet snow is a recent innovation and represents the first time that wet snow mapping is performed on an operational basis. The orbital data are gridded at 25 x 25 km resolution in an equal-area Lambert azimuthal grid. The day on which each grid cell is updated changes with latitude. High latitudes, with frequent overpasses, are updated daily while midlatitude regions are updated every 2-4 days. The number of days since a grid cell was last updated is provided as an additional data layer in the NISE product (see Figure 8).

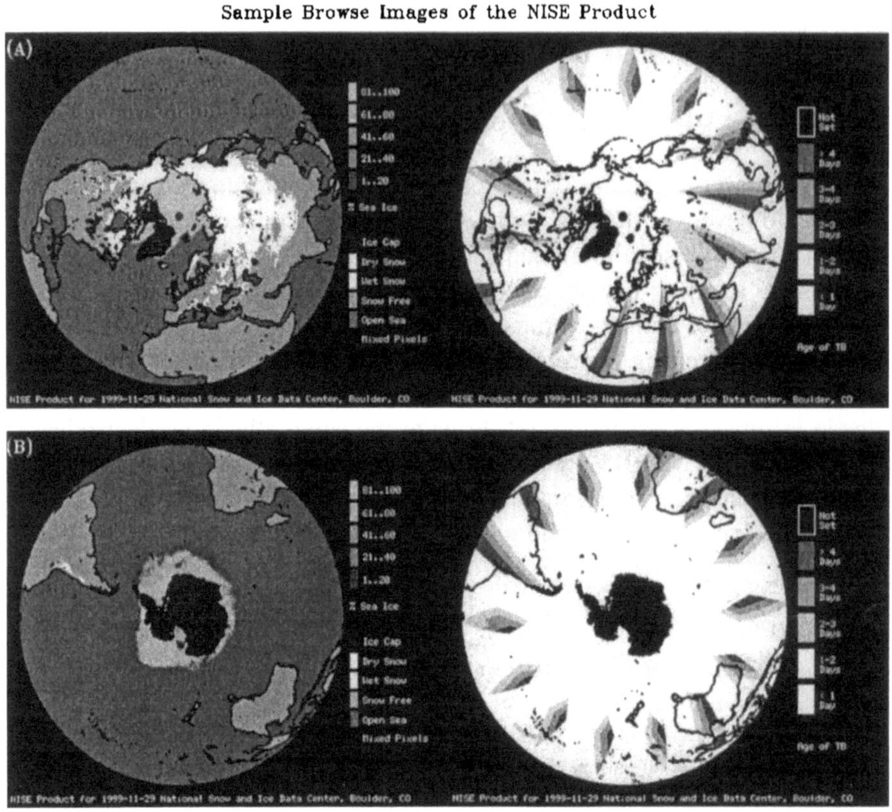


*Figure 8.* An example of browse images accompanying the NISE data. Here, dry snow is shown as white and wet snow as gray. Sea ice concentrations (not discussed) are shown in shades of light gray

Dry snow is identified using brightness temperature data in the 19 GHz and 37 GHz channels including both vertical and horizontal polarizations (Goodison, 1989). The algorithm was originally designed to map snow water equivalent (SWE). Here, any grid cell having a positive value of SWE is designated as dry snow. As discussed earlier, warm storms (clouds composed of liquid water droplets) can be mistaken for snow cover. These effects are partially removed by using a snow mask derived from 30 years of snow cover data from the NOAA weekly snow charts. Any grid location with a 1% chance of snow ever having occurred may be mapped as snow. Since snow in the mid-to-high latitudes is typically associated with ice clouds, warm storms are no considered a major problem. However, no definitive study has been performed and the frequency of these errors of commission remains an open question.

When liquid water appears in the snowpack, vertical polarization brightness temperatures become much greater than horizontal polarization brightness temperatures. This polarization difference is the basis for the wet snow mapping algorithm developed by Walker and Goodison (1993). The NISE product includes a modification of this algorithm requiring wet snow pixels to be colder than 255 K in the 37 GHz channels, thus enabling the identification of melt onset.

For modelers, information on melt onset can be useful for albedo parameterization and model validation. In the absence of ancillary information, albedo values can be assigned on the basis of wet and dry snow. In addition, more sophisticated parameterizations that include the number of days a grid cell has been mapped as wet snow (thereby further decreasing the albedo) are possible.

The NISE product can also help validate modeled snow hydrology. Because its 25-km resolution is smaller than most model grid cells, NISE snow data can be used to compute fractional snow cover. The NISE product shows snow extent, identifies the onset of melt, and maps the regional ablation of snow. An example of this process is shown in Figure 9 where Eurasian snowcover is displayed at two-week intervals from March 1998 through June 1998. Snow cover on March 1$^{st}$ is extensive, covering the mountains of Europe, Scandinavia, the mountains of Turkey and Iran, the Himalayan plateau, and virtually all of northern Asia. As spring progresses, one can see the appearance of wet snow in various areas and its subsequent disappearance. In 1998, the proportion of wet snow in the Northern Hemisphere increases from about 1% in mid-winter to over 65% in early summer. Wet snow increases from about 3% in late March to over 20% in late April. During the spring season, the transition from an extensive cover of bright, dry snow to incomplete, lower albedo, wet snow can occur within a few days.

With it's near-real time availability, 25-km spatial scale, and ability to identify melting snow, the NISE product is appropriate for mapping and characterizing snow extent and melt onset. Furthermore, the distinction between wet snow and dry snow provides some additional information helpful for characterizing snow albedo. Until a daily or weekly snow albedo product becomes available, a dry/wet snow indicator is a satisfactory substitute. Access to the NISE product is available through the NSIDC Data Catalog (http://www-nsidc.colorado.edu/NSIDC/CATALOG/ENTRIES/nsi-0056.html).

Improved passive microwave snow mapping products are planned to begin in 2001 following the launch of the Advanced Microwave Sounding Radiometer (AMSR-E) on NASA's Aqua satellite. AMSR-E has higher spectral and spatial resolution than SSM/I and will map snow covered area, snow water equivalent, and wet snow with improved accuracy and precision.

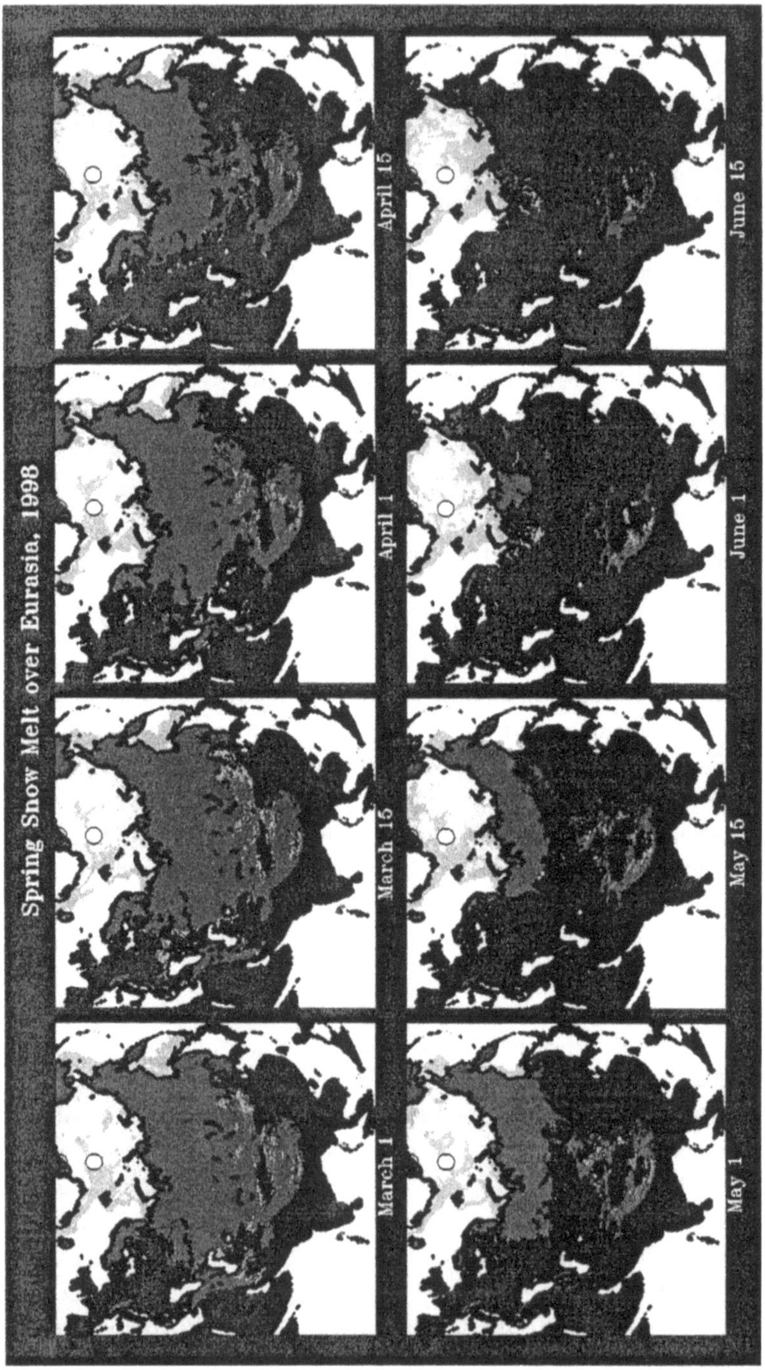


*Figure 9.* Spring 1998 ablation of Eurasian snowcover at two week intervals from the NISE product. Dry snow is shown in black and wet snow is shown in dark gray. Light gray and gray areas around the North Pole region are sea ice

In addition, snow surface temperature and the onset of melt over ice sheets will be mapped as a special product in the post-launch phase (Chang and Rango, 1997).

# 5. CONCLUDING DISCUSSION

In this chapter we demonstrate how snow albedo parameterizations can significantly affect modeled surface-atmosphere fluxes. Grain metamorphism provides an important control on albedo and is directly affected by changes in snowpack energy balance. Furthermore, there is a positive feedback effect between grain size and energy balance whereby increased grain size causes decreased albedo and a subsequent increase in absorption of shortwave energy. Parameterization of snow albedo as a function of snow grain size includes the effects of energy balance changes on surface layer grain size. More simple constant albedo parameterizations can result in individual flux differences of nearly 40 $Wm^{-2}$. This is especially critical during the spring melt season, when increased sensitivity of $\Delta Q$ to the albedo parameterization directly affects the melt rate.

Moreover, similar estimates of $\Delta Q$ can be derived from both constant and variable albedo parameterizations even at times when the individual flux estimates are different. This is because the differences in estimated latent heat flux tend to be of opposite sign and similar magnitude to the differences in estimated sensible heat flux. These differences in estimates of the individual turbulent fluxes may be unimportant from the point of view of snowpack energy balance modelers. However, from the perspective of the atmospheric modeler, the values of the individual fluxes are important because they are treated differently in the models. Therefore, an accurate representation of snowpack albedo is key to accurate computation of surface fluxes.

The accurate detection of snow cover is crucial for quantifying the impacts of snow cover dynamics on local, regional and hemispherical land surface-atmosphere interactions. Moreover, trends in the spatial extent of snow cover may be a key indicator climate change. Snow cover and snow albedo data are needed for accurate land surface representation in climate models as well as for climate model validation. There is currently no snow observational data set that provides all the mapping capabilities required by climate models. Historical products such as the NOAA weekly snow charts (dating back to 1972) and SMMR/SSM/I (dating back to 1978) provide longer term records of snowcover useful for climatological analyses. New products, although lacking in record length, provide improved and additional observations. A global snow albedo product from the MODIS instrument is planned to begin production in year 2001. Currently, the NISE product, derived from passive microwave data, maps global distributions of both dry and wet snow on a daily basis at a spatial resolution that is useful for climate models. Another product scheduled to begin production in 2001 will come from the AMSR-E instrument (also a passive microwave instrument). This product will map global daily snow water equivalent and snow covered area at 12.5 km spatial resolution. Climate models will benefit from improved snow parameterizations and observations and the incorporation of more sophisticated snow albedo parameterizations. In addition, remote sensing algorithm developers should consider the needs of the climate modeling community when designing new snow mapping products.

## 6. ACKNOWLEDGEMENTS

Support for this research is provided through NASA grant NAG5-7543 and NSF grants ATM-9900687 and ATM-9818098.

## 7. REFERENCES

Barnett, T. P., L. Dumenil, U. Schlese, E. Roeckner, and M. Latif (1989), The effect of Eurasian snow cover on regional and global climate variations. J. Atmos. Sci., 46:661-685.Basist , A., N. C. Grody, T. C. Peterson and C. N. Williams (1998), Using the Special Sensor Microwave Imager to monitor land surface temperatures, wetness and snow cover. J. Appl. Meteorol., 37:888-911.

Basist, A., D. Garrett, R. Ferraro, N. Grody and K. Mitchell (1996), A comparison between snow cover products derived from visible and microwave satellite observations. *J. Appl. Meteorol.,* 35:163-177.

Bonan, G. B. (1996), A land surface model (LSM version 1.0) for ecological, hydrological, and atmospheric studies: Technical description and user's guide. NCAR Tech. Note, NCAR/TN-417+STR, Boulder, CO, 150 pp.

Caroll, S. S., T. R. Carroll and R. W. Poston (1999), Spatial modeling and prediction of snow water equivalent using ground-based, airborne, and satellite snow data, *J. Geophys. Res.,* 104:19623-19629.

Carroll, T. R. (1990) Operational airborne and satellite snow cover products of the National Operational Hydrologic Remote Sensing Center, Proceedings of the 1990 Eastern Snow Conference, CRREL Special Report 90-44, 87-98.

Chang, A. T. C., J. L. Foster, and D. K. Hall (1990), Satellite sensor estimates of Northern Hemisphere snow volume. *Int. J. Remote Sensing,* 11:167-171.

Chang, A. T. C. and A. Rango (1997), Algorithm Theoretical Basis Document (ATBD) for the AMSR Snow Water Equivalent Algorithm, Vers. 2.0, NASA EOS publication, available online at http://wwwghcc.msfc.nasa.gov/AMSR/html/amsratbd.html, 26pp.

Clark, M. P. (1998), The Role of Snow Cover in the Climate System. Ph.D. Dissertation, Dept. of Geog., Univ. CO, Boulder, 108 pp.

Cohen, J. and D. Entekhabi (1999), Eurasian snow cover variability and Northern Hemisphere climate predictability. *Geophys. Res. Lett.,* 26:345-348.

Cohen, J. and D. Rind (1991), The effect of snow cover on climate. *J. Climate,* 4:689-706.

Colbeck, S. C. (1979), Grain clusters in wet snow. *J. Colloid Interface Sci.*, 72: 371-384.

Colbeck, S. C. (1982), An overview of seasonal snow metamorphism. *Rev. Geophys. and Space Phys.*, 20:45-61.

Dewey, K. F. (1977), Daily maximum and minimum temperature forecasts and the influence of snow cover. *Mon. Wea. Rev.*, 105, 1594-1597.

Dickson, R. R. and J. Namias (1976), North American influence on the circulation and climate of the North Atlantic sector. *Mon. Wea. Rev.* 104:1255-1265.

Ellis, A. W. and D. J. Leathers (1998), The effects of a discontinuous snow cover on lower atmospheric temperature and energy flux patterns. Geophys. Res. Lett., 25:2161-2164.

Frei, A. and D. A. Robinson (1999), Northern hemisphere snow extent: Regional variability 1972-1994. Int. J. Climate, 19:1535-1560.

Gates, W. L. (1992), AMIP: The Atmospheric Model Intercomparison Project. Bull. Amer. Meteor. Soc., 73:1962-1970.

Goodison, B. E. (1989), Determination of areal snow water equivalent on the Canadian prairies using passive microwave satellite data. IGARSS '89, Proceedings 3:1243-1246.

Goodison, B. E. and A. E. Walker (1993), Use of snow cover derived from satellite passive microwave data as an indicator of climate change. Ann.Glaciol. 17:137-142.

Grody, N. C. (1991), Classification of snow cover and precipitation using the special sensor microwave imager. J. Geophys. Res.96:7423-7435.

Grody, N. C. and A. N. Basist (1996), Global identification of snowcover using SSM/I measurements. IEEE Trans. Geosci. Remote Sens,. 34:237-249.

Groisman, P. Y., T. R. Karl, and R. W. Knight (1994), Observed impact of snow cover on the heat balance and the rise of continental spring temperatures. Science, 263:198-200.

Gutzler, D. S. and J. W. Preston (1997), Evidence for a relationship between spring snow cover in North America and summer rainfall in New Mexico. Geophys. Res. Lett., 24:2207-2210.

Hall, D. K., A. B. Tait, G. A. Riggs, V. V. Salomonson (1998), Algorithm Theoretical Basis Document (ATBD) for the MODIS Snow-, Lake Ice-, and Sea Ice- Mapping Algorithm, Version 4.0. NASA EOS publication, available online at http://ltpwww.gsfc.nasa.gov/MODIS_Snow/modis.html, 50 pp.

Jordan, R. (1991), A one-dimensional temperature model for a snow cover: technical documentation for SNTHERM.89. Special Report 91-16, US Army Corps of Engineers, Cold Regions Research and Engineering Laboratory, Hanover, NH, 49 pp.

Klein, A. G., D. K. Hall, and G. Riggs (1998), Improving snow-cover mapping in forests through the use of a canopy reflectance model. Hydrological Processes, 12:1723-1744.

Loth, B., H-F. Graf, and J. M. Oberhuber (1993), Snow cover model for global climate simulations. J. Geophys. Res., 98:10451-10465.

Marks, D. (1988), Climate Energy Exchange, and Snowmelt in Emerald Lake Watershed, Sierra Nevada. Ph.D. Dissertation, Dept. of Geog., Univ. Calif., Santa Barbara, 149 pp.

Marshall, S. (1989), A physical parameterization of snow albedo for se in climate models. NCAR Cooperative Thesis no. 123, Boulder, CO.

Matson, M. (1986), The NOAA satellite-derived snow cover data base: Past, present, and future. Proceedings of Snow Watch '85, World Data Center for Glaciology, Report GD-18, 115-124.

Nolin, A. W., and J. C. Stroeve (1997), The changing albedo of the Greenland ice sheet: Implications for climate change, Ann. Glaciol., 25: 51-57.

Robinson, D. A., J. D. Tarpley, and B. H. Ramsay (1999), Transition from NOAA weekly to daily hemispheric snow charts. AMS, Tenth Symp. on Global Change Studies, Dallas, TX, p. 487-490.

Schweiger, A. J., R. Armstrong, and R. G. Barry (1987), Snow cover parameter retrieval from various data sources in the Federal Republic of Germany. In Large Scale Effects of Seasonal Snow Cover, Proceedings of he Vancouver Symposium, August 1987, IAHS Publ. no. 166, Oxford: International Association of Hydrometeorological Sciences, pp. 353-364.

Stamnes, K., S. C. Tsay, W. Wiscombe, and K. Jayaweera (1988), Numerically stable algorithm for discrete-ordinate-method radiative transfer in multiple scattering and emitting layered media. Appl. Opt., 27:2502-2509.

Stroeve, J. C., A. W. Nolin, and K. Steffen (1997), Comparison of AVHRR-derived and in situ surface albedo over the Greenland ice sheet. Remote Sens. Environ., 62:262-276.

Tait, A. and R. Armstrong (1996), Evaluation of SMMR satellite-derived snow depth using ground-based measurements, *Int. J. Remote Sensing,* 17:657-665.

Verseghy, D. L. (1991), CLASS – A Canadian land surface scheme for GCMs, I: soil model, *J. Climatol.,* 11:111-133.

Walker, A. E., and B. E. Goodison (1993), Discrimination of a wet snow cover using passive microwave satellite data. *Ann Glaciol.* 17:307-311.

Walland, D. J. and I. Simmonds (1997), Modelled atmospheric response to changes in Northern Hemisphere snow cover. *Climate Dynamics,* 13:25-34.

Walsh, J. E. (1993), Observational and modeling studies of the influence of snow anomalies on the atmospheric circulation, in *Prediction of the interannual climate variations, NATO ASI Ser. 6,* edited by Shukla, Kluwer, Norwell, Mass., pp. 89-105.

Walsh, J. E. and B. Ross (1988), Sensitivity of 30-day dynamical forecasts to continental snow cover. *J. Climate,* 1:737-754.

Wanner, W., A. Strahler, B. Hu, P. Lewis, J.-P Muller, X. Li, C. Barker-Schaaf, and M. Barnsley (1997), Global retrieval of BRDF and albedo over land from EOS MODIS and MISR data: Theory and algorithm. *J. Geophys. Res.,* 102:17143-17162.

Warren, S. G., and Wiscombe, W. J. (1980), A model for the spectral albedo of snow, II, Snow containing atmospheric aerosols, *J. Atmos. Sci.*, 37: 2734-2745.

Wiscombe, W. J., and Warren, S. G. (1980), A model for the spectral albedo of snow, I, Pure snow, *J. Atmos. Sci.*, 37:2712-2713.

# Using the Special Sensor Microwave Imager to Monitor Surface Wetness and Temperature

ALAN BASIST and CLAUDE WILLIAMS
*National Climatic Data Center, Asheville, USA*

Abstract: The current network of in situ stations is inadequate for monitoring regional temperature and moisture anomalies across the land surface, leaving the climate monitoring community insufficient information to identify spatial structure and variations over many areas of the world. Therefore, we need to blend satellite observations with in situ data to obtain global coverage. In order to accomplish this task, we have calibrated and independently validated an algorithm that derives land surface temperatures from the Special Sensor Microwave Imager (SSMI). The goal of this exercise is to blend both the in situ and satellite data sets into one superior product, then merge this product with an sea surface temperature anomaly field form the same base period. The value of the global product has extremely valuable applications to climate modeling community, since it can serve as a validation tool and/or direct input to the surface parameterization, allowing the radiation feed back to be realistically grounded on surface temperature and humidity observations.

# 1. INTRODUCTION

Historically, global land surface temperatures and wetness have been obtained from *in situ* point sources located mainly in populated and industrialized regions. Unfortunately, these stations are neither located evenly nor densely around the globe. Specifically, observations are sparse over large regions of Africa, tropical South America, southeastern and central Asia, and large sections of the Arctic and Antarctic. Therefore, we developed a technique to derive the global distribution of land surface temperature and wetness from satellite observations. In order to derive global surface temperature anomalies at 1 degree resolution, we merged that land values (derived from



*M. Beniston and M.M. Verstraete (eds.),*
*Remote Sensing and Climate Modeling: Synergies and Limitations,* 181-202.

the procedure described below) with sea surface values provided by Reynolds and Smith (1994).

This technique uses channel measurements from the SSM/I sensors on three separate Defense Meteorological Satellite Program (DMSP) polar orbiting satellites (F08, F11, and F13) from 1987 to 1997. Background information on the SSM/I instrument can be obtained from the world wide web at: http://www.ngdc.noaa.gov/dmsp/descriptions/doc_ssmi.html. These DMSP satellites provide sun synchronized overpasses at 6 A.M. and 6 P.M. These twice daily satellite overpasses are processed into 1/3 x 1/3 degree "pixels" by NESDIS and archived at the National Climatic Data Center (NCDC) in near real time. From August 1988 to the end of 1991, erratic signals from the F08 85GHz channels forced the removal of the data from our analysis. The SSM/I instrument measures the brightness temperature at four frequencies: 19, 37, and 85 GHz with vertical and horizontal polarization and 22 GHz with only vertical polarization. All of these frequencies are in atmospheric window regions with the 22 and 85 GHz channels having weak water vapor absorption. Various signatures among the seven channel measurements were used to identify surface types and calculate dynamic emissivity adjustments. In this paper will distinguish between the various SSM/I channels by their frequency in GHz, where V stands for vertical and H stands for horizontal polarization (i.e. the 37 GHz horizontally polarized channel will be denoted as the 37H channel.

Observations of passive microwave radiation by polar orbiting satellites can be used to measure many of the Earth's atmospheric and geophysical properties. In particular, brightness temperatures from the Special Sensor Microwave Imager (SSM/I) have been used over land to derive: surface wetness (Basist et al. 1998), snow cover (Grody and Basist 1996), surface emissivities (Prigent et al. 1997), precipitation (Ferraro and Marks 1994), and soil moisture (Vinnikov et al. 1999). Ferraro et al. (1996) give an excellent overview of numerous surface and atmospheric products developed from the SSM/I instrument. Land surface temperature has been derived having different accuracies depending on surface conditions (McFarland et al. 1990, Neale et al. 1990, Njoku 1994, Weng and Grody 1998). Basist et al. (1998) developed a technique that dynamically adjusts the SSM/I algorithm coefficients for the effect of liquid water on surface emissivity, resulting in improved temperature estimates.

The primary difficulty in deriving surface temperature from passive microwave measurements is the variable emissivity associated with different surfaces. For the microwave spectrum the emissivity of soil depends on its water and/or mineral content, as well as the effects of vegetation and surface roughness. Since the microwave emissivity is variable, the brightness temperature is not a function of surface temperature alone. Therefore, any algorithm that attempts to estimate surface temperature must first infer the particular surface condition for a microwave measurement, and either make appropriate emissivity adjustments to the microwave measurement, or filter the measurement if reliable adjustments are not currently possible. The approach used here assumes no a priori information about the surface conditions, allowing the satellite observations to provided a dynamic assessment of the surface type and current emissivity. The technique will be explained more fully in the methodology section below.

The Basist Wetness Index (BWI) is simply the emissivity adjustment associated with water in the radiating surface. However, wetness values also comes from the magnitude of precipitation, which can is derived directly from the SSM/I instrument (Ferraro et al. 1996). The wavelength of microwave frequencies are near the diameter of precipitation. Consequently, large hydrometeors (i.e. large raindrops and snow flakes) exceed the wavelength at 85 GHz, allowing them to scatter the high frequencies channel measurements, while the low frequency channel measurements (19and 22 GHz) remain transparent to these hydrometeors. This reduction of upwelling radiation at high frequencies is known as a scattering signature, where the greater the difference between the low and high frequencies, the greater the magnitude of precipitation. Moreover, large ice grains in the anvil tops of deep convection can also scatter emission at the lower microwave frequencies (i.e. 19 GHz wavelengths). Since all channel measurements are affected by deep convection, there is no base line for removing for the influence of hydrometeors at the higher frequency channels, forcing us to remove these observations from the temperature product. However, the gradient in scattering across the frequencies still provides as a reliable magnitude of precipitation. Therefore, when the temperature is above freezing (Grody and Basist 1996), we associate the scatter index to a wetness value. Unfortunately, these scattering values do not have the same scaling as the surface wetness measurements; none-the-less, they are included in the BWI, after weighting the consequence of not associating a rainfall event with an underlying wet surface.

## 2. DATA

This paper describes how we identify various surface types, calibrate the emissivity adjustment for each unique SSM/I signature, verify of the accuracy of our approach to estimate shelter height temperatures from the passive microwave observations, and validate the utility of the wetness index with independent precipitation measurements. Hourly first order stations served as the reference for emissivity adjustments, and least absolute difference techniques were used to minimize the standard error for each surface type. A dense U.S. cooperative network of monthly mean *in situ* temperature anomalies provided validation of the temperature anomalies derived from the SSM/I channels.

The Global Precipitation Climate Program (GPCP) provides a reference or validation (Huffman et al. 1995) of the BWI. The GPCP precipitation values have been compiled over the same six year period, thereby allowing for a direct comparison of anomalies over the same base period. The GPCP values are derived from numerous sources: *in situ* point measurements, as well as infrared and microwave observations. The final product uses an interpolation scheme to produce global fields at 2.5° monthly resolution for each month. Where *in situ* data are available, they get considerably more weight than satellite observations. Since it would be inappropriate to compare the BWI in locations where the microwave observations have significant weight in the GPCP product, (i.e. the two products would not be independent), the comparisons were made over areas where *in-situ* observations were adequate to dominate the final product, thereby allowing us to assume that the two data sets are largely independent. The fact that these data are largely independent is substantiated by a map of weight for all data sources in the GPCP analysis and by the analysis performed is this study. If there was a lot covariance (cross talk) due to dependence, the correlations would not sharply increase when the precipitation is allowed to lag the wetness anomalies. A major reason the GPCP product is used in this study rests on the fact that it is probably the most comprehensive global analysis of precipitation, and that the same quality control and interpolation schemes were used across national borders and throughout all six study areas. A comparison of the two monthly anomalies provide a venue to validate the utility of the BWI, and its ability to serve as a proxy for precipitation over an area.

## 3. METHODOLOGY

The eastern half of the U.S. was chosen as the validation site for the satellite derived surface temperature product. This area spans from 100°W to the eastern seaboard and from the Canadian border to the Gulf of Mexico. The complex topography of the western half of the U.S. excludes that region because rapid elevation changes cause large discrepancies between the point source *in situ* data and the areal integrated SSM/I data. Our study area contains a moderate density of quality *in situ* stations with temperature measurements (Fig. 1). Over a year, the surface conditions within this region consist of wet and dry land covered by trees, grass, cities, snow, ice and dead vegetation in a moderate elevation range (sea level to 1600 meters).

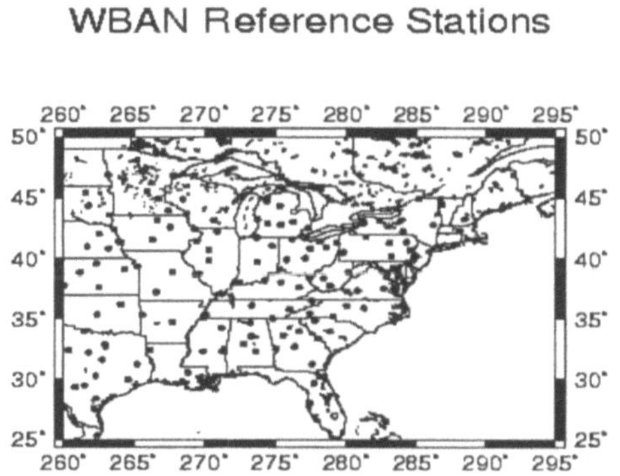


*Figure 1.* The study area and the First Order stations used for calibration of the SSM/I derived surface temperature

Water surfaces are excluded due to their very low emissivity. This low emissivity allows other physical parameters to increase their contribution in the signal, and introduces noise in the final product. For instance, wind across a water surface increases the emissivity and therefore the observed temperature. The concentration of atmospheric water vapor also alters the observed temperature over a water surface. For these reasons, the high polarization difference of water at 19 GHz was used to identify surfaces that were primarily water and remove them from the data sets. Due to the low resolution of the instrument: rivers, ponds, and small lakes all contribute to the integrated signal. As long as the majority of the radiating surface is the dry earth or a vegetated surface, those observations remain in the analysis. However, coastal margins, large lakes and seas are filtered out either through a moderate resolution land-sea mask or the polarization difference at 19 GHz.

As mentioned above, no a priori information is used to identify surface type or the background emissivity. For each observation of every orbit, the seven SSM/I channel measurements provide all the information used to identify the surface type and make an emissivity correction. It took a sophisticated expert to make a dynamic determination of the land and/or atmospheric conditions for each time and place, and identify the surface emissivity. This system calculates the change of surface emissivity due the percentage of liquid water on the radiating surface, at the time of the satellite over flight.

The *in situ* stations monitor the microclimate in a shelter 2 meters above a grass surface. Tests have shown that these instruments have an accuracy of 0.1°C (Nadolski 1992) at a specific point. On the other hand, the SSM/I integrates the entire footprint (12.5 to 60 km) at each of the four frequencies into an average brightness temperature. The SSM/I measurements mainly emanate from the radiating surface, which can be barren and/or vegetated though there is a small atmospheric contribution at the 22 and 85 GHz water vapor frequencies. For the most accurate calibration between each SSM/I signal and the shelter height temperature at a given location, only hourly surface observations within one hour of the SSM/I overpass of 1/3° square occupied by the *in situ* station were used. The following steps elucidate the methodology and procedures to simulate the shelter height temperatures from the SSM/I channel measurements.

We begin by identifying a surface type (e.g. desert, wet surface, snow cover) from the theoretical and empirical relationships between the seven channel measurements. This analysis also allows us to identify which microwave signatures (i.e. scattering, emission and/or polarization) should be used in a regression equation to minimize the standard error between the *in situ* and SSM/I derived surface temperature. A regression equation for each surface type uses the unique relationship between the SSM/I channel measurements to make dynamic emissivity adjustments.

Neglecting atmospheric effects, the emissivity $\epsilon_s$, at a particular frequency v, is the ratio of the brightness temperature at that frequency $T_b(v)$, to the actual surface temperature, $T_s$.

$$\epsilon_s(v) = T_b(v)/T_s \qquad (1)$$

When the ground is dry and vegetated, the vertically polarized SSM/I channels have a nominal emissivity $e_o$ of 0.95. However when the surface is other than dry and vegetated, the emissivity must be implied through a function of scattering and polarization characteristics of the radiating surface. The function can be written as:

$$\epsilon_s(v) = \beta_0\,[f_0(v)] + \beta_1\,[f_1(v)] + \ldots + \beta_n\,[f_n(v)] \qquad (2)$$

where $f_0$ ,$f_1$,..., $f_n$ are functions that contain brightness temperatures, scattering over frequency ranges, or polarization difference at given frequencies. The proportionality constants ($\beta$ through $\beta_n$) account for the different contributions of the various characteristics. Furthermore, the fields of view at the various frequencies range from 60 Km at 19 GHz to 12.5 Km at 85 GHz, and the proportionality constants in (2) partially account for this variation as well.

Substituting (2) into (1) and solving for $T_s$ we obtain

$$T_s = T_b(v)\ /\ (\ \beta_0\,[f_0()] + \beta_1\,[f_1(v)] + \ldots + \beta_n\,[f_n(v)]\ ) \qquad (3)$$

It is evident that the functional form of $e_s(v)$ and the accuracy of $T_s$ will improve as we learn more about the emissivity characteristics of different surfaces. Currently the specific functional form for each adjustable surface is empirically derived from SSM/I measurements and surface temperatures from the eastern half of the United States. The results of this study introduces significant improvements over the previous paper (Basist et al. 1998), since the additional analyses expanded the number of surface types for which emissivity adjustments are derived. The new proportionality coefficients have increased the explained variance and reduced the error of the SSM/I derived temperature product.

Hourly temperature measurements from the 'First Order' network were compared to the corresponding SSM/I derived temperature at the time of the DMSP satellite overpass. The SSM/I channel measurements for these comparisons came from the descending (morning) pass of the F11 satellite because this satellite has the most stable orbit of the three satellites (see section 4) and the near surface boundary layer temperature approaches equilibrium with the skin temperature in the early morning (Betts and Ball 1995).

The functional form of $\Delta\ \varepsilon$ (change of emissivity) was empirically determined from global SSM/I measurements and surface temperatures over wet and dry land. These measurements include a myriad of climatological conditions, e.g., river valleys adjacent to dry ground, irrigated regions adjacent to non-irrigated areas, melting snow next to snow free ground, and recent rain soaked ground near dry ground. To minimize internal factors, we conscientiously chose sites where the surface characteristics are internally homogeneous and where there is nominal topographic gradient. To maximize external factors, we used sites from all over the world, at all times of the year, from both morning and afternoon passes. A scatter diagram between the expected and predicted surface temperature had a correlation coefficient of 0.95 and a standard error less 2.5°C (Basist et al. 1998). Since these are individual pixel errors, the standard error would further diminish when mean monthly values over 1x1 degree boxes are determined. In addition to computing the surface temperature, a new parameter called the BWI is also derived. This parameter is defined as $\Delta\varepsilon{*}T_s$ which is based on equation (4),

$$\mathrm{BWI} = \Delta\varepsilon{*}T_s = \beta_0[(T_b(v_2) - T_b(v_1)] + \beta_1[(T_b(v_3) - T_b(v_2)] \quad (4)$$

Figure 2 presents a histogram of the error characteristic of the derived surface temperature once emissivity is corrected for water in the radiating surface. The distribution, based on 44,619 observation, is approximately guassian, with a average error 0.02°C. The kurtosis is extremely small, meaning that the distribution is bunched near zero error, and there is nominal skewness, indicating that the distribution is symmetrical around the mean. Figure 3 shows the calculated wetness index over the globe for July 1997. Note the high values over the Figure 3. A map of of the Basist wetness index (BWI) for the globe during the month of July 1999. The highest wetness values occur over tundra, swamps, and broad river valleys, moderate values in the moist areas of the globe, and values below 2 in the dry regions or where dense vegetation hides the ground. When these adjustments are added directly to the global temperature, they effectively remove the low bias associated with wet ground (Williams et al. 1999), and provide an excellent approximation of the actual surface temperature.

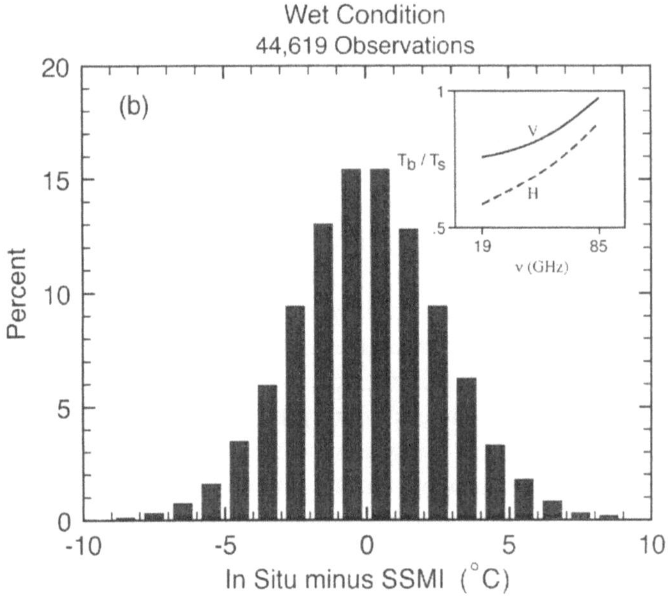


*Figure 2.* A histogram of the differences between in situ and SSM/I derived surface temperatures when wetness is detected on the surface. The right corner of each figure provides an idealized depiction of the SSM/I spectral signature for a range of wet surfaces

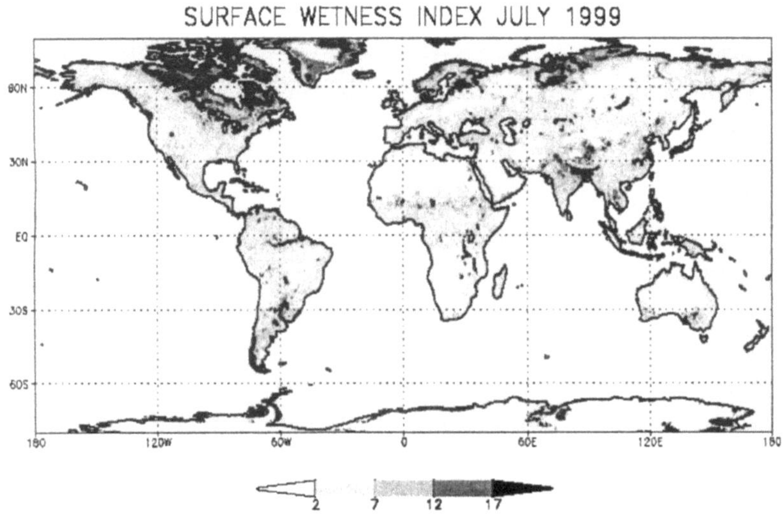


*Figure 3.* A map of the Basist Wetness Index (BWI) for the globe during the month of July 1999

# 4. RESULTS

Figure 4 shows the distribution of the residuals which is the result of *in situ* minus satellite derived monthly temperature anomalies in a 1°x1° grid. This analysis of over 10,000 data points with a standard deviation of 0.76°C indicates that the satellite derived monthly temperature anomalies have a strong correspondence to the *in situ* based anomaly. It is important to note that these differences can be ascribed to several factors: errors in the derived temperatures, *in situ* measurement errors, variations in atmospheric contributions, differences between point measurements and spatially integrated values, and discrepancies between shelter height temperature anomalies and those at the radiating surface. An analysis of the spatial structure in the satellite-derived field consistently demonstrates strong coherence well within 0.5°C.

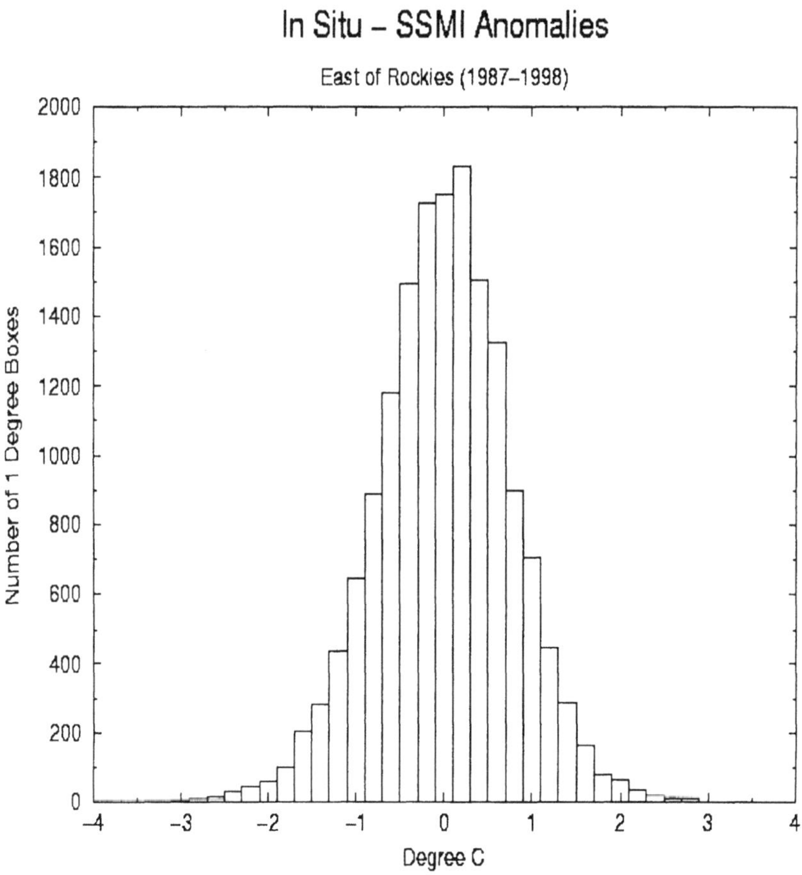


*Figure 4.* Distribution of the residuals between each 1°x1°-grid of *in situ* minus SSM/I monthly temperature anomalies. Each monthly grid box value is a separate observation providing over 10,000 data points over the study period

The time series of monthly anomalies across the study area over the period 1987 through 1997 is presented in Fig. 5; the period between 1989-1991 removed due to erroneous satellite measurements at 85 GHz. The regional anomalies generally fluctuate between -2°C and 2°C and the average fluctuation from the mean is 1.15°C. The two data sets share most of the variance, i.e. have a high signal to noise ratio, with a mean difference of 0.34°C, while the spatial correlation coefficient (r) over the time series is 0.92. Fig. 6 shows a time series of the difference between the two data sets. There are two occasions when the differences exceed 1°C. During almost every month the differences were far less than the actual temperature anomaly. The difference series indicates some temporal auto-correlation, which we will investigate in a future analysis. Since there are some unanswered questions on the second or third order drift characteristics of the satellite, we may be seeing an error associated with changes in the radiometer's precision over time.

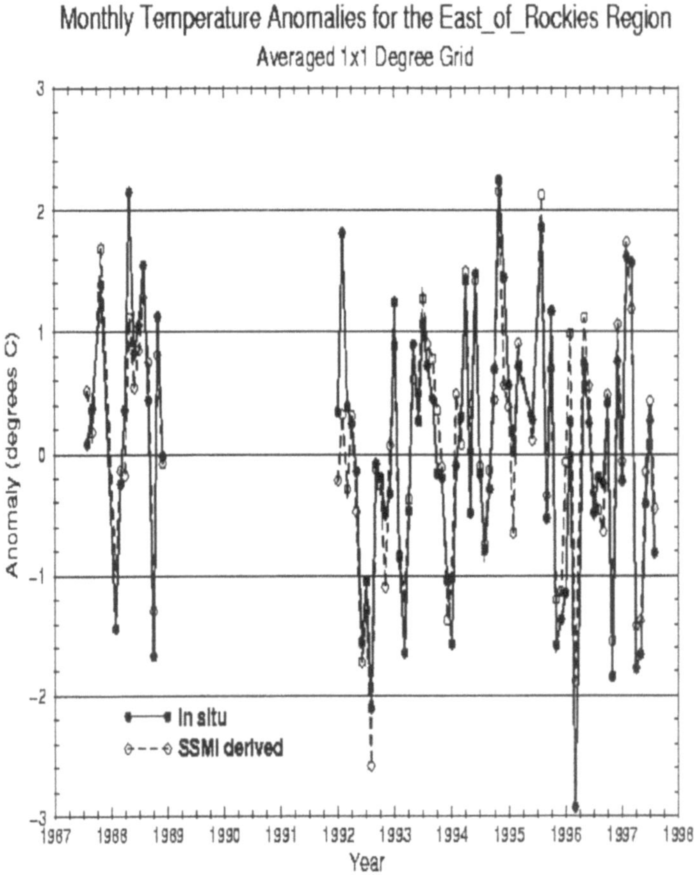


*Figure 5.* Time series of monthly anomalies across the study area over the period 1987 through 1997, with a period between 1989-1991 removed (as discussed in the text). Anomalies during the record generally fluctuate between -2°C and 2°C, with average fluctuation from the mean averaging 1.15°C

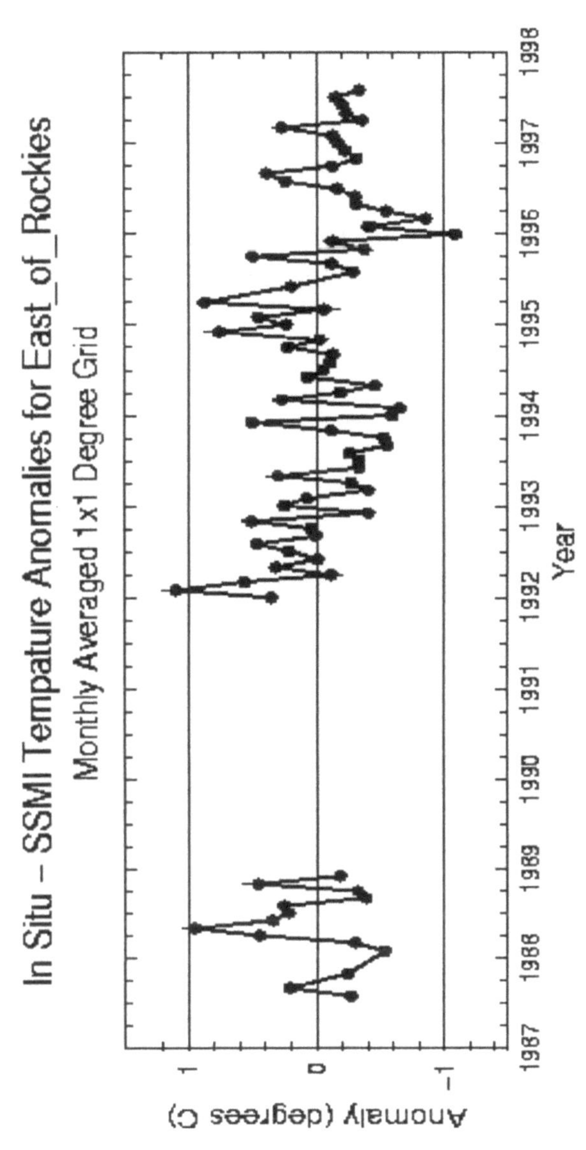


*Figure 6.* Time series on the difference between the two data sets presented in Fig. 5

The temporal correlation in each grid box of the study area is mapped on Fig. 7. There is a significant area where the correlation coefficient (r) exceeds 0.90, this areas covers portions of the Northern Plains, Midwest and Tennessee Valley. The majority of the study area has correlation coefficients greater than 0.80, which means that over two thirds of the variance is shared between the *in situ* and satellite derived temperature anomalies. There are some limited areas where correlation coefficients drop below 0.70. One of these are areas is western Texas where we have investigated potential causes to the lower correlation over this area and have not been able to identify any particular source of the problem. Another area with a lower correlation is the lower Mississippi Valley. This area experiences extreme flooding during part of the year. When there is extensive surface water, the relationship between channel measurements becomes non-linear and introduces noise in the derived product. In general the two fields have both high spatial and temporal correlation throughout the study.

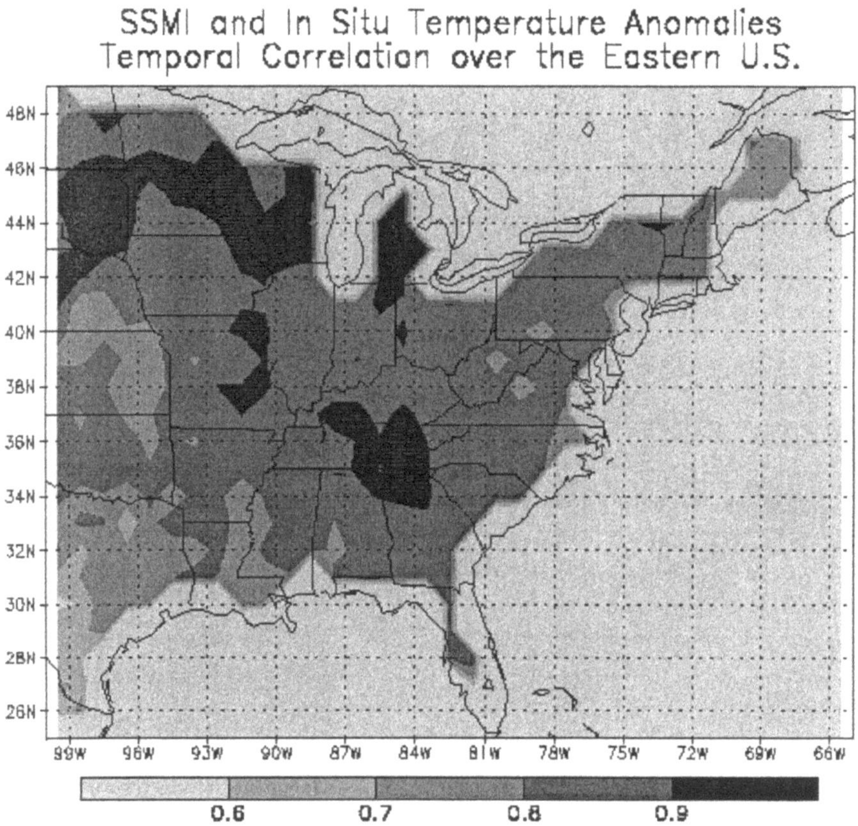


*Figure 7.* Temporal correlation in each grid box of the study area. There is a significant area where the correlation coefficient exceeds 0.90, this areas covers portions of the northern plains, Midwest and Tennessee Valley. The majority of the study area has correlation coefficients greater than 0.80, although some isolated areas do drop below 0.70

For land surface temperatures, these satellite-derived analyses were blended with and anchored to *in situ* mean monthly temperature observations, provided by the Global Historical Climate Network (GHCN). Then the land and ocean observations were merged to produce near-global coverage (peterson et al. 2000). The spatial resolution of the data set was defined by the existing 1x1 degree SST data base. The SSM/I data could provide 1/3rd degree resolution but averaging up to the SST's 1x1 degree made the data set more robust by helping remove outliers (the averaging being done with a biweight approach; Lanzante 1996). The temporal resolution, monthly, was defined by high-quality internationally exchanged *in situ* climate reports which are more reliable than calculating, for example, weekly means from incomplete synoptic reports. And the period of record was governed by data availability from the passive microwave satellite, i.e. the period 1992-present. The GHCN uses climate data from approximately 1,400 stations are exchanged internationally each month. A station's monthly mean temperature is calculated by the source station or country before being transmitted. While thousands more stations transmit synoptic messages over the GTS, monthly means derived from synoptic reports have serious data quality problems primarily because synoptic reports are seldom complete (e.g., Schneider, 1992). Therefore, no data derived from synoptic sources are used in GHCN.

Our approach to blending the data focused on minimizing the impact of significant though rare errors in each of the data sets rather than producing a data set with a slightly more accurate mode. Toward this end, we (a) transformed the data into anomalies to the 1992-present base period so all interpolation or merging was done in anomaly space; (b) interpolated GHCN data out only a modest distance where the interpolated values are most reliable, covering a 5°x5° square centered on the station; (c) weighted all GHCN-derived values equally, that is, an anomaly value interpolated out 2 degrees is given the same weight as a station in that 1x1 degree grid box; and (d) gave the SSM/I anomaly value for a 1x1 degree grid box the same weight as a single *in situ* station. This means that for land areas with many *in situ* observations, the final product primarily represents *in situ* anomalies. If there is only one station in the vicinity, the product reflects both the interpolated *in situ* and SSM/I-derived anomalies equally. Where no *in situ* data are present, the final product is purely SSM/I-derived anomalies (e.g., in much of the tropics) or missing (e.g., snow-covered areas).

To ensure that changes in satellite drift or degradation of the SSM/I instrument doesn't impart a bias to the product, each month's SSM/I analysis was adjusted to anchor it on the *in situ* observations. The difference between every GHCN station monthly temperature anomaly and the SSM/I anomaly for the 1x1 degree grid box containing the station, if available, was calculated. The monthly anomaly used in this calculation was based on a mean of only those years that have both GHCN and collocated SSM/I temperatures. The mean difference for each calendar month was determined. Next, the monthly time series of these differences was smoothed with a three month running average filter to allow more data points to contribute to the calculation of the adjustment. This smoothed difference value was then added to every 1x1 degree SSM/I grid box anomaly value.

The final 1x1 degree product is not fully global. Snow and ice-covered regions are limited to *in situ* observations. Note the excellent coverage in Figures 4d and 5d in the *in situ* data sparse tropics. Clearly there is a spatial coherence to the structure of the monthly climate signal captured by this data set. While it is not error free, it does clearly define the regions of significant climate anomalies and is therefore very useful for monitoring the climate in regions that were previously data sparse. It also shows the spatial structure and cohesion far better than any of the individual data set could provide.

We compared the BWI directly to precipitation derived from the Global Precipitation Climate Program (GPCP), using anomalies from the 1992 through 1997 base period. Analysis were performed over six regions of the globe, each corresponding to an agricultural area on a different continent with spatial dimensions of 5° by 5°. The background characteristics of each region are used to understand the unique relationship between precipitation and the BWI. Correlations between the two fields were derived in the monthly anomalies over the 6 year period. The precipitation anomalies were also aggregated over multiple months, in order to determine the memory of the BWI and its correspondence to upper level soil moisture. Memory in the BWI implies that it retains information on precipitation anomalies during the earlier period, and that it can show how these cumulative values impact the upper level soil moisture.

Southeastern Australia (32.5°N - 37.5°N and 145°E - 150°E) is one of the most important

agricultural areas in Australia, a region where wheat, corn, and oats is grown in abundance. Precipitation is fairly evenly distributed throughout the year, with a slightly higher concentration in the summer. The strong correspondence between precipitation and the BWI anomalies is illustrated by them sharing 65% of their variance (fig. 8). When the precipitation anomaly was aggregated between the previous and concurrent month, the explained variance drops to 44%. The weak memory is associated with the inability of the wetness index to see below the surface, large porosity in the soil, and moderate precipitation amounts that fall over the areas. Therefore rarely does the water pool near the surface for any extended period, and surface moisture is usually depleted within a two month period.

We compared the anomalies over central (45°N -50°N and 00°E - 5°E), where the major crops include: wheat, grapes, oats, barley, sugar beets, fruit trees, and grazing land. Annual precipitation in the region averages between 600 to 800 mm, while the mountains in the southern section have an orographic influence, where some up-slope areas receive over 1200 mm, while downwind areas receive less than 500 mm. Precipitation is fairly evenly distributed throughout the year over this region. Across the time series two data sets share 22% of their variance (fig. 9). When the wetness anomalies were correlated with the precipitation anomalies aggregated between the concurrent and previous month, the explained variance rose to 40%. Furthermore, it rose to 63% when the precipitation anomaly from the previous two months was added with the concurrent month. The strong memory implies that the BWI is a good proxy to the upper level soil moisture, which frequently has a two month memory of precipitation in this region.

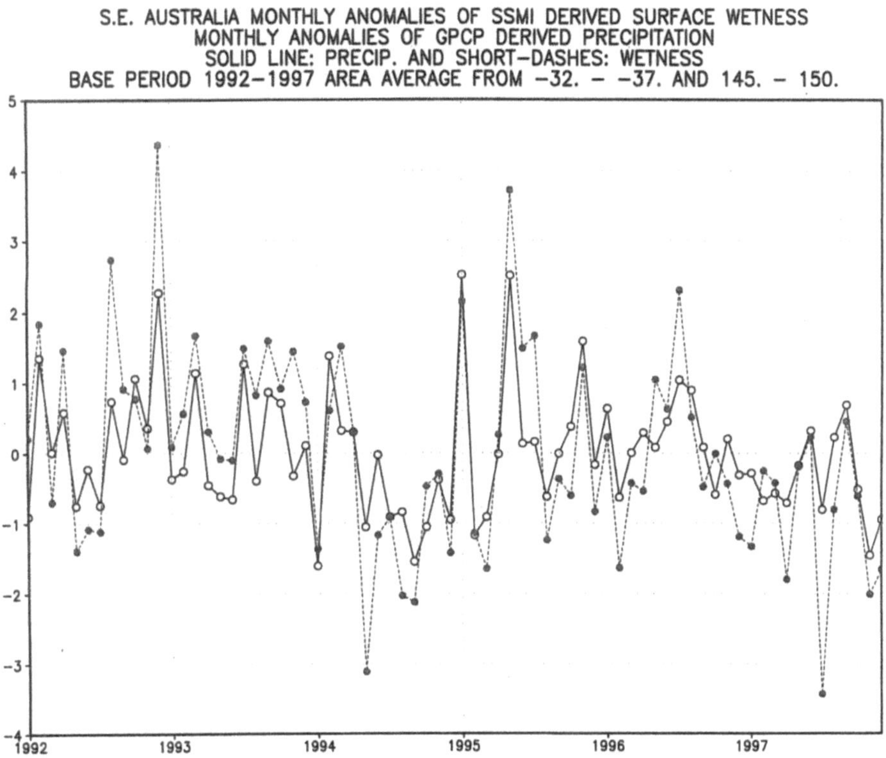


*Figure 8.* A time series of the surface wetness and precipitation anomalies for Southeastern Australia over the period 1992- 1997. Unit for the precipitation anomalies are in millimeters per day, while surface wetness index is unit-less

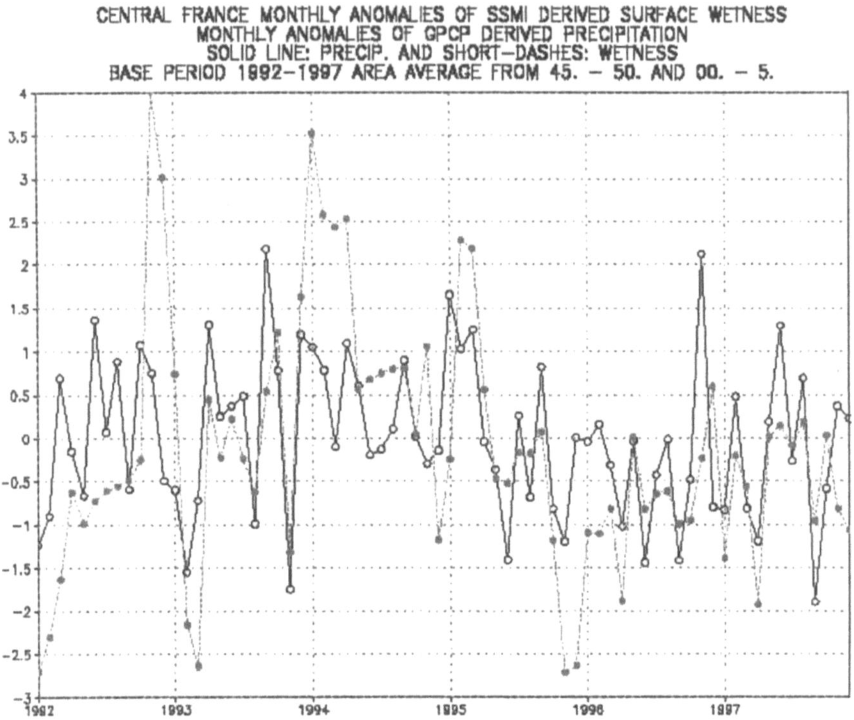


*Figure 9.* The same as Figure 8, except for Central France

Northern Argentina (37°S - 32°S and 63°W - 58°W) is an extremely important source of wheat, corn oats, millet, potatoes, flaxseed, and soybean. The precipitation across the region averages around 700 mm a year, and is slightly drier to the west. Rainfall is most abundant in summer, it is lowest in winter, and the seasonality increases towards to the west. The two data sets share 29% of their variance across the monthly time series (fig. 10). When the precipitation anomalies were aggregated between the concurrent and previous month, the explained variance rose to 49%, and it rose to 59% when the precipitation anomaly from the previous two months was added with the concurrent month. This result indicates that the BWI has memory that corresponds with upper level soil moisture.

The region in the western Sahel (5°W - 0° and 10°N - 15°N) supports grazing land, millet, sorghum, peanuts, soybeans, corn, and cotton. The climate is semiarid with the rainy season primarily restricted from June to September. Annual precipitation averages around 100 cm with a clear gradient of wetter to the south and drier to the north. An 8 months from October to May generally accounts for less than 25% of the annual precipitation. The two data sets share 45% of their variance across the time series (fig. 11). When the wetness values was correlated with the precipitation anomalies aggregated between the concurrent and previous month, the variance rose to 61%, and it rose to 78% when the precipitation anomaly was aggregated from the two previous months. This relationship supports the fact that the wetness anomaly corresponds with the upper level soil moisture, although an early or late start to the rainy season can confuse the signal.

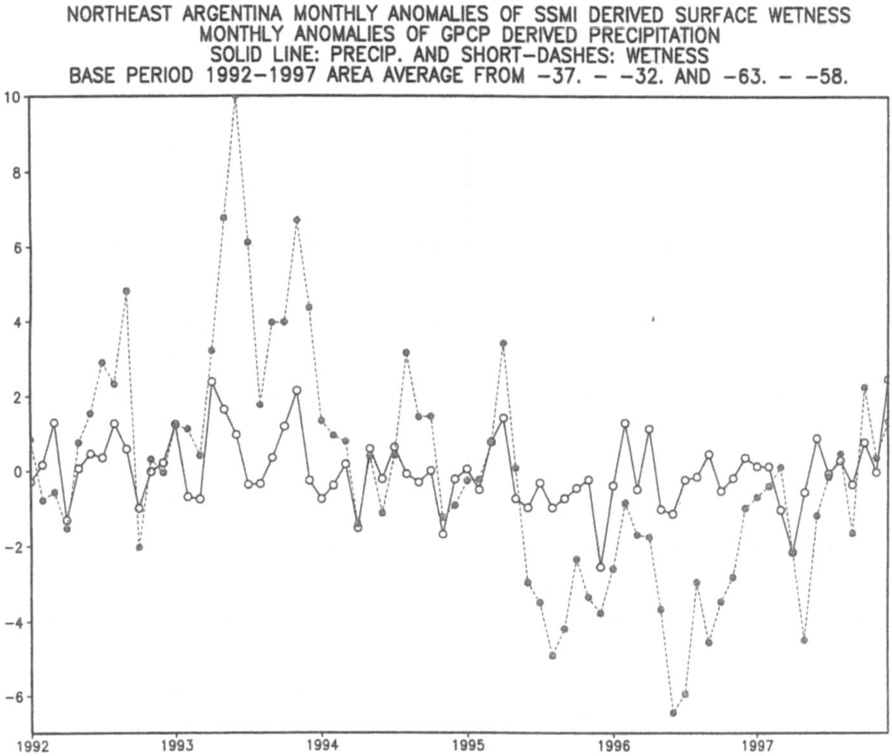


*Figure 10.* The same as Figure 8, except for northern Argentina

The region in the east central China (112.5°E - 117.5°E and 30°N - 35°N) primarily grows rice and cotton. Precipitation in the area ranges from 1300 mm in the south to 700 mm to the north. Precipitation has a summer maximum followed by autumn and spring, while the winter season receive less than 10% of the annual precipitation total. The two variables share 40% of their variance in common (fig. 12). The figure indicates that the correlation would be much higher except for a few notable outliers. Another reason for the moderate correlation is the widespread irrigation of rice pattics over the area, since the source of water covering the surface does not correspond with precipitation. The correlation slightly drops as precipitation anomalies from the previous month were aggregated with the concurrent month. This drop may correspond to the density and type of vegetation, which intercepts rainfall, and partially hides soil moisture from the satellite.

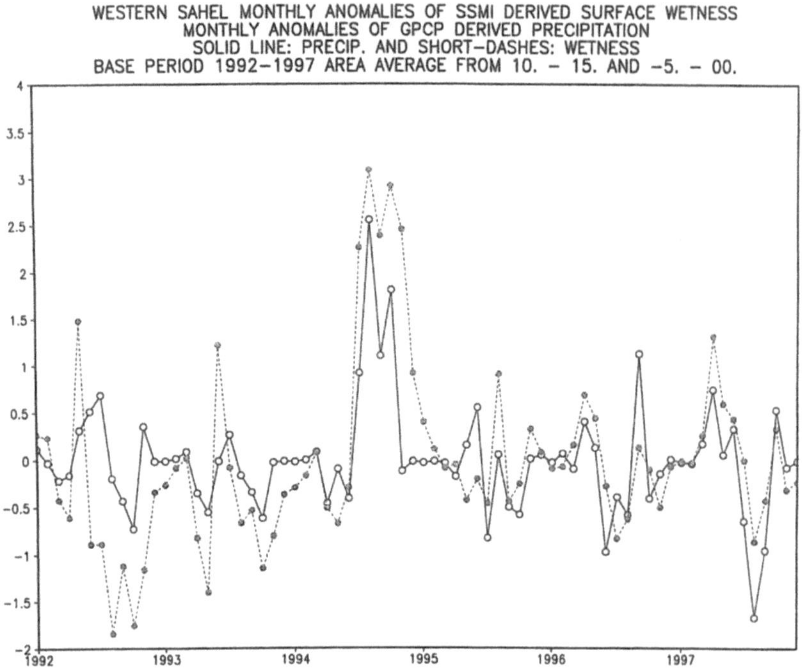


*Figure 11.* The same as Figure 8, except for the Western Sahel

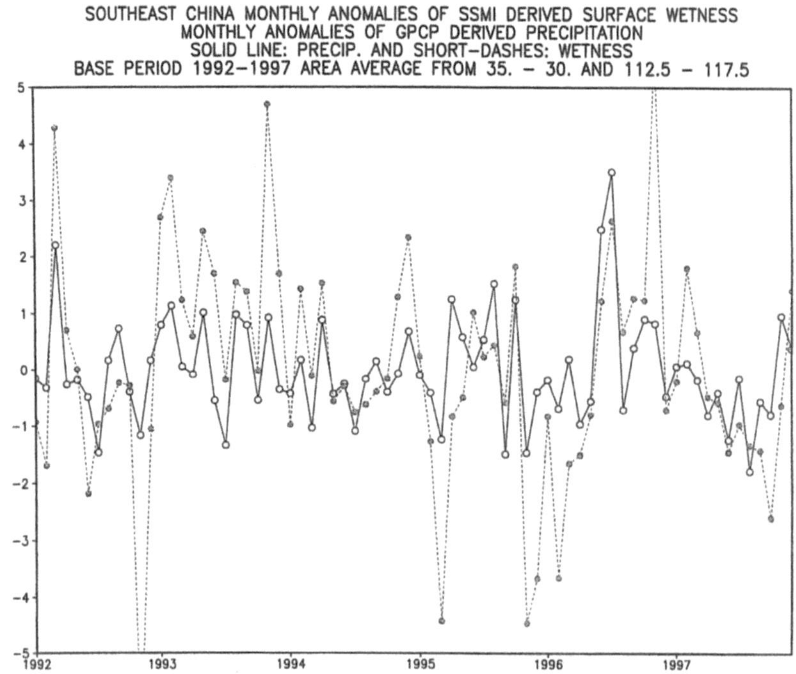


*Figure 12.* The same as Figure 8, except for Central China

The region in the southeastern United States (30°N -35°N and 90°W - 95°W) primarily grows corn, soybean, cotton, and pine trees. Annual precipitation across the area averages around 1300 mm a year. Over 100 days of year receive more than 0.25 mm , and about 35 of those days receive more than 35 mm, accounting for 80 percent of the total precipitation. The two variables share 10% of their variance in common (fig. 13). This weak relationship is caused by dense vegetation (pine and hardwood trees) throughout thc year, and crops with a high leaf area index. Therefore, under densely vegetated conditions, the wetness signal observed by the satellite does not have a strong correspondence to the soil moisture. None-the-less, the shared variance nearly doubles, rising to 17% when the precipitation anomaly from the previous month is added to the anomaly from the concurrent month. This indicates that a appreciable signal does still come from the ground, during at least a portion of the year, and that the wetness index has memory of precipitation from the previous month.

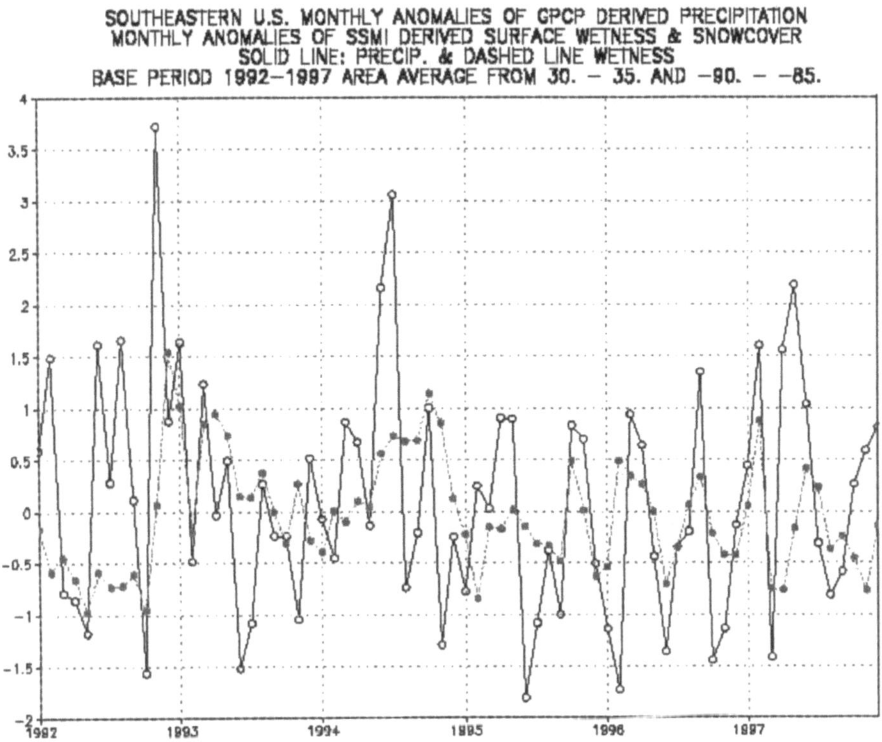


*Figure 13.* The same as Figure 8, except for the southeastern U.S

## 5. SUMMARY AND CONCLUSIONS

*In situ* temperature observations do not provide adequate global coverage; therefore, additional data sources are needed to supplement these traditional observations. Satellite borne infrared observations have been tried for years (Davis and Tarpley 1983), but cloud contamination continues to resist filtering. Although passive microwave measurements have considerably lower resolution than the infrared or visible observations, they provide a viable option. Fortunately, clouds are not much problem in the microwave spectrum; although there is the issue of variable emissivity of the radiating surface. We addressed this variation by understanding the radiating characteristics of numerous surfaces and made dynamic emissivity adjustments as a consequence of the surface type of each observed signal. Spatially and temporally coincident SSM/I observations and *in situ* data were used to develop regression equations to estimate the emissivity adjustments. Each adjustable surface type within the SSM/I data has a different set of parameters and goodness of fit with respect to the shelter height temperatures.

An independent high-resolution data set was used to validate the accuracy and precision of the satellite derived surface temperature anomalies. Both the satellite and reference time series had similar distributions around their mean, and the difference between the two anomaly fields had a standard variance of 0.76°C with low kurtosis and skewness. The spatial correlation over the time series was 0.92 and temporal correlation over the study area generally exceeded 0.80. There was some spatial and temporal auto-correlation in the residual, but it is at an acceptable level.

The present blended global surface temperature product is clearly better than any one of the sources alone. For example, insights into the climate can be obtained by seeing the transition between SST and continental air temperatures which in some instances is very smooth and in other cases have abrupt changes. The SSM/I-derived temperatures provide dramatically improved coverage over *in situ* data in some climatologically important areas and at the same time the SSM/I temperatures are improved by anchoring the monthly anomalies on *in situ* observations. Blended together, the three sources of data provide near-global coverage.

The BWI had strong correspondence with precipitation measurements through many important agricultural areas of the globe. The relationship between the two variables generally increased as precipitation was aggregated across the concurrent and previous months. This indicates that the BWI measures upper level soil moisture, and it provides valuable information across many areas of the world where *in situ* are not available. The BWI does not represent soil moisture with much accuracy over densely vegetated surfaces, since the satellite can not see the ground through the vegetative cover.

Products of global surface temperature anomalies and surface wetness can be obtained from the world wide web at http://www5.ncdc.noaa.gov:7777/plwebapps/plsql/ssmimain. These are available in near real time by the 15th day of the following month. The historical anomaly fields are also available from the January 1992 to the present. The data also be accessed free of charge from the web by going through the site listed above.

# 6. ACKNOWLEDGMENTS

The authors gratefully acknowledge the technical assistance of Michael Burgin. We would also like to acknowledge the support and funding of the NOAA Climate and Global Change Programs' Climate Change Data and Detection element and the National Climatic Data Center.

# 7. REFERENCES

Basist, A. N., N. C. Peterson, T. C. Peterson, and C. N. Williams, 1998: Using the Special Sensor Microwave Imager to monitor land surface temperature, wetness, and snow cover. J. Appl. Meteor., 37, 888-911.

Betts, A. K. and J. H. Ball, 1995: The FIFE surface diurnal cycle climate. J. Geophys. Res., 100, 25,679-25,693.

Davis, P. A. and J. D. Tarpley, 1983: Estimation of shelter temperatures from operational satellite sounder data. Climate and Appl. Meteor., 22, 369-376.

Ferraro, R. R., F. Weng, N. C. Grody, and A. N. Basist, 1996: An eight-year (1987-1994) time series of rainfall, clouds, water vapor, snow cover, sea ice derived from SSM/I measurements. Bull. Amer. Meteor. Soc., 77, 891-905.

Ferraro, R. R. and G. F. Marks, 1994: Effects of surface conditions on rain identification using the SSM/I. Remote Sens. Rev., 11, 195-209. Grody, N. C. and A. Basist, 1996: Global identification of snow cover using SSM/I measurements. IEEE Trans. Geosci. Remote Sens., 34, 237-249.

Grody, N.C. and A. Basist, 1996: Global identification of snowcover using SSM/I measurements. IEEE Trans. On Geoscience and Rem. Sensing. Vo. 34, No.1, 237-249.

Huffman, G. J., R. F. Adler, B. Rudolf, U. Schneider, P. R. Keehn, 1995: Global precipitation estimates based on a technique for combining satellite-based estimates, rain gauge analysis, and NWP model precipitation information. J. Climate., 8, 1284-1295.

McFarland, J.M., R.L. Miller, and C.M.U. Neale, 1990: Land surface temperature derived from the SSM/I passive microwave brightness temperatures. IEEE Transactions Geoscience and Rem. Sens., 28, 839-845.

Neale, C.M.U., M.J. McFarland, K. Chang, 1990: Land-surface-type classification using microwave brightness temperatures from the Special Sensor Microwave/Imager. IEEE Trans. On Geoscience and Rem. Sensing, 28, 829-238.

Nadolski, V. (program manager), 1992: Automated Surface Observing System User's Guide. NOAA Dept. of Commerce publication. 12.

Njoku, E. G., 1994: Surface temperature estimation over land using satellite microwave radiometry. Remote Sensing of Land-Atmosphere Interactions, St. Laury, France. 509-530.

Peterson, T.C., A. N. Basist, C. N. Williams and N. C. Grody. A Blended Satellite–*In situ* Near-Global Surface Temperature Data Set. Bulletin of Amer. Meteor. Soc., accepted.

Prigent, C., W. B. Rossow, E. Matthews, 1997: Microwave land surface emissivities estimated from SSM/I. J. Geophys. Res., 102, 21867-21890.

Vinnikov, K. Y., A. Robock, S. Qui, J. K. Entin, M. Owe, B. J. Choudhury, S. E. Hollinger, and E. G. Njoku., 1999: Satellite remote sensing of soil moisture in Illinois, United States. J. Geophys. Res., 104, 4145-4165.

Weng. F. and N.C. Grody, 1998: Physical retrieval of land surface temperature using the Special Sensor Microwave Imager. J. Geophys. Res., 103, 8839-8848.

Williams, C., A. Basist, T. C. Peterson, and N. Grody, 1999: Calibration and validation of land surface temperature anomalies derived from the SSM/I. Bull. Amer. Meteorol. Soc., accepted.

# Snow Cover Fraction in a General Circulation Model

A. ROESCH, M. WILD and A. OHMURA
*Swiss Federal Institute of Technology, Zurich, Switzerland*

Abstract: Snow cover fraction (SCF) has a significant influence on the surface albedo and thus on the radiation balance and surface climate. Long-term three dimensional simulations with General Circulation Models (GCMs) showed that the SCF greatly affects the climate in the Northern Hemisphere.

By means of both ground observations and remotely sensed data, several deficiencies in the SCF parameterization used in the current ECHAM4 GCM were identified: over mountainous areas a substantial overestimation in the SCF was found whereas flat areas showed a distinctly underestimated SCF. This paper proposes a new parameterization of the SCF for use in GCMs. Evaluations illustrate that it is beneficial to include the effects of (i) flat, non-forested areas, (ii) mountainous regions and (iii) forests.

A new SCF parameterization for flat, non-forested areas was derived by using global datasets of ground-based snow depth and remote sensing observations of snow cover data. A 3-dimensional ECHAM4 simulation showed that this modification raises the SCF by up to approximately 20%, mainly in areas with a relatively thin snow cover.

The comparison between remotely sensed and simulated mean monthly surface albedo revealed a significant overestimation of the surface albedo in snow covered mountainous areas. The extension of the current SCF parameterization in ECHAM4, according to the French climate model Arpège, yielded a close agreement with satellite-derived surface albedo.

Using remotely-sensed SCF data in ECHAM4 over forested areas produced unrealistic results due to the masking of snow cover on the ground underlying the canopy. Therefore, we adopted the submodel for snow albedo as used in the Canadian Land Surface Scheme (CLASS) to simulate the SCF of snow-covered canopies. This model combined with a newly-developed simple snow interception model demonstrated the ability to capture the main physical processes of snow covered canopies, including the albedo. This modification has a beneficial impact on the delayed snow melt in spring, a well-known problem in many current GCMs: The simulated surface albedo over the boreal forests decreases by approximately 0.1 during winter and spring, which is in better agreement with ground-based observations. This induces a significant rise in the surface temperature over extended parts of Eurasia and North America in



*M. Beniston and M.M. Verstraete (eds.),*
*Remote Sensing and Climate Modeling: Synergies and Limitations,* 203-232.

late spring, which subsequently yields a faster snowmelt and an accelerated retreat of the snow line.

# 1. INTRODUCTION

Of all the surface conditions, the extent of snow shows the largest spatial and temporal fluctuations. Over 50% of Eurasia and North America can be seasonally covered by snow (Robinson et al., 1993). Snow cover extent exhibits a strong seasonal cycle with a distinct interannual variability in the middle latitudes.

Snow cover extent has a strong influence on the Earth's heat balance, since a large part of the incoming shortwave radiation is reflected by the high snow albedo. The induced radiative cooling is reinforced by the high thermal emissivity of the snow cover which increases static stability in the boundary layer and consequently reduces turbulent fluxes. This effect is enhanced by a reduced roughness of snow covered vegetation when compared to snow free conditions.

Many studies have shown the importance of snow cover extent for weather forecasts as well as for climate simulations. For example, the sensitivity of the Indian monsoon to the extent of the Eurasian snow cover has been confirmed by several numerical experiments (e.g., Barnett et al., 1989). Walsh and Ross (1988) tested the sensitivity of 30-day forecasts to the presence of continental snow cover and found considerable sensitivity over Eurasia.

Snow extent is related to a number of feedbacks (Randall et al., 1994), the most obvious being the snow albedo feedback: A positive temperature bias leads to larger snow melt and favours rain over snowfall which leads to a decrease of surface albedo. This allows more absorption of solar radiation and therefore reinforces further warming.

Snow cover extent is measured by the snow cover fraction (SCF), which is the fraction of a surface element covered by snow. SCF data are obtained from snow depth (SDH) or snow water equivalent (SWE) data. For relatively thick snow covers, the SCF is obtained easily, for relatively thin snow covers, the computation of the SCF from SDH or SWE is more difficult: When the SDH decreases the exposed patches of ground and the transparency of the snow cover increase. Both cause the reflective properties of the underlying ground to affect the albedo (Kung et al., 1964). In climate models, SCF is diagnostically derived from the SWE, which is a prognostic variable in most models. A correct simulation of the snow cover fraction (SCF) is crucial for the computation of surface albedo during the winter season and the literature presents several parameterizations for use in GCMs

(Dickinson et al., 1993; Marshall et al., 1994; Sellers et al., 1996a; Yang et al, 1997).

In many GCMs, the parameterization of SCF is oversimplified. As a typical deficiency in GCMs, it is assumed that SCF is independent of the vegetation type. However, it is known that SCF over forests, e.g., depends on a number of parameters such as the density of the foliage and the snow intercepted by the canopy. The more closed the canopy, the less incoming solar radiation directly reaches the ground without being reflected at tree elements. In order to realistically compute SCF, it is thus crucial to account for varying canopy densities and the snow amount which is intercepted by the canopy.

As another typical deficiency, SCF in GCMs does not depend on orography, despite the fact that snow spreads more homogenously in flat areas compared to mountainous regions where steep slopes encourage redistribution of snow by wind and avalanches. Moreover, southern faced slopes (northern faced in the Southern Hemisphere) yield more rapid snow melt due to higher insolation when compared to horizontal plains. In addition, SCF depends on the height variation within the grid. As temperature usually decreases with height, the form of the precipitation and snow melting can differ within a single GCM grid element. The inclusion of the subgrid scale orography, i.e., the deviations of height within the grid square from the mean grid-box height, might thus yield more realistic SCF parameterization (Walland et al., 1996; Roesch, 2000).

For the above reasons, the SCF should include the effects of forests and mountainous regions in their parameterization. This paper provides a detailed investigation of SCF over forests as well as flat and mountainous areas and derives parameterizations which are tested in 3-D climate simulations (Section 4). As the main result, a compact formula for SCF is proposed in Section 4.4. The new SCF parameterizations are tested within the framework of 3-dimensional model simulations using the ECHAM4 GCM at T42 horizontal resolution (Section 5). Section 2 describes the model and Section 3 the data.

## 2. MODEL DESCRIPTION AND EXPERIMENTAL DESIGN

### 2.1 ECHAM4

The SCF parameterization is analyzed in the framework of the ECHAM4 GCM of the Max Planck Institute for Meteorology, Hamburg. The structure of this GCM is described in detail by Roeckner et al. (1992) and Roeckner et al. (1996). It has evolved from the spectral numerical weather forecasting model of the European Centre for Medium-Range Weather Forecasts, and has been extensively modified for climate applications. For the present study, the processes affecting the surface albedo are particularly interesting and are summarized below.

In ECHAM4, the background surface albedo $\alpha_{sb}$ for every grid element is constant in time. $\alpha_{sb}$ is calculated on the basis of three blended data sets as described in Claussen et al. (1994). In snow covered regions, the surface albedo $\alpha_{sfc}$ is modified according to

$$\alpha_{sfc} = (1 - f_s) \cdot \alpha_{sb} + f_s \cdot \alpha_s \tag{1}$$

where $\alpha_s$ is the snow albedo and $\alpha_{sb}$ the background albedo. The snow cover fraction $f_s$ is calculated according to Equation 2:

$$f_s = \frac{S_n}{S_n + S_n^*} \tag{2}$$

where $S_n$ is the water equivalent of snow in metres and $S_n^*$ is the critical snow depth (=0.01 m). The albedo of snow and ice covered surfaces $\alpha_s$ is a function of the surface type ($t_s$), surface temperature ($T_S$), and the fractional forest area ($a_f$) over land. A maximum value $\alpha_{smax}$ is assumed for temperatures below 263.15 K and a minimum value $\alpha_{smin}$ for temperatures above the freezing point. Snow albedo for a surface temperature between $T_0$ = 263.15 K and $T_m$ = 273.15 K is obtained by linear interpolation:

$$\alpha_s = \alpha_{smax} - (\alpha_{smax} - \alpha_{smin}) \cdot \frac{T_S - T_0}{T_m - T_0} \tag{3}$$

where $\alpha_{smin}$ and $\alpha_{smax}$ are as given in Table 1 and further depend on the forest fraction ($a_f$) as follows:

$$\begin{aligned} \alpha_{min}(a_f) &= a_f \cdot \alpha_{smin}(a_f = 1) + (1 - a_f) \cdot \alpha_{smin}(a_f = 0) \\ \alpha_{max}(a_f) &= a_f \cdot \alpha_{smax}(a_f = 1) + (1 - a_f) \cdot \alpha_{smax}(a_f = 0) \end{aligned} \tag{4}$$

taking into account the fraction $a_f$ of the grid-cell covered with forest with $\alpha_{smax}$ and $\alpha_{smin}$ as given in Table 1.

*Table 1.* Minimal ($\alpha_{min}$) and maximal ($\alpha_{max}$) surface albedos used in Eq. 4. $a_f$ is the fractional forest area

| surface type | $\alpha_{smin}$ | $\alpha_{smax}$ |
|---|---|---|
| sea ice | 0.5 | 0.75 |
| land ice | 0.6 | 0.8 |
| snow on land $a_f$=0 | 0.4 | 0.8 |
| snow on land $a_f$=1 | 0.3 | 0.4 |

## 2.2 Experimental design

All model simulations are performed with the 3-dimensional ECHAM4 GCM at T42 resolution and were carried out at the Swiss Scientific Computing Center (CSCS) in Manno, Switzerland, on a NEC SX4 vector computer. Initialization is made by using atmospheric data from the ECMWF analysis. Sea surface temperature and sea-ice coverage are prescribed from the atmospheric intercomparison project (AMIP) dataset (Gates, 1992).Each simulation covers a period of eleven years and three months. Since these simulations require a spin-up time of about one year in order to reach an equilibrium climate, the first 15 months were discarded. Therefore, all climatological means refer to a ten-year period.

# 3. DATA

## 3.1 Snow water depth

In this study, the model outputs are validated against the global snow depth (SDH) climatology of the U.S. Air Force Environmental Technical Application Center (USAF/ ETAC) as documented in Foster and Davy (1988). This data set provides a mid-monthly mean SDH climatology with the highest spatial resolution (1° x 1° equal-angle grid) currently available, which was compiled from a comprehensive set of station data for the months of September through to June. The USAF data is generally considered to be one of the most reliable and accurate snow depth climatologies available (Douville et al., 1995b) and is used in several studies dealing with the validation of snow models (Douville et al., 1995b; Marshall et al., 1994; Foster et al., 1996). Throughout the United States, Canada and Eurasia, there is high confidence in the observations, as they generally contain more than 5 years of data and are of good coverage. In areas of sparse data coverage, SDHs are estimated using precipitation and satellite analyses of snow extent.

These areas are generally assumed to have a lower confidence level (e.g., Arctic and Antarctica).

## 3.2 Snow cover fraction

Northern hemispheric monthly SCF data is averaged from weekly values of the "weekly digital Northern Hemisphere snow and ice product" compiled by the National Oceanic and Atmospheric Administration (NOAA) and National Environmental Satellite, Data and Information Service (NESDIS) from 1973 - 1996. NOAA charts are based on a visual interpretation of photographic copies of visible satellite imagery by trained meteorologists. The data are given on a regular 1° x 1°-grid. In general, the NOAA charts are considered to be the most accurate means of obtaining snow extent information on large regional to hemisphere scales. Furthermore, they comprise the longest satellite-based record available and has been intensively used in former studies (Gutzler and Rosen, 1992; Iwasaki, 1991; Kukla and Robinson, 1981; Masuda et al., 1993; Robinson et al., 1993).

The principal shortcomings in using visible satellite imagery to chart snow cover are (i) the inability to detect snow cover when solar radiation is small, (ii) difficulties in discriminating snow from clouds, and (iii) the underestimation of snow cover where dense forests mask the underlying snow. Moreover, problems arise when the snow cover is unstable or rapid changes occur. These deficiencies should be taken into account when interpreting the results in later sections.

## 3.3 Albedo data

The Surface Radiation Budget (SRB, Whitlock et al., 1995) dataset provides incoming and reflected shortwave radiation and thus albedo at the surface. In order to derive surface radiation fluxes and surface albedo from top-of-atmosphere (TOA) radiation fluxes, the algorithm developed by Staylor (Darnell et al., 1992) at the NASA Langley Research Center is used. The SRB dataset is computed from data which covers the period 1984 - 1990. They are given on the International Satellite Cloud Climatology Project (ISCCP) equal-area grid which comprises 6596 gridboxes. Close to the equator,its resolution is 2.5° x 2.5°.

The comparison between simulated and observed data requires the interpolation from a 1° x 1°-grid (snow water equivalent and snow cover fraction) and from the ISCCP-grid (surface albedo) on the T42 grid or T106 grid. The necessary interpolations are performed using an area-weighted interpolation. Reliable results are expected for small-scale grids while the ISCCP-grid raises some problems due to coarse resolution, primarily in the higher latitudes. This should be considered when comparing modelled and observed surface albedos in northern Eurasia or Canada/Alaska.

# 4. RELATIONSHIP BETWEEN SNOW COVER FRACTION AND SNOW WATER EQUIVALENT

In ECHAM4, SCF is calculated from the prognostic SWE without taking into account the vegetation and the orography (Section 2).

The major goal is to develop a more sophisticated yet still compact relationship between SCF and SWE which accounts for different vegetation types and orography. The new parameterization should be simple to save computing time and, furthermore, use only surface boundary fields which are typically available in GCMs such as roughness length, forest fraction and leaf area index.

Before developing a new parameterization, it is useful to review available relationships between SCF and SWE: In most land-surface schemes, SCF is derived from the prognostic SWE. Several approaches account for varying vegetation roughness. Only a few expressions, however, take into account the reduction of surface albedo due to subgrid-scale orography, e.g., over rough mountainous regions (e.g., Douville et al., 1995a and Walland and Simmonds, 1996).

Quantitatively, large differences between the existing parameterizations are found. For example, in Fig. 1, the relationship between the SCF and the SDH used in five climate models is shown. For a snow depth of 5 cm, the 2nd version of the NCAR climate model (CCM2, Marshall et al., 1994) computes snow cover fraction $f_S = 0.20$ while in the "Europa-Modell" of the German Weather Service (Edelmann et al., 1995), the grid box is almost completely snow covered ($f_S = 0.998$). Part of the variations may be due to the fact that they are used in models with different resolution. Nevertheless, Fig. 1 suggests that the SCF is calculated with large uncertainties in climate models.

In the following sections, new parameterizations for the calculation of the snow cover fractions over (i) flat, non-forested areas, (ii) mountainous, non-forested areas and (iii) forests are presented. This partition reflects the fundamentally different characteristics of snow cover in forests and non-forested areas,as well as between flat and mountainous regions, respectively. Finally, it is shown how these parameterizations are blended into a short, compact formula for use in land surface schemes.

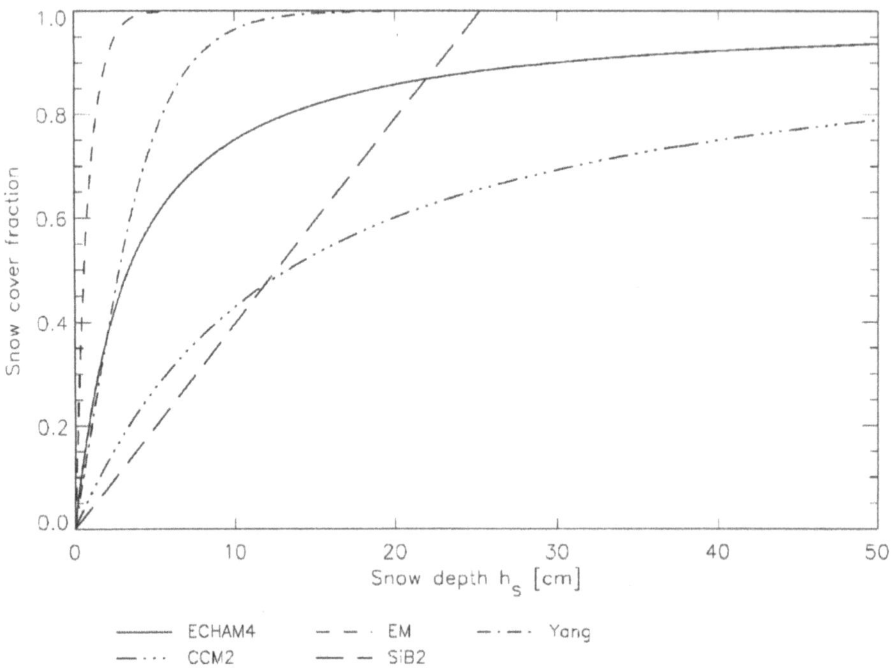


*Figure 1.* Relationship between snow depth and snow cover fraction assuming $z_0 = 2$ cm (grass) and a snow density $\rho_s = 300$ kgm$^{-3}$ using various parameterizations. EM: "Europa-Modell" (Edelmann et al., 1995); Yang: Yang et al. (1997); CCM2: 2nd version of NCAR climate model (Marshall et al., 1994); SiB2: Simple Biosphere (Model Sellers et al., 1996a)

## 4.1 Flat, non-forested areas

For flat surfaces without forests, the relationship between SCF and SWE (fig. 2) was derived by using satellite-based and ground-based observations. On one hand, the remote sensing data are used for estimating SCF, since SCF cannot be obtained in areas comparable to the size of GCM grid cells from ground observations. On the other hand, ground observations are used for SWE, since from satellites, snow depth and density cannot be measured. As both satellite- and ground-based observations are involved, the equation which applies only for non-forested areas for the following reasons: (i) Satellite-based observations of SCF are not well defined in forested areas due to masking of snow on the ground below the canopy and, (ii) ground observation are rare in densely forested regions such as the boreal forests.

The improved parameterization for the SCF was established in four steps. In the first step, SDH $h_s$ from the USAF climatology was transformed into snow water equivalents $S_n$ (in millimetres) according to Verseghy (1991), since snow depth is not simulated in ECHAM4:

$$\rho_s = 188.82\,\mathrm{kgm}^{-3} + 0.419\,\mathrm{m}^{-1} \cdot S_n \leq 450\,\mathrm{kgm}^{-3}, \tag{5}$$

where $\rho_S$ is the density of snow. Following Verseghy (1991), snow density does not exceed 450 $\mathrm{kgm}^{-3}$, but remains constant for $S_n > 0.623$ m. Equation 5 accounts only for changes in snow density due to mechanical compaction but neglects metamorphism. Temperature induced metamorphism leads to significant density changes at the end of the snow season. Therefore, Eq. 5 probably underestimates the snow density in spring since the density increases as the snow cover melts. However, Robock et al. (1995) validated Eq. 5 using daily observed values of snow depth and SWE at six Russian stations between 1978 - 1983.

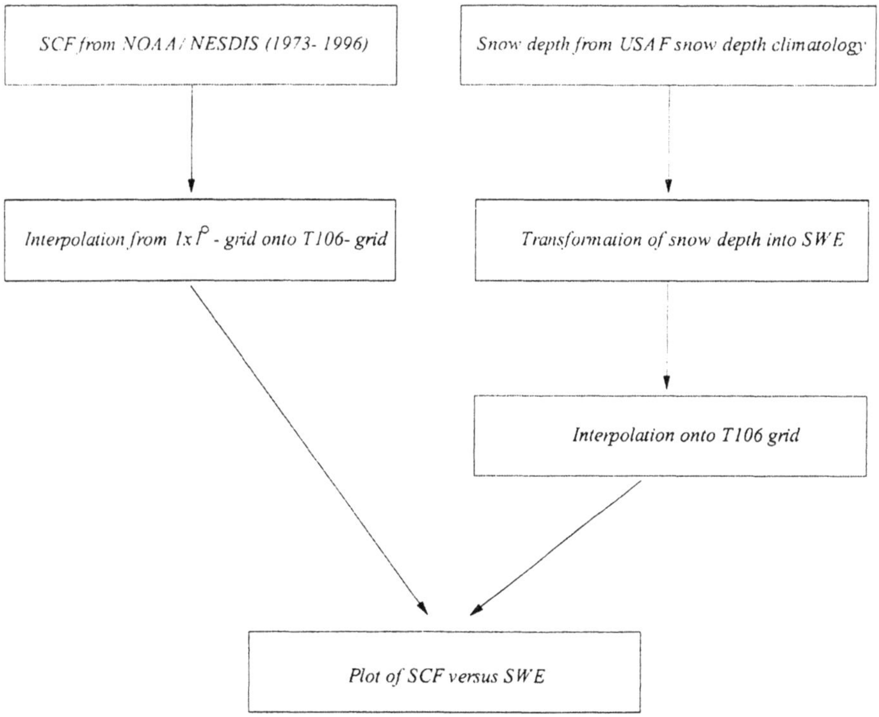


*Figure 2.* Flow chart for derivation of the relationship between SWE and SCF

In the second step, SWE obtained with Eq. 5 from the USAF snow depth climatology (Section 3) was transformed onto the T106 grid (1.1° resolution) by using an area-weighted interpolation. The error introduced by interpolation is negligible as the old and new mesh have similar resolution.

In the third step, a global snow cover climatology grid was compiled on the T106 grid from the NOAA/NESDIS data (Section 3). This compilation involved the calculation of a frequency of occurrence from the weekly present/absent data and an area-weighted interpolation onto the T106 model grid.

In the fourth step, the SCF was plotted as a function of the SWE in Fig. 3. Data measured in regions with forests and mountains were excluded using data from grid-cells with less than 10% forest cover (as compiled by Claussen et al., 1994) and a standard deviation of the sub-grid scale orography smaller 300 m. The few data measured in Greenland and Antarctica are not considered.

A *tanh*-function, proposed by Yang et al. (1997) proved to be most adequate for the relationship between $f_s$ and $S_n$. The parameters a and b in Eq. 6 were estimated with a non-linear least-squares fit (Marquardt, 1963). These parameters were calculated separately for each month from December to March, yielding a = 1.05 $cm^{-1}$ and b = 0.96 for December, a = 0.98 $cm^{-1}$ and b = 0.96 for January, a = 1.02 $cm^{-1}$ and b = 0.94 for February, a = 1.03 $cm^{-1}$ and b = 0.95 for March. Since the differences are small, a = 1.0 $cm^{-1}$ and b = 0.95 were chosen for the whole season with snow cover. Thus, the equation for SCF is:

$$f_s = b \cdot \tanh(aS_n) \quad \text{with } a = 1.0\ \text{cm}^{-1};\ b = 0.95. \tag{6}$$

Further investigations including the roughness length for vegetation ($z_{0,veg}$) as a new independent variable in the parameterization of SCF led to no further improvement in the parameterization. This might be partly due to large uncertainties and inaccuracies in determining the roughness lengths for GCM grid boxes. In addition, it is hazardous to use grid-mean roughness lengths rather than computing roughness lengths separately over each vegetation type within a grid-cell.

From Fig. 3, it is evident that the original parameterization in ECHAM4 substantially underestimates the snow cover fraction in flat, non-forested grid boxes, while the parameterization used in the "Europa-Modell" is an upper envelope of the point cloud displayed in Fig. 3. Both old parameterizations are a poor description of the relationship between SCF and SWE for flat surfaces without forests.

These results are in line with the findings of Yang et al. (1997). They argued, based on measurements of daily mean surface albedo and snow depth by Baker et al. (1991), that there are two distinct stages in the relationship between surface albedo and snow albedo. In the first stage, albedo rapidly increases with increasing SDH until it reaches a critical depth. For larger SDHs, during the second stage, the surface albedo increases minimally as SDH increases. They emphasize that the *tanh*-form pertains to grasslands and agricultural lands, where vegetation slumps under the snow burden.

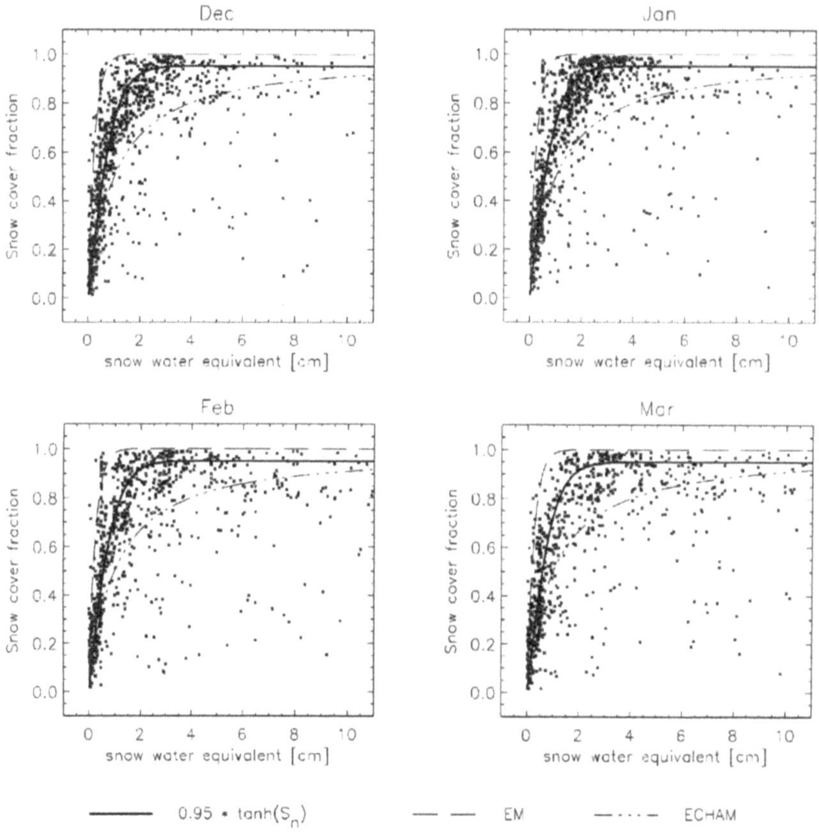


*Figure 3.* SCF as a function of SWE in flat, non-forested T106 grid cells. SWE interpolated from USAF snow depth climatology; SCF from NOAA data (1973 - 1996). Curves are displayed for the new parameterization(solid thick line, Eq. 6), "Europa-Modell" (EM, long-dashed) and ECHAM (dash-dotted, Eq. 2)

## 4.2 Mountainous, non-forested areas

In regions with mountains, ECHAM4 overestimates the surface albedo in winter (Fig. 4). Since the snow albedo was shown to be realistic (Roesch et al., 1999), the albedo bias is primarily due to an overly large SCF. Therefore, in order to reduce the SCF over mountainous regions, the parameterization 7, which is applied in the Météo-France climate model, was tested in detail. Eq. 7 was proposed by Douville et al. (1995a). This parameterization is based on the standard deviation of the subgrid orography ($\sigma_Z$ in [m]):

$$f_s = \frac{S_n}{S_n + S_n^*} \cdot \sqrt{\frac{S_n}{S_n + \max(1.0\text{m}, 0.15\sigma_z)}}, \tag{7}$$

where $S_n^* = 0.01$ m. Eq. 7 accounts for snow patches remaining on the northern slopes of the mountains, whereas the slopes exposed to solar radiation are snow free during the melt period. In addition, on steep mountains, the bare rock can be exposed on account of snow sliding, even during winters with heavy snowfall. This might markedly reduce the SCF and thus surface albedo. Fig. 5a displays how SCF is significantly reduced over rough mountainous regions. The SCF is about 70% over steep mountains ($\sigma_z \geq 1000\text{m}$) for $S_n = 0.2$ m, compared to more than 90% over flat land areas.

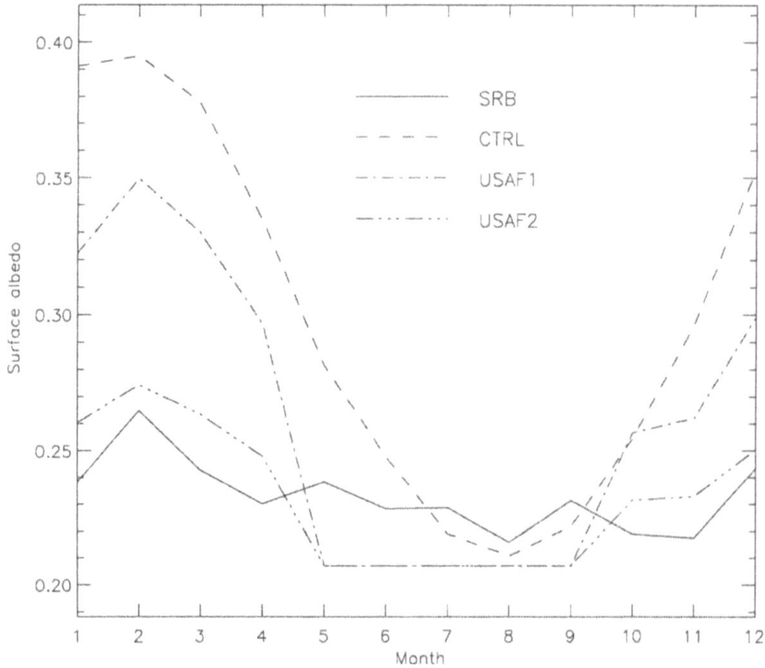


*Figure 4.* Monthly mean surface albedos for the Himalayan region. SRB: remotely sensed surface albedo of the SRB Project (1984 - 1990); CTRL: simulation of the current climate with ECHAM4/T42; USAF1 and USAF2: modified albedos based on the USAF snow depth climatology and SCFs determined with Eqs. 2 and 7, respectively

The calculation of $\sigma_z$ is based on a global digital elevation model, compiled by the U.S. Geological Survey's EROS data center in Sioux Falls, South Dakota (Bliss and Olson, 1996). The horizontal grid spacing is 30-arc-seconds (which corresponds to about 1 km). This global height distribution dataset provides a resolution ten times better than the US Navy dataset with a resolution of 5' x 5', as used in the ECHAM4. A 1-km-resolution is better adapted for resolving the characteristic dimension of valleys and the steepness of mountain slopes which are relevant to processes such as snow sliding and snow melt.

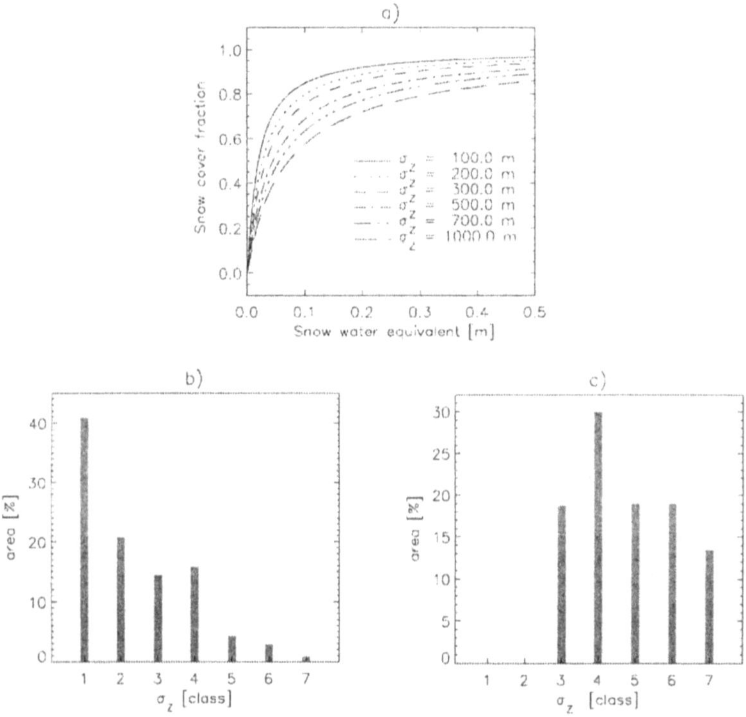


*Figure 5.* Snow cover fraction as implemented in the Météo-France climate model (Eq. 7). b): Histogram of $\sigma_Z$ for all T42 grid boxes over land. The computation of $\sigma_Z$ is based on the global high-resolution topographical dataset compiled by the U.S. Geological Survey's EROS data center Bliss1996}. d): As c), but limited to the Himalayan area

Figures 5b and 5c show the distribution of $\sigma_Z$ for all land grid points and the Himalayan area, respectively. The classes '1' to '7' embrace the following ranges for $\sigma_Z$: class 1: $\sigma_z < 100\text{m}$; class 2: $100\text{m} \le \sigma_z < 200\text{m}$; class 3: $200\text{m} \le \sigma_z < 300\text{m}$; class 4: $300\text{m} \le \sigma_z < 500\text{m}$; class 5: $500\text{m} \le \sigma_z < 700\text{m}$; class 6: $700\text{m} \le \sigma_z < 1000\text{m}$; class 7: $\sigma_z \ge 1000\text{m}$. More than 50% of the Himalayas is covered with grid boxes with $\sigma_z \ge 500\text{m}$ at T42 resolution. Approximately 25% of the Earth's land surface has a standard deviation of the subgrid orography larger than 300 m at T42 resolution, thereby pointing out the relevance of incorporating $\sigma_Z$. In addition, these regions are mostly in mountainous (and, therefore, higher and colder) areas that are usually snow covered during winter.

## 4.3 Forests

The SCF of forests is difficult to assess since snow on the ground is strongly masked by the overlying canopy. In addition, snow intercepted on the canopy further complicates the SCF parameterization.

A detailed representation of the radiation fluxes within the canopy is a prerequisite for the computation of the total forest albedo. Numerous canopy models which focus on radiation transfer have been developed in the past (Dickinson, 1983; Sellers, 1985; Yamazaki et al., 1992; Joseph et al., 1996). They are usually based on the two-stream approximation and thus reduce the complex radiation transfer through canopies to a one-dimensional problem. The main deficiency of these rather sophisticated models arises from the need for further assumptions, such as randomly distributed leaves, and from the necessity to make those input data available on a GCM-grid.

Therefore, a second approach involving less assumptions and only available input data is used for the development of a new SCF parameterization. Various studies (Otterman, 1984; Barker et al., 1994; Yang et al., 1997) demonstrate that the SCF over forests is likely to differ from that over grass and agricultural lands. These studies suggest that over forests, an exponential relationship between SCF and SWE is more appropriate. However, these simple approaches neglect the structure of the forests. They consider neither variations in the leaf area index nor do they distinguish broad- and needleleaf trees. They do not account for openings in the canopy where solar radiation reaches the (possibly highly reflective snow covered) ground without being reflected at tree elements. Furthermore, the process of snow intercepted by the canopy is neglected. Most of the above processes are considered in the Canadian Land Surface Scheme (CLASS, detailed in Verseghy, 1991). This parameterization is kept as simple as possible and does not require further surface boundary fields. As CLASS does not include the relevant processes leading to variations in the snow mass intercepted by forests, a simple snow interception model was developed. The relevant equations adopted from CLASS are briefly discussed in the next subsection.

## 4.4 Albedo of snow covered forests as in CLASS

CLASS computes the albedo of snow-covered forests using Eqs. 8, 9, 10, 11 and 12. Note that no other CLASS parameterizations (e.g., the computation of the snow albedo or the structure and heat conduction of the snowpack) were adopted in ECHAM4.

CLASS allows for snow on the forest canopy. It is based on a simple algorithm, yet is dedicated to capture the principal relationships between the canopy albedo and snow water equivalent on both the underlying ground and the canopy.

The key parameter for the computation of the albedo of snow covered forests in CLASS is the sky view factor (SVF) which describes the degree of canopy closure. The SVF is related to the leaf are index by an exponential function:

$$SVF = e^{-0.5LAI} \qquad \text{(needleleaf trees)}$$
$$SVF = e^{-1.5LAI} \qquad \text{(broadleaf trees).} \tag{8}$$

The total surface albedo of forests is computed as

$$\alpha = SVF \cdot \alpha_g + (1 - SVF) \cdot \alpha_c , \tag{9}$$

where $\alpha_g$ is the albedo of the ground underneath the canopy and $\alpha_c$ is the albedo of the closed canopy. The snow albedo on the ground is assumed to be the same as in the open area, which is in line with the findings of Pomeroy and Dion (1996). $\alpha_c$ is given by

$$\alpha_c = f_{sc} \cdot \alpha_{sc} + (1.0 - f_{sc}) \cdot \alpha_{c,snowfree} , \tag{10}$$

where $\alpha_{sc}$ is the albedo of closed canopy with a maximal snow interception, and $\alpha_{c,\ snowfree}$ is the snowfree canopy albedo. $\alpha_{sc}$ is set to 0.20, from values given in the literature (e.g., Verseghy, 1991; Harding and Pomeroy, 1996; Pomeroy and Dion, 1996). The fraction of the canopy covered by snow, $f_{sc}$, is defined as

$$f_{sc} = S_{nc} / S_{nc,max} , \tag{11}$$

where $S_{nc}$ is the water equivalent of snow intercepted by the canopy and

$$S_{nc,max} = 0.2\,mm \cdot LAI. \tag{12}$$

Verseghy (1991) reports that Eq.12 works well for both rain and snow and for a wide variety of vegetation types and precipitation events. This means that for LAI = 5, an amount of snow equal to 1 $kgm^{-2}$ (or 1 cm of fresh snow) is sufficient to fill the canopy storage capacity.

The computation of the albedo of snow covered forests using Eqs. 8 - 12 requires $S_{nc}$, the snow water equivalent of the snow intercepted by the canopy. In CLASS, snow on the canopy is removed by sublimation only. Therefore, a model for $S_{nc}$ must be developed. This will be assessed in the next section.

## 4.5 Snow interception model

CLASS does not allow for the processes which are relevant for the snow mass intercepted by the canopy. In particular, it does not account for down-loading of snow triggered by wind and temperature close to or above free-zing point. Hence, as the current version of ECHAM4 sustains no reservoir for snow intercepted by the canopy ($S_{nc}$), a simple snow interception model was developed.

The prognostic variable $S_{nc}$ evolves according to the following equation:

$$\frac{\partial S_{nc}}{\partial t} = P_{sc} - S_{nc} \cdot f(T_1) \cdot f(v) - E_{sc} , \qquad (13)$$

with

$E_{sc}$ Evaporation rate from the skin reservoir for intercepted snow [$kgm^{-2}s^{-1}$]
$P_{sc}$ Snowfall rate per unit area intercepted by the canopy [$kgm^{-2}s^{-1}$]
$S_{nc}$ Snow water equivalent of the snow intercepted by the canopy [$kgm^{-2}$]
$f(T_1)$ Function describing unloading of intercepted snow per time step caused by temperature $T_1$ at the lowest model level [$s^{-1}$]
$f(v)$ Function describing wind-induced downfall of intercepted snow per time step [$s^{-1}$]

Eq. 13 includes the major processes affecting the amount of snow intercepted by the canopy, i.e.:

(1) snowfall rate
(2) unloading due to temperature (melt/drip and slipping) (Eq. 14)
(3) unloading due to wind (Eq. 15)
(4) sublimation of intercepted snow.

It is assumed that snow sublimation is at its potential rate.

In order to keep Eq. 13 as simple as possible, $f(T_1)$ is specified as a linear function in $T_1$ (°C)

$$f(T_1) = \frac{3^{\circ}C + T_1}{1.87 \cdot 10^5 \,{}^{\circ}C \cdot sec} . \qquad (14)$$

such that $f(-3°C) = 0.0$. The unloading rate due to melting is assumed to vanish for temperatures below $T_l = -3°C$. The value of the denominator allows for the unloading of half the intercepted snow during 12 hours at $T_l = 0°C$. For a temperature of 2°C, approximately 70% of the intercepted snow is unloaded during 12 hours. The considerable increase in unloading for air temperature above 0°C agrees with a rapid decrease of snow on trees after a snowfall caused by slipping and melt (Nakai et al., 1994). Eq. 14 is also in line with an exponential decrease of the crown-snow ratio with time, depending on the air temperature and wind speed (Yamazaki et al., 1996). According to this study, the response time $\tau$ (reduction to ≈37% of initial value) of the crown-snow is about 1/2 day when the air temperature is below 0°C, and 1 - 5 hours when it is above 0°C. This is in reasonable agreement with the newly-developed model.

After snowfall, release of intercepted snow, triggered by branch movement due to wind influence, is generally observed. Unloading may also be caused by the atmospheric shear stress exerted by wind on the branches and snow. Betts and Ball (1997) analysed Boreal Ecosystem-Atmosphere Study (BOREAS) measurements from 1994 and 1995. They found that (winter) forest albedos above 0.3 correspond to days with low wind speeds of less than approximately 3 $ms^{-1}$. Miller (1962) reports that snow interception considerably decreases when the wind speed during snowfall is larger than 2 $ms^{-1}$. The equation for the wind induced unload of intercepted snow was assumed to be similar to that of temperature. In Eq. 15, v [$ms^{-1}$] represents the wind speed at 10 m above the ground, which corresponds, to a first approximation, to the mean canopy height.

$$f(v) = \frac{v}{1.56 \cdot 10^5 \, m}. \tag{15}$$

With this denominator 50% of the intercepted snow is unloaded within 6 hours for $v = 5\ ms^{-1}$. This interception model does not presume that the intercepted snow load approaches zero between each snowfall event as most simple interception models do (Hedstrom and Pomeroy, 1998).

## 4.6 Compact formula for the surface albedo of the entire grid-box

In the previous sections, improved SCF parameterizations have been separately derived for both non-forested and forested areas. In order to determine a closed formula for the surface albedo of an entire grid box, the formulae for SCF of non-forested areas will be combined.

Applying both Eqs. 6 and 7 produces an inhomogeneity as Eq. 7 approaches the current ECHAM parameterization (Eq. 2) for $\sigma_z \rightarrow 0$. In order to fulfil the requests of both parameterizations, Eqs. 6 and 7 were merged:

$$f_{s,nf} = \tanh(S_n) \cdot \sqrt{\frac{S_n}{S_n + \max(1.0\mathrm{m}, 0.15\sigma_z)}}. \tag{16}$$

Eq. 16 applies to non-forested (subscript ``nf") areas.

The total surface albedo $\alpha_{sfc}$ can now be determined as an area-weighted sum of the surface albedo over non-forested (1-$a_f$ [1-SVF]) and forested ($a_f$ [1-SVF]) areas, respectively. The non-forested part consists of the two terms (1-$a_f$) and ($a_f \cdot$SVF) as the canopy openings must be counted to the forest-free part.

$$\alpha_{sfc} = [1 - a_f(1 - SVF)] \cdot [f_{s,nf} \cdot \alpha_s + (1 - f_{s,nf}) \cdot \alpha_{snfr}] + a_f(1 - SVF) \cdot \alpha_c \tag{17}$$

where

$a_f$ fractional forest area
$a_c$ albedo of closed forest (Eq. 10)
$a_s$ snow albedo
$a_{snfr}$ albedo of unforested and snow free surfaces
$a_{sfc}$ surface albedo for the entire grid box.

All other abbreviations are as in Eqs. 9 and 16.

## 5. EFFECT OF MODIFIED SCF PARAMETERIZATIONS IN 3-D CLIMATE SIMULATIONS

The effects of the modifications presented in Section 4 are studied in 3-D ECHAM4/T42 experiments described in Chapter 2. For a detailed discussion of the response, the modifications for

(i) flat, non-forested areas (Experiment MOD1)
(ii) mountainous, non-forested areas and (Experiment MOD2)
(iii) forests (Experiment MOD3)

were separately implemented and tested in three independent model runs.

## 5.1 Flat areas

The implementation of the *tanh*-function for computing the snow cover fraction from the SWE in the model experiment MOD1 (Eq. 6) substantially raises the SCF compared to the 10-year control experiment. This is shown in Fig. 6 for the winter season (DJF) in the Northern Hemisphere. The notable increase in SCF is mainly limited to flat regions with relatively thin snow cover and few forests. These characteristics apply to areas such as the vast lowlands in Ukraine and Kazakhstan with predominantly grass or Steppe vegetation (Fig. 6b). Further significant differences in the simulated SCFs are found south of Lake Baikal in Mongolia. In North America, substantial differences in the SWE are simulated at the Great Lakes and west of Lake Winnipeg with their generally flat landcapes.

The SCF is only marginally affected in highly forested areas (boreal forests) as well as mountainous regions (Rockies, Himalayas or Alps) since the new parameterization is only applied to flat, non-forested areas. Small differences between CTRL and MOD1 are also found in regions with thick snow cover such as Arctic Russia and northern Canada with a mean (modeled) SCF above 80% during winter (Fig. 6a). The largest differences in the SCF, using Eq. 6, occur for thin snow cover. This difference is larger than the interannual DJF variability given in Fig. 6c.

The maximum difference in the SCF is associated with a snow water equivalent of 1.6 cm and amounts to more than 25%. For thicker snow cover, the difference between the two expressions decreases rapidly and is small for SWE larger than 10 cm.

This modification of the SCF parameterization for flat regions without forests (MOD1) produces only a few significant changes in the surface climate, as summarized by the sensitivities in Table 2. These sensitivities were averaged from land grid boxes with measurable snow ($S_n > 0.1$ cm) in February. Annual differences larger than 3% are found for snow cover fraction, surface albedo, reflected shortwave radiation and sensible heat flux. The percentage for the sensible heat flux is large because annual means are close to zero. The larger snow cover fraction increases the reflected shortwave radiation and thus the heating of the ground is reduced, which implies lower surface temperatures. Lower surface temperatures yield a higher fraction of snow in the total precipitation and less snowmelt which again increases the SCF in a positive feedback. However, the cooling of the surface and the increase of snow water equivalent are small, being statistically insignificant. The lower amount of available radiation near the ground may also reduce the magnitude of the hydrological cycle which is supported by a decrease in total precipitation. It should be noted that the above discussion is purely one-dimensional, thus precluding any advective processes. For example, it cannot be excluded that changes in the moisture convergence, rather than lower net radiation, yield the simulated decrease in the annual precipitation.

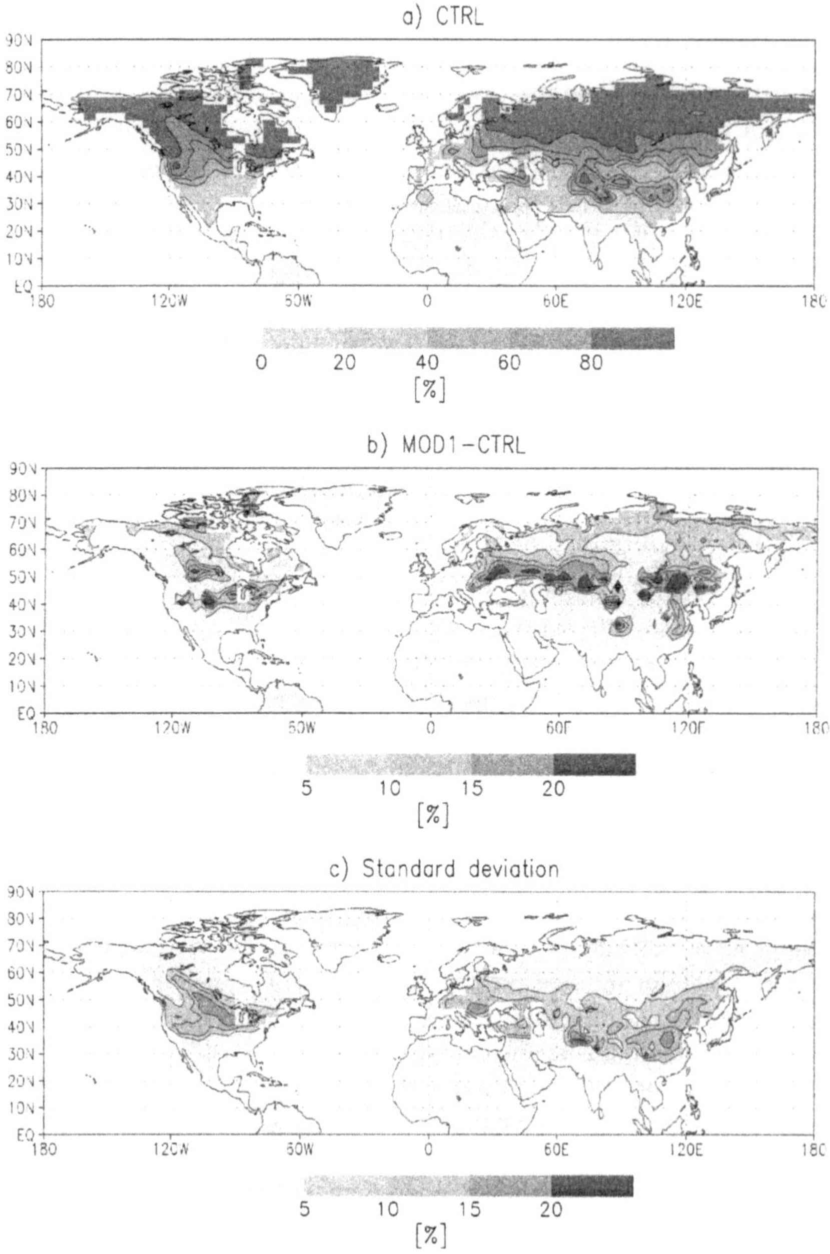


*Figure 6.* Simulated SCF for the winter season (DJF). a) mean winter SCF for the control simulation; b) difference between MOD1 and CTRL; c) standard deviation as calculated from the 10 available winter means simulated in MOD1

*Table 2.* Comparison of 10-year-means between the modified simulation (MOD1) and the control climate. Figures refer to averages over all land points with $S_n$> 0.1 cm in February (according to the USAF snow depth climatology), excluding ice covered grid boxes. Last column: percentage differences. Sign convention: downward fluxes are positive

| Parameter | Unit | MOD1 | CTRL | Diff. | Diff. (%) |
|---|---|---|---|---|---|
| Snow cover | % | 35.9 | 33.6 | 2.3 | 6.9 |
| Snow water eq. | cm | 3.24 | 3.23 | 0.01 | 0.3 |
| Surface albedo | | 0.32 | 0.31 | 0.01 | 3.4 |
| Net SW, surface | $Wm^{-2}$ | 95.1 | 95.8 | -0.7 | -0.7 |
| Global radiation | $Wm^{-2}$ | 127.5 | 127.2 | 0.3 | 0.2 |
| SW, up, surface | $Wm^{-2}$ | -32.4 | -31.4 | -1.0 | 3.3 |
| Net LW, surface | $Wm^{-2}$ | -51.0 | -51.2 | 0.2 | -0.4 |
| 2-m temperature | K | 272.86 | 272.95 | -0.09 | |
| Latent heat flux | $Wm^{-2}$ | -31.3 | -31.7 | 0.4 | -1.2 |
| Evapotranspiration | mm/day | 1.06 | 1.07 | -0.01 | -1.2 |
| Sensible heat flux | $Wm^{-2}$ | -1.4 | -1.6 | 0.2 | 12.5 |
| Rel. soil moisture | % | 73.7 | 73.4 | 0.3 | 0.4 |
| Precipitation | mm/day | 1.67 | 1.69 | -0.02 | -1.1 |
| 10 m windspeed | $ms^{-1}$ | 3.59 | 3.62 | -0.01 | -0.3 |

The impact of increasing surface albedo on land evapotranspiration E is negative (Table 2), which is in line with other sensitivity studies with 3-dimensional GCMs (Garratt, 1993). In the current experiment, E decreases by 0.01 mm/day for an increase in surface albedo by 0.1, which compares well with the value determined by Mylne and Rowntree (1991). However, a number of studies quoted in Garratt (1993) suggest a distinctly higher impact of surface albedo on evapotranspiration. The sensitivity of total precipitation to surface albedo is approximately 1.7 times larger than the response of the surface albedo on E, in good agreement with the sensitivity studies reviewed in Garratt (1993). The response of evapotranspiration to changes in surface albedo is consistent with results from off-line experiments using atmospheric forcing from the Cabauw site in the Netherlands (Roesch et al., 1997).

To summarize, the change in directly affected variables such as snow cover fraction and surface albedo is significant, whereas other surface variables do not considerably change in MOD1 compared to the control simulation.

## 5.2 Mountainous areas

In experiment MOD2, the new SCF parameterization (Eq. 7) for regions with mountains was compared with the old one given in Eq. 2.

The new parameterization considerably improves the calculation of the surface albedo for mountains with snow cover, which predominate in regions as the Himalayan or Rockies. For the Himalayan example in Fig. 4, a marked difference between the satellite derived (SRB) and simulated (CTRL)

monthly mean surface albedos is evident. While the ECHAM4/T42 experiment of the current climate simulates a pronounced annual cycle with an amplitude of about 0.2, the SRB data varies only slightly.

In Fig. 4, the curve labelled USAF1 was computed from the USAF snow depth climatology and the following algorithms from the ECHAM4: (i) the parameterization of the snow albedo, (ii) the calculation of the background albedos and surface temperatures, (iii) the transformation equation to compute the SCF and (iv) the forest fraction. The USAF1 albedo is approximately 0.05 lower than the control simulation in winter. This is due to the fact that SWE in the control simulation exceeds the observation.

The curve labeled USAF2 in Fig. 4 is calculated as USAF1 except that the SCF is calculated using the new parameterization (Eq. 7). The difference is due to the subgrid orography not accounted for by the old parameterization.

During summer, where the USAF climatology provides snow free conditions, the SRB albedo is significantly higher than in USAF1 and USAF2. This may be due to the problem that the measurements sites used for compiling the snow depth climatology do not represent the very high mountain regions with permanent snow cover.

The influence of the new SCF parameterization for regions with mountains is summarized by the sensitivities in Table 3. To exclude regions that are not of interest, the averaging was again limited to land grid boxes with snow cover in February. The decreased SCF leads to a lower albedo and stronger heating of the surface due to increased shortwave and, thus, net radiation. The higher surface temperature and higher wind speeds lead to enhanced turbulent heat fluxes and, therefore, to higher precipitation. This may be related to an intensification of the hydrological cycle.

*Table 3.* As Table 2, but for simulation MOD2

| Parameter | Unit | MOD2 | CTRL | Diff. | Diff. (%) |
|---|---|---|---|---|---|
| Snow cover | % | 28.6 | 33.6 | -5.0 | -14.9 |
| Snow water eq. | cm | 3.14 | 3.23 | -0.09 | -2.9 |
| Surface albedo | | 0.29 | 0.31 | -0.02 | -8.0 |
| Net SW, surface | $Wm^{-2}$ | 97.5 | 95.8 | 1.7 | 1.8 |
| Global radiation | $Wm^{-2}$ | 126.3 | 127.2 | -0.9 | -0.7 |
| SW, up, surface | $Wm^{-2}$ | -28.7 | -31.4 | 2.7 | -8.5 |
| Net LW, surface | $Wm^{-2}$ | -51.5 | -51.2 | -0.3 | 0.6 |
| 2-m temperature | K | 273.23 | 272.95 | 0.32 | |
| Latent heat flux | $Wm^{-2}$ | -32.4 | -31.7 | 0.6 | 2.0 |
| Evapotranspiration | mm/day | 1.09 | 1.07 | -0.02 | -1.9 |
| Sensible heat flux | $Wm^{-2}$ | -2.4 | -1.6 | -0.76 | 46.7 |
| Rel. soil moisture | % | 73.3 | 73.4 | -0.09 | -0.1 |
| Precipitation | mm/day | 1.70 | 1.69 | 0.01 | 0.8 |
| 10 m windspeed | $ms^{-1}$ | 3.65 | 3.62 | 0.03 | 0.8 |

It is of some interest to compare surface climate sensitivities of MOD2 and MOD1. The responses are of opposite sign but overall changes in MOD2 are two to three times larger than in MOD1. This indicates that the combined effect of the modification introduced in MOD1 and MOD2 is to reduce the snow cover fraction and thus the surface albedo in the average.

## 5.3 Forested areas

The model run with modified surface albedos of snow-covered forests as described in Sections 4.3, 4.4 and 4.5, MOD3, significantly changes the surface climate. Annual means of the Northern Hemisphere (with measurable snow cover in February) are given in Table 4. Considerable differences are found for SWE, SCF, surface albedo and surface temperature. The decreasing albedo leads to higher temperatures and, thus, to less snowfall and earlier snow melt in spring when the higher temperatures reduce stability and, therefore, produce larger turbulent heat fluxes. This increases the vertically integrated water vapour (and cloud water) by about 2%, which leads to a slight increase in precipitation. Nevertheless, the snowfall rate decreases due to higher temperatures, mainly during the transient seasons. Again, the above discussion is based on 1-dimensional considerations. Local processes are, however, also affected by large-scale advective processes such as moisture convergence.

*Table 4.* As Table 2, but for simulation MOD3

| Parameter | Unit | MOD3 | CTRL | Diff. | Diff. (%) |
|---|---|---|---|---|---|
| Snow cover | % | 32.4 | 33.6 | -1.2 | -3.5 |
| Snow water eq. | cm | 3.10 | 3.23 | -0.13 | -4.1 |
| Surface albedo | | 0.29 | 0.31 | -0.02 | -7.5 |
| Net SW, surface | $Wm^{-2}$ | 98.2 | 95.8 | 2.4 | 2.6 |
| Net LW, surface | $Wm^{-2}$ | 126.6 | 127.2 | -0.6 | 1.1 |
| 2-m temperature | K | 273.55 | 272.95 | 0.6 | |
| Evapotranspiration | mm/day | 1.09 | 1.07 | 0.02 | 2.3 |
| Sensible + latent heat flux | $Wm^{-2}$ | -35.2 | -33.7 | -1.5 | 4.3 |
| Rel. soil moisture | % | 72.6 | 73.4 | -0.8 | -1.2 |
| Snow fall | mm/day | 0.636 | 0.650 | -0.014 | -2.3 |
| Precipitation | mm/day | 1.71 | 1.69 | 0.02 | 1.0 |
| Cloud cover | % | 62.4 | 62.7 | -0.3 | -0.5 |

Higher deviations between the surface climate in MOD3 and CTRL are found on smaller scales as, e.g., the boreal forests. The differencc between MOD3 and CTRL can be compared with the effect of deforestation on the (surface) climate. Their responses show equal tendencies on the surface climate: Thomas and Rowntree (1992) showed that the removal of the boreal forests increases the land surface albedo and snow depth but decreases air temperature, surface net radiation, sensible heat flux, latent heat flux and precipitation during the months of March, April and May. The impact caused by changes in the surface albedo is likely to be larger when including the oceanic feedback instead of prescribing the sea surface temperature (Bonan et al., 1995): colder winter climates increase the extent of sea ice, thereby reinforcing the cooling caused by higher ocean albedos.

The largest deviations between MOD3 and CTRL occur in spring: Fig. 7 shows the long-term differences for the surface albedo and 2-m temperatures in March. The pronounced decrease in the surface albedo leads to substantially higher shortwave net radiation and, consequently, to enhanced surface temperatures. Fig. 7c indicates that the differences are statistically significant on the 95% level, using the statistical t-test. The substantial warming leads to an accelerated snow melt in late spring which reduces the overestimated snow cover in late spring in ECHAM4. Based on daily model output, it is found that the retreat of the simulated snow line is approximately 5 days later between mid-April and mid-June than in CTRL (not shown). The largest albedo differences are found over the boreal forests in both the higher latitudes of Eurasia and North America. This feature is mainly attributed to Eq. 9 which assumes the albedo of snow covered forests to be 0.2. This value has been confirmed as realistic by several authors (Verseghy, 1991; Harding and Pomeroy, 1996; Pomeroy and Dion, 1996, and others). A crude estimate of the maximum surface albedo of snow covered evergreen forests, as calculated with CLASS, leads to a value of approximately 0.25, which is in agreement with observational studies (Pomeroy and Dion, 1996 and Betts and Ball, 1997). This estimate is based on the following assumptions: (i) sky view factor SVF = 5 - 6% (Eq. 8 for needleleaf trees with LAI = 6), and (ii) snow on the ground with $\alpha_S = 0.8$. The albedo of 0.25 obtained with CLASS is distinctly lower than the albedos suggested in ECHAM4, being 0.4 for boreal forests in winter and surface temperatures below -10°C.

The impact of changes in the forest albedo on the absorbed shortwave radiation and surface temperature from December through February is small due to low sun and, hence, low global radiation.

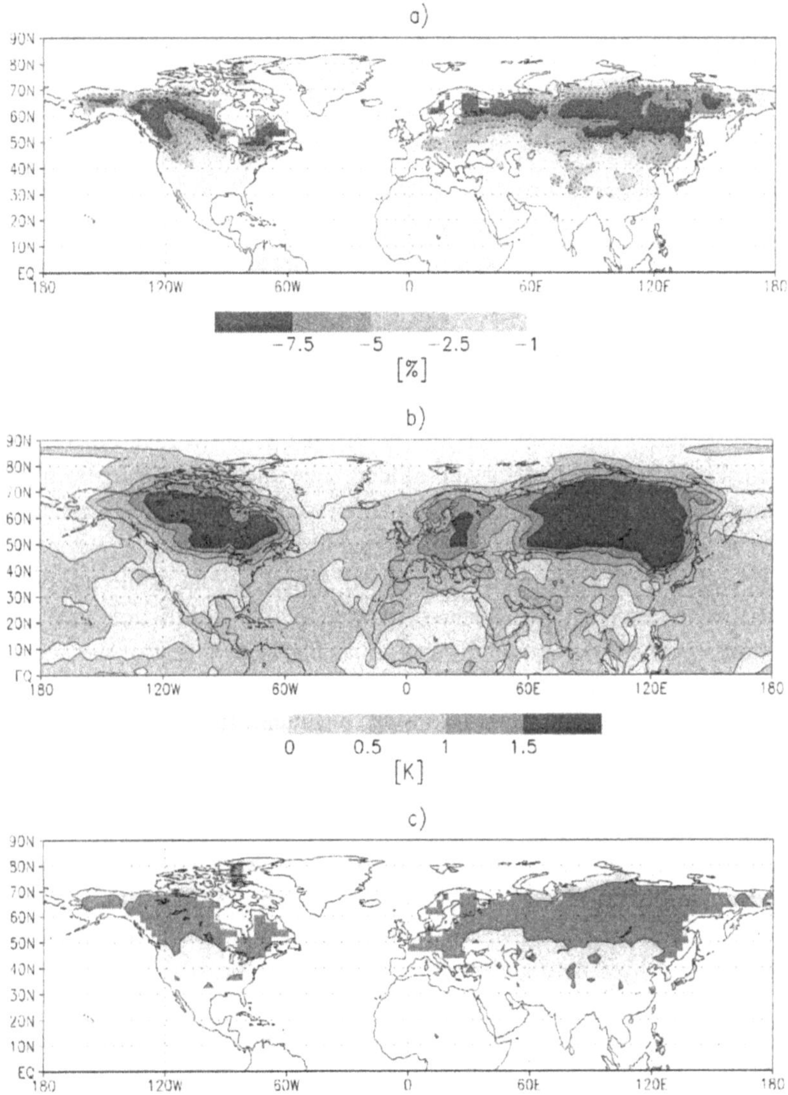


*Figure 7.* Mean differences between MOD3 and CTRL for March (10-year means). a) Surface albedo, b) 2-m-temperature, c) t-statistic test for surface albedo - dark shaded areas differ on the 95% level. Areas shaded in bright grey depict the area with measurable snow in February

# 6. SUMMARY AND OUTLOOK

This paper describes a new parameterization of SCF in ECHAM4. However, the more general results of the investigation of surface processes are also of interest for other GCMs.

Comparisons with observations revealed that it is beneficial to parameterize the SCF for three surface types:

(i) flat, non-forested lands
The simulated SCF is in best agreement with satellite-based measurements when using a *tanh*-function to relate SWE to SCF.

(ii) mountainous, non-forested land
When the subgrid orography is included in the calculation according to Douville et al. (1995a), the simulation of surface albedo is improved.

(iii) forests
The use of the CLASS (Canadian Land Surface Scheme for GCMs) submodel for snow albedo combined with a simple interception model, considerably improves the simulation of the snow-covered forest albedo. For example, the surface albedo over the boreal forests in Siberia and Canada decreases by up to 0.1 in winter, which is in better agreement with observations. The subsequent rise in surface temperature over extended parts of Eurasia and North America, due to the increased radiative heating of the surface, is statistically significant and yields a more rapid spring snowmelt and an accelerated retreat of the snow line. This reduces the overestimated snow amount as simulated in CTRL during late spring, a well-known problem in many current GCMs (Foster et al., 1996).

Two remote-sensing data sets (surface albedo from SRB, NOAA/NESDIS for SCF) were instrumental in the development of the new parameterizations in the ECHAM4 surface scheme. In general, remote sensing data products are vital when surface schemes in GCMs are enhanced, as shown by the following examples.

For the validation of simulated SWE, the USAF snow depth climatology is deemed to be the most reliable of the limited data sets available (Foster et al., 1996). However, since models simulate SWE rather than snow depth, a transformation from snow depth to SWE is required, which can be a significant source of error. In addition to visible satellite imagery, passive microwave data from the Scanning Multichannel Microwave Radiometer (SMMR) can also be used to estimate snow cover extent and snow depth (Chang et al., 1987). However, SMMR significantly underestimates snow mass during winter, especially in North America (Foster et al., 1996, Yang et al., 1999). Remote-sensed SWE data therefore need increased accuracy in order to be used for model validation.

However, remote sensing is the most suitable technique for deriving large consistent data sets of surface albedos and snow cover (Hall et al., 1995). High quality surface data at high resolution can be expected from the EOS moderate resolution imaging spectroradiometer (MODIS) launched in December 1999. Surface albedo, e.g., will be available daily at 250 m resolution (Strahler et al., 1999).

The leaf area index (LAI) in ECHAM4 is constant for each grid square. This deficiency could be easily eliminated by using monthly LAIs retrieved from the International Satellite Land Surface Climatology Project (ISLSCP, Sellers et al., 1996b). Furthermore, better boundary conditions (e.g., fraction of grid square covered with forest and vegetation) could be provided from ISLSCP Initiative I global datasets (Sellers et al., 1996b).

In summary, development and validation of ECHAM4 can benefit from the rapidly growing amount of accurate global remote-sensed surface data.

# 7. ACKNOWLEDGEMENTS

The research reported herein was sponsored by the NF project CLIMATE-2000 Grant 20-50533. The author is indebted to USAF/ ETAC, NOAA and the U.S. Geological Survey's EROS data center who made the necessary datasets available and appreciates the valuable comments and critism offered by numerous individuals from the Institute for Climate Research.

# 8. REFERENCES

Baker, D., R. Skaggs, and D. Ruschy, 1991: Snow depth required to mask the underlying surface. *J. Appl. Meteor.* **30**, 387-392.

Barker, H. and J. Davies, 1989: Surface albedo estimates from Nimbus-7 ERB data and a two-stream approximation of the radiative transfer equation. *J. Climate* **2**, 409-418.

Barker, H., Z. Li, and J.-P. Blanchet, 1994: Radiative characteristics of the Canadian Climate Centre second-generation general circulation model. *J. Climate* **7**, 1070-1091.

Barnett, T., L. Dümenil, U. Schlese, E. Roeckner, and M. Latif, 1989: The effect of Eurasian snow cover on regional and global climate variations. *J. Atmos. Sci.* **46**, 661-685.

Betts, A. and J. Ball, 1987: Albedo over the boreal forest. J. *Geophys. Res.* **102**, 28901-28909.

Bliss, N. and L. Olson, 1996: Development of a 30-arc-second digital elevation model of South America. In *Pecora thirteen, human interactions with the environment-perspectives from space*. Sioux Falls, South Dakota.

Bonan, B., F.C. Ill, and S. Thompson, 1995: Boreal forest and tundra ecosystems as components of the climate system. *Clim. Change* **29**, 145-167.

Chang, A., J. Foster, and D. Hall, 1987: Nimbus-7 SMMR derived global snow cover parameters. *Annals of Glaciology* **9**, 39-44.

Claussen, M., U. Lohmann, E. Roeckner, and U. Schulzweida, 1994: *A global data set of land-surface parameters* Max Planck Institute for Meteorology, Report No. 135, MPI for Meteorology, Hamburg.

Darnell, W., W. Staylor, S. Gupta, N. Ritchey, and A. Wilber, 1992: Seasonal variation of surface radiation budget derived from International Satellite Cloud Climatology Project C1 Data. *J. Geophys. Res.* **92**, 15741-15760.

Dickinson, R., 1983: Land surface processes and climate-surface albedos and energy balance. In B. Saltzman (Ed.), *Theory of climate*, Volume 25 of *Advances in Geophysics*, pp. 305-353. Academic press.

Dickinson, R., A. Henderson-Sellers, and P. Kennedy, 1993: Biosphere-Atmosphere Transfer Scheme (BATS) Version 1e as coupled to the NCAR Community Climate Model. In *NCAR/TN-387+STR*. National Center for Atmospheric Research, Boulder, Colorado.

Douville, H., J.-F. Royer, and J.-F. Mahfouf, 1995a: A new parameterization for the Météo-France climate model, Part 1: Validation in stand-alone experiments. *Climate Dynamics* **12**, 21-35.

Douville, H., J.-F. Royer, and J.-F. Mahfouf, 1995b: A new parameterization for the Météo-France climate model, Part 2: Validation in a 3-D GCM experiment. *Climate Dynamics* **12**, 37-52.

Edelmann, W., D. Majewski, E. Heise, P. Prohl, G. Doms, B. Ritter, M. Gertz, and T. Hanisch, 1995, *Dokumentation des EM/DM-Systems*. DWD.

Foster, D. and R. Davy, 1988: Global snow depth climatology. In *USAF publication USAFETAC/TN-88/006*, pp. 48. Scott Air Force base, Illinois.

Foster, J., G. Liston, R. Koster, R. Essery, H. Behr, L. Dümenil, D. Verseghy, S. Thompson, D. Pollard, and J. Cohen, 1996: Snow cover and snow mass intercomparison of general circulation models and remotely sensed datasets. *J. Climate* **9**, 409-426.

Garratt, J. 1993: Sensitivity of climate simulations to land-surface and atmospheric boundary-layer treatments - a review. *J. Climate* **6**, 419-449.

Gates, W.L., 1992: AMIP: The atmospheric model intercomparison project. *Bull. Amer. Meteor. Soc.* **73**, 1962-1970.

Gutzler, D. and R. Rosen, 1992: Interannual variability of wintertime snow cover across the Northern Hemisphere. *J. Climate* **5**, 1441-1447.

Hall, F., J. Townshend, and E. Engman, 1995: Status of remote sensing algorithms for estimation of land surface state parameters. *Remote Sens. Environ.* **51**, 138-156.

Harding, R. and J. Pomeroy, 1996: The energy balance of the winter boreal landscape. *J. Climate* **9**, 2778-2787.

Hedstrom, N. and J. Pomeroy, 1998: Measurements and modelling of snow interception in the boreal forest. *J. Hydrol. Processes* **12**, 1611-1625.

Iwasaki, T., 1991: Year-to-year variation of snow cover area in the Northern Hemisphere. *J. Meteor. Soc. Japan* **69**, 209-217.

Joseph, J., J. Iaquinta, and B. Pinty, 1996: The use of two-stream approximations for the parameterization of solar radiative energy fluxes through vegetation. *J. Climate* **9**, 2326-2336.

Kukla, G. and D. Robinson, 1981: *Accuracy of snow and ice monitoring*. Snow watch 1980, Glaciological Data, Report GD-5.

Kung, E., R. Bryson, and D. Lenschow, 1964: Study of a continental surface albedo on the basis of flight measurements and structure of the earth's surface cover over North America. *Monthly Weather Review* **92**, 543-564.

Marquardt, 1963: An algorithm for least-squares estimation of nonlinear parameters. *J. Soc. Ind. Appl. Math.* **11**, 431-441.

Marshall, S., J. Roads, and G. Glatzmaier, 1994: Snow hydrology in a general circulation model. *J. Climate* **7**, 1251-1269.

Masuda, K., Y. Morinaga, A. Numaguti, and A. Abe-ouchi, 1993: *The annual cycle of snow cover extent over the Northern Hemisphere as revealed by NOAA/NESDIS satellite data*, Volume 28 of *Geographical Reports of Tokyo Metropolitan University*.

Miller, D., 1962: Snow in trees - where did it go? In *Proc. of the 30th annual meeting of the Western Snow Conference*, Volume 30, pp. 21-29.

Mylne, M. and P. Rowntree, 1991: Deforestation of Amazonia - modelling the effects of albedo change. Technical Report Clim. Res. Tech. Note CRTN 7, Hadley Centre, UK Met. Office, Bracknell.

Nakai, Y., T. Sakamoto, T. Terajima, H. Kitahara, and T. Saito, 1994: Snow interception by forest canopies:weighing a conifer tree, meteorological observation and analysis by the Penman-Monteith formula. IAHS Publ. No. 223, pp. 227-236.

Otterman, J., 1984: Albedo of a forest modeled as a plane with dense protrusions. *J. Clim. Appl. Meteor.* **23**, 297-307.

Pomeroy, J. and K. Dion, 1996: Winter radiation extinction and reflection in a boreal pine canopy: measurements and modelling. J. *Hydrol. Processes* **10**, 1591-1608.

Randall, D., R. Cess, J. Blanchet, S. Chalita, R. Colman, D. Dazlich, A.D. Genio, E. Keup, A. Lacis, H.L. Treut, X.-Z. Liang, B. McAvaney, J. Mahfouf, V. Meleshko, J.-J. Morcrette, P. Norris, G. Potter, L. Rikus, E. Roeckner, J. Royer, U. Schlese, D. Sheinin, A. Sokolov, K. Taylor, R. Wetherald, I. Yagai, and M.-H. Zhang, 1994: Analysis of snow feedbacks in 14 general circulation models. *J. Geophys. Res.* **99**, 20757-20771.

Robinson, D., K. Dewey, and R. Heim, 1993: Global snow cover monitoring: an update. *Bull. Amer. Meteor. Soc.* **74**, 1689-1696.

Robock, A., K. Vinnikov, C. Schlosser, N. Speranskaya, and Y. Yue, 1995: Use of midlatitude soil moisture and meteorological observations to validate soil moisture simulations with biosphere and bucket models. *J. Climate* **8**, 15-35.

Roeckner, E., K. Arpe, L. Bengtsson, M. Christoph, M. Claussen, L. Dümenil, M. Esch, M. Giorgetta, U. Schlese, and U. Schulzweida, 1996: *The atmospheric general circulation model ECHAM-4: Model description and simulation of present day climate*. Max Planck Institute for Meteorology Report No. 218. MPI for Meteorology, Hamburg.

Roeckner, E., K. Arpe, L. Bengtsson, L.D.S. Brinkop, M. Christoph, M. Esch, E. Kirk, F. Lunkeit, M. Ponater, B. Rockel, R. Sausen, U. Schlese, S. Schubert, and M. Windelband, 1992: *Simulation of the present day climate with the ECHAM model: impact of model physics and resolution*. Max Planck Institute for Meteorology Report No. 93. MPI for Meteorology, Hamburg.

Roesch, A., 2000: *Assessment of the land surface scheme in climate models with focus on surface albedo and snow cover*. Zürcher Klima-Schriften No. 78. ETH Zürich.

Roesch, A., H. Gilgen, M. Wild, and A. Ohmura, 1999: Assessment of GCM simulated snow albedo using direct observations. *Climate Dynamics* **15**, 405-418.

Roesch, A., J.-P. Schulz, and M. Wild, 1997: *Comparison and sensitivity studies of the land-surface schemes in the ECHAM General Circulation Model and the Europa-Modell*. Max Planck Institute for Meteorology Report No. 244. MPI for Meteorolgy, Hamburg.

Sellers, P., 1985: Canopy reflectance, photosynthesis and transpiration. *Int. J. Remote Sensing* **6**, 1335-1372.

Sellers, P., D. Randall, G. Collatz, J. Berry, C. Field, D. Dazlich, C. Zhang, G. Collelo, and L. Bounoua, 1996a: A revised land surface parameterization (SiB2) for atmospheric GCMs, Part I: Model formulation. *J. Climate* **9**, 676-705.

Sellers, P., B. Meeson, J. Closs, J. Collatz, F. Corprew, D. Dazlich, F. Hall, Y. Kerr, R. Koster, S. Los, K. Mitchell, J. McManus, D. Myers, K.-J. Sun, and P. Try, 1996b: The ISLSCP Initiative global datasets: Surface boundary conditions and atmospheric forcings for land-atmosphere studies. *Bull. Amer. Meteor. Soc.* **77**, 1987-2005.

Strahler, A., J.-P. Muller, W. Lucht, C. Schaaf, T. Tsang, F. Gao, X. Li, P. Lewis, and M. Barnley, 1999: MODIS BRDF/ Albedo Product: Algorithm theoretical basis document, pp. 53.

Thomas, G. and P. Rowntree, 1992: The boreal forests and climate. *Quart. J. R. Met. Soc.* **118**, 469-497.

Verseghy, D., 1991: CLASS - A Canadian Land Surface Scheme for GCMs. I Soil model. *Int. J. Climatology* **13**, 11-133.

Walland, D. and I. Simmonds, 1996: Sub-grid-scale topography and the simulation of Northern Hemisphere snow cover. *Int. J. Climatology* **16**, 961-982.

Walsh, J. and B. Ross, 1988: Sensitivity of 30-day dynamical forecasts to continental snow cover. J. *Climate* **1**, 739-754.

Whitlock, C., T. Charlock, W. Staylor, R. Pinker, I. Laszlo, A. Ohmura, H. Gilgen, T. Konzelmann, R. DiPasquale, C. Moats, S. LeCroy, and N. Ritchey, 1995: First global WCRP surface radiation data set. *Bull. Amer. Meteor. Soc.* **76**, 905-922.

Yamazaki, T., K. Fukabori, and J. Kondo, 1996: Albedo of forest with crown snow. *J. Jpn. Soc. Snow and Ice* **58**, 11-18.

Yamazaki, T., J. Kondo, and T. Watanabe, 1992: A heat-balance model with a canopy of one or two layers and its application to field measurements. *J. Appl. Meteor.* **31**, 86-103.

Yang, Z.-L., R. Dickinson, A. Hahmann, G.-Y. Niu, M. Shaikh, X. Gao, R. Bales, S. Sorooshian, and J. Jin, 1999: Simulation of snow mass and extent in general circulation models. *Hydrol. Process.* **13**, 2097-2113.

Yang, Z.-L., R. Dickinson, A. Robock, and K. Vinnikov, 1997: Validation of the snow sub-model of the bioshere-atmosphere transfer scheme with Russian snow cover and meteorological observational data. *J. Climate* **10**, 353-373.

# Boreal Forest Fire Regimes and Climate Change

B.J. STOCKS[1], B.M. WOTTON[1], M.D. FLANNIGAN[2], M.A. FOSBERG[3], D.R. CAHOON[4], and J.G. GOLDAMMER[5]
[1]*Canadian Forest Service, Sault Ste. Marie, Ontario, Canada*
[2]*Canadian Forest Service, Edmonton, Alberta, Canada*
[3]*IGBP-BAHC Core Project Office, Potsdam, Germany*
[4]*NASA Langley Research Center, Hampton, Virginia, USA*
[5]*Max Planck Institute for Chemistry/University of Freiburg, Freiburg, Germany*

Abstract: Stretching in two broad transcontinental bands across Eurasia and North America, the global boreal zone covers approximately 12 million square kilometres, two-thirds in Russia and Scandinavia and the remainder in Canada and Alaska. Situated generally between 45 and 70 degrees north latitude, with northern and southern boundaries determined by the July 13°C and July 18°C isotherms respectively, the boreal zone contains extensive tracts of coniferous forest which provide a vital natural and economic resource for northern circumpolar countries. The export value of forest products from global boreal forests is ca. 47% of the world total (Kusela 1990, 1992).

The boreal forest is composed of hardy species of pine (Pinus), spruce (Picea), larch (Larix), and fir (Abies), mixed, usually after disturbance, with deciduous hardwoods such as birch (Betula), poplar (Populus), willow (Salix), and alder (Alnus), and interspersed with extensive lakes and organic terrain. This closed-crown forest, with its moist and deeply shaded forest floor where mosses predominate, is bounded immediately to the north by a lichen-floored open forest or woodland which in turn becomes progressively more open and tundra-dominated with increasing latitude. To the south the boreal forest zone is succeeded by temperate forests or grasslands.

Forest fire is the dominant disturbance regime in boreal forests, and is the primary process which organizes the physical and biological attributes of the boreal biome over most of its range, shaping landscape diversity and influencing energy flows and biogeochemical cycles, particularly the global carbon cycle since the last Ice Age. The physiognomy of the boreal forest is therefore largely dependent, at any given time, on the frequency, size and severity of forest fires. The overwhelming impact of wildfires on ecosystem development and forest composition in the boreal forest is readily apparent and understandable. Large contiguous expanses of even-aged stands of spruce and pine dominate the landscape in an irregular patchwork mosaic, the result of periodic



*M. Beniston and M.M. Verstraete (eds.),*
*Remote Sensing and Climate Modeling: Synergies and Limitations,* 233-246.

severe wildfire years and a testimony to the adaptation of boreal forest species to natural fire over millennia. The result is a classic example of a fire dependent ecosystem, capable, during periods of extreme fire weather, of sustaining the very large, high intensity wildfires which are responsible for its existence. This chapter presents data on recent and current trends in circumpolar boreal fire activity, with particular emphasis on Canada, Russia and Alaska, and describes the physical characteristics of boreal fires in terms of fuel consumption, spread rates, and energy release rates. The potential impact of a changing climate on boreal fire occurrence and severity, with resultant impacts on atmospheric chemistry and the global carbon budget, is discussed in detail, with reference to current research activities in this area.

# 1. FOREST FIRE ACTIVITY IN THE BOREAL ZONE

Human settlement and exploitation of the resource-rich boreal zone has been accomplished in conjunction with the development of highly efficient forest fire management systems designed to detect and suppress unwanted fires quickly and efficiently. Over the past century people throughout northern forest ecosystems have, at times somewhat uneasily, coexisted with this important natural force, as fire management agencies attempted to balance public safety concerns and the industrial and recreational use of these forests, with costs, and the need for natural forest cycling through forest fires. Canadian, Russian, and American fire managers have always designated parts of the boreal zone, usually in northern regions, as "lower priority" zones that receive little or no fire protection, since fires occurring there generally have little or no significant detrimental impact on public safety and forest values. This policy has become more widely accepted with the realisation that total fire exclusion is neither possible nor ecologically desirable, which initiated a gradual move toward the widespread adoption of fire management strategies that priorize protection of high-value resources while permitting natural fire in more remote areas. This is particularly true in the boreal forest regions of Canada, Russia, and Alaska where lower population densities and forest use allow more flexible fire management strategies.

A detailed examination of forest fire statistics from northern circumpolar countries shows that, while humans have had an influence on the extent and impact of boreal fires, fire still dominates as a disturbance regime in the boreal biome, with an estimated 5-10 million hectares burning annually in this region (Stocks 1991). Canada and Alaska, despite progressive fire management programs, still regularly experience significant, resource-stretching fire problems. In contrast, Scandinavian countries do not seem to have major large fire problems, probably due to the easy access resulting from intensive forest management over virtually all of the forested area of these countries.

Russian fire statistics are available over the past four decades but, until recent years, these statistics are considered very unreliable. The following boreal zone fire statistics are summarised from Stocks 1991, and updated using published UN-ECE/FAO statistics (e.g. UN-ECE/FAO 1997) and summaries from International Forest Fire News (United Nations FAO/ECE, Geneva).

## 1.1 Alaska

Forest fire statistics are available for the past half-century, and generally indicate that the area burned in this northernmost US state has decreased steadily while fire incidence has increased. During the 1940s Alaska recorded an annual average of 114 fires, which burned over an annual average area of 502,000 hectares. By comparison, the 1990-96 period saw annual averages of 670 fires and 383,000 hectares. Increased accessibility has influenced both fire incidence and area burned. Road and rail access meant both an increase in forest use, which resulted in increased fire occurrence, but also a corresponding enhanced detection capability and a shortened response time. Faster initial attack, particularly using smoke jumpers, coupled with aerial detection, are the major contributors to the reduction in area burned. Lightning fires, generally occurring in areas where response intervals are longer, account for a large percentage of the area burned in Alaska (38% of Alaska fires are lightning-caused and these fires account for 80% of the area burned). In addition, many fires in Alaska are fought on a priority basis, with extensive zones of limited protection, resulting in recent area burned statistics being somewhat inflated as a result of selective fire suppression.

## 1.2 Scandinavia

Somewhat limited fire statistics are available for this region, with Finland having the only continuous records from 1952. Fire statistics from Sweden are available only from 1950 to 1980 and post-1982, while Norwegian fire statistics have been recorded only since 1980. In general, considering Scandinavia as a whole, fire incidence is relatively constant, with ~4000 fires occurring annually. Area burned also varies slightly, averaging under 5000 hectares a year. Unlike Canada, the Russia, and Alaska, Scandinavian countries do not appear to experience large forest fires. As mentioned earlier, this is likely attributable to the high degree of accessibility in these smaller countries as a result of intensive forest utilisation and management. In addition, lightning fires, a higher proportion of which tend to occur in remote areas, are not a major factor in Scandinavia where they account for less than 10% of all fires.

## 1.3 Russia

Although northern Russia and Siberia have long been noted as areas where extensive forest fire activity is common (Lutz 1956), no documented statistics were ever published by the former Union of Soviet Socialist Republics (USSR) which would allow accurate quantification of the magnitude of the problem in that country. Documentary accounts from the early 1900s describe enormous forest fire losses covering thousands of square kilometres in Siberia, and giving the impression that it was difficult to find areas where evidence of recent fire was not present. In the particularly dry year of 1915, an estimated total of 14,000,000 hectares burned in Siberia (Shostakovitch 1925). Periodically some qualitative accounts of the role of fire in the Siberian forests were published, but these contained only partial statistics at best, which did not permit even rudimentary analysis. 1987 was a particularly severe fire year in Inner Mongolia and Siberia. The well-publicised Great China Fire burned in excess of one million hectares near the China-USSR border during the early spring of that year (Stocks and Jin 1988, Cahoon et al. 1991). NOAA AVHRR satellite imagery revealed that a much larger area was burning in central Siberia during the same period. Analysis of this low-resolution imagery revealed 40-50 fires, ranging in size from 20,000 to 2,000,000 hectares, had burned over a total of approximately 10,000,000 hectares in this part of the USSR (Cahoon et al. 1994). While the absolute accuracy of this estimate may be questionable due to the coarse resolution of the NOAA imagery, it still provides, in the absence of any official statistics from the USSR, a reasonable indication of the enormous forest fire problems that existed in this region in 1987, and is supported in a recent paper by Rylkov (1996). While fire activity in the USSR can be assumed to fluctuate dramatically from year to year, as is the case in other countries, the 1987 scenario is strong evidence that a major proportion of the earth's large boreal forest fires occur in Siberia.

With the dissolution of the USSR in the early 1990s, western and Russian fire managers and scientists began to work cooperatively, and this has resulted in a more accurate representation of forest fire impacts in Russia Korovin (1996) presented fire statistics for the 1956-1990 period, which indicated that, on average, 16,500 fires burned over ~650,000 hectares annually in the former USSR, with very little annual variation. Russian fire managers agree, however, that these numbers are a gross underestimation of the actual extent of boreal fire in Russia, primarily due to an incomplete reporting structure that emphasised under-reporting actual fire statistics. Recent fire statistics for Russia (1990-96) show an annual average of ~22,000 fires burning approximately 1.12 million hectares. While these statistics still appear to be largely incomplete, new NOAA AVHRR satellite down links established in Siberia are expected to solve this problem. All of Russia, including remote regions previously unmonitored, will be covered under the new satellite-based system of fire detection and monitoring.

The strongly continental climate of Russia, and in particular Siberia, produces fire weather and fire danger conditions that match ,or even exceed, those observed in Canada and Alaska (Stocks and Lynham 1996) over a much larger land base. It seems likely then that Russian fire statistics should show significant annual variation in area burned, with periodic major fire years, as is the case in both Canada and Alaska. Given the importance of Russia's boreal forests in a global context, it is critical that an accurate representation of fire activity in that major part of the boreal zone be obtained, and extensive satellite monitoring should provide that information in the near future.

## 1.4 Canada

As in the Russia, the forested area of Canada is dominated by the boreal forest zone, which extends in a broad belt from the Atlantic to the Pacific Oceans, and lies immediately to the north of the heavily populated region along the Canada/United States border. Over the past century the use of the boreal forest zone in Canada, for both industrial and recreational purposes, has increased dramatically, and this has resulted in a concurrent increase in both forest fire incidence and the fire management capability mobilised to deal with this problem. Provincial and territorial agencies in Canada have progressed to the point where state-of-the-art centralised and highly computerised fire management systems are common, yet forest fires continue to exert a tremendous influence on the forest resource in this country. Periods of extreme short-term fire weather, in combination with a recognition of both the ecological desirability of natural fire and the economic impossibility of controlling all fires, have resulted in the realisation that forest fires in Canada are a problem that cannot, and should not, be eliminated.

Detailed forest fire statistics have been archived since 1920 in Canada and, within limits, this extensive record permits a general analysis of trends in this country. It is recognised that the Canadian fire record prior to the early 1970s (when satellite coverage began) is incomplete, as various parts of the country were not consistently monitored during this period. It is expected that this problem increases as one goes back in time, likely being more of a problem in the earlier part of the century than during the mid-1900s. Keeping this uncertainty in mind, annual fire occurrence in Canada, without fluctuating greatly on a year-to-year basis, has increased rather steadily from approximately 6,000 fires annually in the 1930-1960 period, to almost 10,000 fires during the 1980s and 1990s. This is a reflection of a growing population and increased forest use, but is also due to an expanded fire detection capability. The area burned by Canadian forest fires fluctuates tremendously on an annual basis, with the 1980-96 period significant in this regard, due to major fire years in 1981, 1989, 1994 and 1995. While fire occurrence numbers were relatively constant over the 1920-1959 period, and have increased steadily since that time, area burned actually decreased over the first four decades of record only to increase over the last three decades. The most dramatic increase occurred during the 1980s, and 1990s, primarily due to periods of short-term extreme fire weather in western and central Canada. During the 1981-96 period an average of 9,246 fires annually burned over an average of 2,519,105 hectares in Canada, with annual area burned fluctuating by an order of magnitude (0.76 million to 7.28 million hectares). Lightning accounts for 35% of Canada's fires, yet these fires result in 85% of the total area burned, due to the fact that lightning fires occur randomly and therefore present access problems usually not associated with human-caused fires, with the end result that lightning fires generally grow larger, as detection and subsequent initial attack is often delayed.

A recent evaluation of Canadian fire statistics (Stocks 1991) also identified some of the reasons why Canadian fire impact varies significantly. Sophisticated fire management programs are largely successful at controlling the vast majority of forest fires at an early stage, such that only~2% of fires grow larger than 200 hectares in size, but these fires account for ~98% of the area burned across Canada. In addition, the practice of "modified" or "selective" protection in remote regions of Canada results in many large fires in low-priority areas being allowed to perform their natural function. Recent studies comparing fire sizes relative to levels of protection indicate that, on average, fires in the largely unprotected regions of the boreal zone are much larger than fires in intensively protected regions (Stocks 1991; Ward and Tithecotte 1993). An examination of the spatial distribution of all 1980-89 Canadian fires >200 hectares (Stocks et al. 1996) showed that by far the greatest area burned occurred in the boreal region of west-central Canada, and attributed this to a combination of fire-prone ecosystems, extreme fire weather, lightning activity, and reduced levels of protection in this region.

## 2. CHARACTERISTICS OF BOREAL FOREST FIRES

Boreal forest fires may be classified, based on their physical fire behaviour characteristics, into three general categories (Van Wagner 1983): smoldering fires in deep organic layers with frontal fire intensity levels <10 kW/m, surface fires with intensities ranging between 200 and 15,000 kW/m, and crown fires with intensities from 8,000 to >100,000 kW/m (frontal fire intensity is the product of a fire's rate of spread, the amount of fuel consumed in the flaming front, and the latent heat of combustion). Crown fires can be either intermittent (trees torching individually) or active (with solid flame development in the crowns), with active crown fires being by far the most common. Crown fire development depends on a number of interacting factors: the height of the crown layer above he ground, the bulk density of crown foliage, the crown foliage moisture content, and the initial surface fire intensity. In general, surface fires must generate sufficient intensity to involve the crown layer, resulting in ready access to the ambient wind field which largely determines the rate of spread of the fire. The surface and crown phases of the fire advance as a linked unit dependent on each other. The fast-spreading active crown fires that dominate the boreal landscape are primarily the result of strong winds, and are aided by both short- and long-range spotting of firebrands ahead of the flame front.

The frequency of fires in a given area depends on both the climate and the rate at which potential fuels accumulate following each fire. The fire frequency must be in long-term equilibrium with the longevity of the primary tree species and their reproductive ages. The natural fire cycle averages 50-200 years in the boreal forest (Heinselman 1981). However, human use/protection of the boreal zone has created a much wider gap in fire return intervals than would be the case under natural conditions. Stocks et al (1996), based on 1980s data for Canada, showed mean fire return intervals ranging from <100 years in remote, modestly-protected regions of the northern boreal to >500 years in heavily protected boreal zones.

Fire-adapted forests can generally be divided into two categories (Van Wagner 1983): those species able to regenerate although all trees have been killed over a large area, and those species of which some individuals must remain alive to provide seed for the next generation. Species of the first type are either conifers that store seed in insulated serotinous cones that require heat to open, or hardwoods that regenerate through suckering from the root layer following fire. Species of the second type are conifers that release seed every year when the cones mature. Canadian and Alaskan boreal forests are dominated by species (e.g. *Pinus banksiana* (jack pine) and *Picea mariana* (black spruce)) that bear serotinous cones and require lethal fire to regenerate, and the boreal landscape in North America reflects this, consisting almost entirely of large tracts of pure, even-aged stands of fire-origin species resulting from high-intensity, active crown fires. Alternatively, Eurasian boreal forests are dominated by conifer species not generally considered serotinous. Many Eurasian species have adapted to periodic, lower-intensity surface fires (e.g. thicker basal bark), releasing seed annually and creating a much more heterogeneous, uneven-aged forest. It can be assumed then, that active crown fires are far less common in the Eurasian boreal forest, and this is borne out in the Russian fire literature (e.g. Artsybashev 1967) which shows that crown fires account for ~25% of the total area burned in Russia.

Fuel consumption and spread rates can vary considerably, both within and between boreal fires. In general, however, boreal crown fires consume 20-30 tonnes/ha of fuel (Stocks 1991, Stocks and Kauffman 1997) with roughly 2/3 of this total associated with consumption of forest floor (litter, moss, humus layer) and dead woody surface fuels. Crown fuels (needles and fine twigs) account for the remaining 1/3 of the total fuel consumed. Spread rates can vary between ~5 m/min in intermittent (torching) crown fires and >100 m/min in fully-developed crown fires (Stocks and Kauffman 1997). In a recent comparison of the dynamics of boreal and savanna fires, Stocks et al. 1997 showed that boreal fires consume, on average, an order of magnitude more fuel than savanna fires. Despite similar spread rates, this large difference in fuel consumption means boreal fires develop very high energy release rates, and produce towering convection columns that can reach the upper troposphere and lower stratosphere directly. Conversely, savanna fires usually develop less well-defined convection columns, usually only 3-4 kilometres in height. The differing convection column dynamics of boreal and savanna fires are important in terms of the long-range transport of smoke products from biomass burning. Although much larger areas burn in the savannas annually than in the boreal zone (Crutzen and Andreae 1990), smoke transport mechanisms are likely much different. Regionally-generated savanna fire emissions must be transported vertically at the Inter-tropical Convergence Zone (ITCZ) to have a more global impact, whereas boreal fire emissions are injected at much higher atmospheric heights, promoting the likelihood of wider-ranging transport and impacts.

## 3. CLIMATE CHANGE AND BOREAL FOREST FIRE ACTIVITY

Confirming a growing scientific consensus, the Intergovernmental Panel on Climate Change (IPCC) has recently concluded (IPCC 1995) that "the observed increase in global mean temperature over the last century (0.3-0.6°C) is unlikely to be entirely due to natural causes, and that a pattern of climate response to human activities is identifiable in the climatological record". There is also evidence of an emerging pattern of climate response to forcings by greenhouse gases and sulphate aerosols, as evidenced by geographical, seasonal and vertical temperature patterns. In North America and Russia this pattern of observed changes has taken the form of major winter and spring warming in west-central and northwestern Canada, Alaska, and virtually all of Siberia over the past three decades, resulting in temperature increases of 2-3°C over this period (Environment Canada 1995, Hansen et al. 1996).

Numerous General Circulation Models (GCMs) project a global mean temperature increase of 0.8-3.5°C by 2100 AD, a change much more rapid than any experienced in the past 10,000 years. Most significant temperature changes are projected at higher latitudes and over land. In addition, greatest warming is expected to occur in winter and spring, similar to the trends measured recently, although warming is projected for all seasons. While GCM projections vary, in general winter temperatures are expected to rise 6-10°C and summer temperatures 4-6°C over much of Canada and Russia with a doubling of atmospheric carbon dioxide. Global precipitation forecasts under a 2x$CO_2$ climate are more variable among GCMs, but indications are that large increases in evaporation over land due to rising air temperatures will more than offset minor increases in precipitation amounts. In addition, changes in the regional and temporal patterns and intensity of precipitation are expected, increasing the tendency for extreme droughts and floods. Recent transient GCMs, which include ocean-atmosphere coupling and aerosols, and project climate continuously through the next century (e.g. Flato et al. 1999), support these earlier predictions.

Despite their coarse spatial and temporal resolution, GCMs provide the best means currently available to project future climate and forest fire danger on a broad scale. However, Regional Climate Models (RCMs) currently under development (e.g. Caya et al. 1995) and validation (Wotton et al. 1998), with much higher resolution, will permit more accurate regional-scale climate projections. In recent years GCM outputs have been used to estimate the magnitude of future fire problems. Flannigan and Van Wagner (1991) used results from three early GCMs to compare seasonal fire weather severity under a 2x$CO_2$ climate with historical climate records, and determined that fire danger would increase by nearly 50% across Canada with climate warming. Wotton and Flannigan (1993) used the Canadian GCM to predict that fire season length across Canada would increase by 30 days in a 2x$CO_2$ climate. An increase in lightning frequency across the northern hemisphere is also expected under a doubled $CO_2$ scenario (Fosberg et al 1990, 1996; Price and Rind 1994). In two recent studies, Fosberg et al.(1996) used the Canadian GCM, and Stocks et al.(1998) used four current GCMs, along with recent weather data, to evaluate the relative occurrence of extreme fire danger across Canada and Russia, and showed a significant increase in the geographical expanse of severe fire danger conditions in both countries under a warming climate. This increase does not appear to be universal across Canada though, as Flannigan et al. (1998) report results using the Canadian GCM that indicate increased precipitation over eastern Canada could result in a decrease in fire activity in that region. In addition, a dendrochronological analysis of fire scars from northern Quebec indicates a decrease in fire activity during the warming period since the end of the Little Ice Age (ca. 1850). However, most paleoecological studies of lake sediments in North America show fire frequency and intensity have increased in past warmer and drier climates (e.g. Clark 1988, 1990)

In addition to increased fire activity and severity, climate warming of the magnitude projected can be expected to have major impacts on boreal forest ecosystem structure and function in northern circumpolar countries (see Weber and Flannigan 1997). Based on GCM projections large-scale shifting of forest vegetation northward is expected (Solomon and Leemans 1989; Rizzo and Wilken 1992; Smith and Shugart 1993), at rates much faster than previously experienced during earlier climate fluctuations. Increased forest fire activity is expected to be an early and significant result of a trend toward warmer and drier conditions (Stocks 1993), resulting in shorter fire return intervals, a shift in age-class distribution towards younger forests, and a decrease in biospheric carbon storage (Kasischke et al. 1995; Stocks et al. 1996). This would likely result in a positive feedback loop between fires in boreal ecosystems and climate change, with more carbon being released from boreal ecosystems than is being stored (Kurz et al. 1995). Reinforcing this point, a retrospective analysis of carbon fluxes in the Canadian forest sector over the past 70 years (Kurz and Apps 1999) found that Canadian forests have been a net source of atmospheric carbon since 1980, primarily due to increasing disturbance regimes (fire and insects). It has been suggested that fire would be the likely agent for future vegetation shifting in response to climate change (Stocks 1993). Weber and Flannigan (1997) conclude that "Fire regime as an ecosystem process is highly sensitive to climate change because fire behaviour responds immediately to fuel moisture..." and that "interaction between climate change and fire regime has the potential to overshadow the direct effects of global warming on species distribution, migration, substitution, and extinction".

While fossil fuel burning contributes most significantly to increasing atmospheric greenhouse gas concentrations, emissions from biomass burning of the world's vegetation (forests, savannas, and agricultural lands) has recently been recognised as an additional major source of greenhouse gas emissions (Crutzen and Andreae 1990). Recent cooperative international experiments (e.g. Andreae et al. 1994, FIRESCAN Science Team 1996) have confirmed that biomass burning produces up to 40% of gross carbon dioxide and 38% of tropospheric ozone, along with a suite of less common, but equally important greenhouse gases (Levine et al. 1995). While most biomass burning emissions originate from savanna and forest conversion burning in the tropics, there is a growing realisation that boreal and temperate forest fire emissions are likely to play a much larger role under a warming climate. Cofer et al. (1996) recently outlined a number of reasons why the importance of atmospheric emissions from boreal fires may be underestimated: the tremendous fluctuations in annual area burned in the boreal zone, the fact that boreal fires are located at climatically sensitive northern latitudes, the potential for positive feedback between climate warming and boreal fire activity, and the high energy level of boreal fires which traditionally produce smoke columns reaching into the upper troposphere.

The 1997 Kyoto Protocol to the United Nations Framework Convention on Climate Change calls for the "protection and enhancement of sinks and resevoirs of greenhouse gases", and will require all countries to monitor and understand the major factors influencing the exchange of carbon between the biosphere and the atmosphere. With a large amount (37%) of the total global terrestrial carbon stored in boreal forests, boreal countries will be required to be in the forefront of these efforts. As discussed here, fire is the major disturbance regime affecting carbon cycling in the boreal zone and, with the likelihood of significant increases in forest fire activity in this region, predicting future boreal fire regimes is an urgent international research goal. Policy development and adaptation strategies require this information as soon as possible.

## 4. REFERENCES


Andreae, M.O., Fishman, J., Garstang, M., Goldammer, J.G., Justice, C.O., Levine, J.S., Scholes, R.J., Stocks, B.J., Thompson, A.M., and van Wilgen, B.W. 1994. Biomass burning in the global environment: first results from the IGAC/BIBEX field campaign STARE/TRACE-A/SAFARI-92. p. 83-101 **in** *Global Atmospheric-Biospheric Chemistry: The First IGAC Scientific Conference* R. Prinn (ed.), Plenum, NY.

Artsybashev, E. 1967. Achievements of the USSR in the protection of forests from fire. LenNIILKH 1967 (May), pp 1-16.

Cahoon, D.R., Levine, J.S., Minnis, P., Tennille, G.M., Yip, T.W., Heck, P.W., and Stocks, B.J. 1991. The Great Chinese Fire of 1987: a view from space. p. 61-66 **in** *Global Biomass Burning: Atmospheric, Climatic, and Biospheric Implications.* J.S. Levine (ed), MIT Press, Cambridge, MA.

Cahoon, D.R., Stocks, B.J., Levine, J.S., Cofer, W.R., and Pierson, J.M. 1994. Satellite analysis of the severe 1987 forest fires in northern China and southeastern Siberia. J. Geophys. Res. 99(D9): 18627-18638.

Caya, D., Laprise, R., Giguere, M., Bergeron, G., Blanchet, J.P., Stocks, B.J., Boer, G.J., and McFarlane, N.A. 1995. Description of the Canadian Regional Climate Model. p. 477-482 **in** *Boreal Forests and Global Change.* M.J. Apps, D.T. Price, and J. Wisniewski (eds.), Kluwer Acad. Pub., Netherlands.

Clark, J.S. 1988. Effect of climate change on fire regimes in northwestern Minnesota. Nature 334: 233-235.

Clark, J.S. 1990. Fire and climate change during the past 750 years in northwestern Minnesota. Ecol. Monogr. 60: 135-159.

Cofer, W.R., Winstead, E.L., Stocks, B.J., Overbay, L.W., Goldammer, J.G., Cahoon, D.R., and Levine, J.S. 1996. Emissions from boreal forest fires: are the atmospheric impacts underestimated? p. 834-839 **in** *Biomass Burning and Global Change*. J.S. Levine (ed.), MIT Press, Cambridge, MA.

Crutzen, P.J. and Andreae, M.O. 1990. Biomass burning in the tropics: impact on atmospheric chemistry and biogeochemical cycles. Science **250**, 1669-1678.

Environment Canada. 1995. The state of Canada's climate: monitoring change and variability, SOE Report No. 95-1, Ottawa, Canada.

FIRESCAN Science Team. 1996. Fire in ecosystems of boreal Eurasia: the Bor Forest Island Fire Experiment, Fire Research Campaign Asia-North (FIRESCAN). p. 848-873 **in** *Biomass Burning and Global Change*. J.S. Levine (ed.), MIT Press, Cambridge, MA.

Flannigan, M.D., and Van Wagner, C.E. 1991. Climate change and wildfire in Canada. Can. J. For. Res. **21**, 66-72.

Flannigan, M.D., Bergeron, Y., Engelmark, O., and Wotton, B.M. 1998. Future wildfire in circumboreal forests in relation to global warming. J. Veg. Science 9: 469-476.

Flato, G.M., Boer, G.J., Lee, W.G., McFarlane, N.A., Ramsden, D., Reader, M.C., and Weaver, A.J. 1998. The Canadian Centre for Climate Modelling and Analysis Global Coupled Model and its Climate. Climate Dynamics (in press).

Fosberg, M.A., Goldammer, J.G., Rind, D., and Price, C. 1990. Global change: effects on forest ecosystems and wildfire severity. p. 483-486 **in** *Fire in the Tropical Biota: Ecosystem Processes and Global Challenges*, J.G. Goldammer (ed.), Ecological Studies 84, Springer-Verlag, Berlin, Germany.

Fosberg, M.A., Stocks, B.J., and Lynham, T.J. 1996. Risk analysis in strategic planning: fire and climate change in the boreal forest. p. 495-505 **in** *Fire in Ecosystems of Boreal Eurasia*. J.G. Goldammer and V.V. Furyaev (eds.), Kluwer Academic Publ., Netherlands.

Hansen, J., Ruedy, R., Sato, M., and Reynolds, R. 1996. Global surface air temperature in 1995: return to pre-Pinatubo level. Geophysical Research Letters 23: 1665-1668.

Heinselman, M.I. 1981. Fire intensity and frequency as factors in the distribution and structure of northern ecosystems. p. 7-57 **in** *Fire Regimes and Ecosystem Properties*. H. Mooney, J.M. Bonnicksen, N.L. Christensen, J.F. Lotan, and W.A. Reimers (eds), USDA For. Serv. Gen. Tech. Rep. WO-26, Washington, DC.

International Forest Fire News, UN-ECE/FAO, Geneva, Switzerland (J.G. Goldammer (ed.)).

IPCC 1995. Climate change 1995: impacts, adaptations and mitigation of climate change: scientific-technical analysis. R.T. Watson, M.C. Zinyowera, and R.H. Moss (eds.),Cambridge University Press, Cambridge, UK.

Kasischke, E.S., Christensen, N.L., and Stocks, B.J. 1995. Fire, global warming, and the carbon balance of boreal forests. Ecol. Appl. 5(2):437-451.

Korovin, G.N. 1996. Statistics on characteristics and spatial and temporal distribution of forest fires in the Russian Federation. p. 285-302 **in** *Fire in Ecosystems of Boreal Eurasia*. J.G. Goldammer and V.V. Furyaev (eds), Kluwer Academic Publ., Netherlands.

Kurz, W.A., Apps, M.J., Stocks, B.J., and Volney, W.J.A. 1995. Global climate change: disturbance regimes and biospheric feedbacks of temperate and boreal forests. p. 119-133 **in** *Biotic Feedbacks in the Global Climate System: Will the Warming Speed the Warming?* G. M. Woodwell and F. Mackenzie (eds), Oxford Univ. Press, Oxford, UK.

Kurz, W.A. and Apps, M.J. 1999. A 70-year retrospective analysis of carbon fluxes in the Canadian forest sector. Ecol. Appl. (in press).

Kuusela, K. 1990. The Dynamics of Boreal Coniferous Forests. The Finnish National Fund for Research and Development (SITRA), Helsinki, Finland.

Kuusela, K. 1992. Boreal forestry in Finland: a fire ecology without fire. Unasylva 43(170):22

Levine, J.S., Cofer, W.R., Cahoon, D.R., and Winstead, E.L. 1995. Biomass burning: a driver for global change. Env. Sci. And Tech. **29(3)**, 120-125.

Lutz, H.J. 1956. Ecological effects of forest fires in the interior of Alaska. USFS. Tech. Bull. 1133, 121 pp.

Price, C., and Rind, D. 1994. Possible implications of global climate change on global lightning distributions and frequencies. J. Geophys. Res. **99,** 10823.

Rizzo, B., and Wilken, E. 1992. Assessing the sensitivity of Canada's forests to climatic change. Climatic Change **21**, 37-55.

Rylkov, V.F. 1996. Forest fires in the Eastern Trans-Baikal Region and elimination of their consequences. p. 219-226 **in** *Fire in Ecosystems of Boreal Eurasia.* J.G. Goldammer and V.V. Furyaev (eds), Kluwer Academic Publ., Netherlands.

Shostakovitch, V.B. 1925. Forest conflagrations in Siberia. J. For. 23(4):365-371.

Smith, T.M. and Shugart, H.H. 1993. The transient response of carbon storage to a perturbed climate. Nature **361**, 523-526.

Solomon, A.M., and Leemans, R. 1989. Forest dieback inevitable if climate changes. Int. Inst. Appl. Syst. Anal., Luxemburg, Austria. IIASA Options Sept. 1989.

Stocks, B.J., and Jin, J-Z. 1988. The China Fire of 1987: extremes in fire weather and behavior. p. 67-79 **in** Northwest Fire Council Annual Mtg., Victoria, B.C.

Stocks, B.J.: 1991. The extent and impact of forest fires in northern circumpolar countries. p. 197-202 **in** *Global Biomass Burning: Atmospheric, Climatic, and Biospheric Implications.* J.S. Levine (ed.), MIT Press, Cambridge, MA.

Stocks, B.J. 1993. Global warming and forest fires in Canada. For. Chron. 69(3): 290-293.

Stocks, B.J., and Lynham, T.J. 1996. Fire weather climatology in Canada and Russia. p. 481-487 **in** *Fire in Ecosystems of Boreal Eurasia.* J.G. Goldammer and V.V. Furyaev (eds.), Kluwer Academic Publ., Netherlands.

Stocks, B.J., Lee, B.S., and Martell, D.L. 1996. Some potential carbon budget implications of fire management in the boreal forest. p. 89-96 **in** *Forest Ecosystems, Forest Management and the Global Carbon Cycle.* M.J. Apps and D.T. Price (eds.), NATO ASI Series, Subseries 1, Vol. 40 "Global Environmental Change", Springer-Verlag, Berlin, Germany.

Stocks, B.J., and Kauffman, J.B. 1997. Biomass consumption and behavior of wildland fires in boreal, temperate, and tropical ecosystems: parameters necessary to interpret historic fire regimes and future fire scenarios. p. 169-188 **in** *Sediment Records of Biomass Burning and Global Change.* J.S. Clark, H. Cachier, J.G. Goldammer, and B.J. Stocks (eds), NATO ASI Series, Subseries 1, "Global Environmental Change", Vol. 51, Springer-Verlag, Berlin, Germany.

Stocks, B.J., van Wilgen, B.W., and Trollope, W.S.W. 1997. Fire behavior and the dynamics of convection columns in African savannas. p. 47-55 **in** *Fire in Southern African Savannas: Ecological and Atmospheric Perspectives.* B.W. van Wilgen, M.O. Andreae, J.G. Goldammer, and J.A. Lindesay (eds), Wits University Press, Johannesburg, South Africa.

Stocks, B.J., Fosberg, M.A., Lynham, T.J., Mearns, L., Wotton, B.M., Yang, Q., Jin, J-Z., Lawrence, K., Hartley, G.R., Mason, J.A., and McKenney, D.W. 1998. Climate change and forest fire potential in Russian and Canadian boreal forests. Climatic Change 38(1): 1-13.

United Nations ECE/FAO 1997. Forest Fire Statistics 1994-1996. ECE/TIM/BULL/50/4, Volume L (4), Geneva, Switzerland.

Van Wagner, C.E. 1983. Fire behavior in northern coniferous forests. p. 65-80 **in** *The Role of Fire in Northern Circumpolar Ecosystems.* R.W. Wein and D.A. MacLean (eds), SCOPE, John Wiley and Sons, UK.

Ward, P.C. and Tithecott, A.G. 1993. The impact of fire management on the boreal landscape of Ontario. Ont. Min. Nat. Res. AFFMB Pub. No. 305.

Weber, M.G., and Flannigan, M.D. 1997. Canadian boreal forest ecosystem structure and function in a changing climate: impact on fire regimes. Environ. Rev. 5: 145-166.

Wotton, B.M., and Flannigan, M.D. 1993. Length of the fire season in a changing climate. For. Chron. **69**:187-192.

Wotton, B.M., Stocks, B.J., Flannigan, M.D., Laprise, R., and Blanchet, J-P. 1998. Estimating current anf future fire climates in the boreal forest of Canada using a Regional Climate Model. p. 1207-1221 **in** Proc. Third International Conference on Forest Fire Research and Fourteenth Conference on Fire and Forest Meteorology, November 16-20, 1998, Luso, Portugal.

# Specification of surface characteristics for use in a high resolution regional climate model: on the role of glaciers in the swiss alps

STÉPHANE GOYETTE, CLAUDE COLLET and MARTIN BENISTON
*Institute of Geography – University of Fribourg – Switzerland*

Abstract: Certain aspects of the specification of the land cover characteristics for use in high-resolution regional climate models (*RCMs*) are considered in this paper. We demonstrate the importance of specifying the appropriate surface characteristics at high horizontal resolution and discuss their impacts on the simulated surface prognostic variables, on the surface energy flux as well as on the surface winds in the alpine domain of Switzerland, using the Canadian regional climate model (*CRCM*). Fixing lower boundary conditions consists in prescribing primary ground characteristics such as land-use (vegetation and soil types and their relative spatial coverage), and the surface height with respect to mean sea level. In the current version of the *CRCM* land-surface scheme, the land-use serves to fix the surface albedo and the large-scale roughness height, the vegetation type affects the soil water holding capacity, the evapotranspiration efficiency, the snow masking depth, while the soil type determines the soil thermal conductivity and specific heat, thus determining the behaviour of the momentum and sensible heat fluxes, as well as the evapotranspiration at the surface. This in turn may have significant effects on mesoscale circulations. The sensitivity of certain simulated surface fields in the *CRCM* is assessed through an appropriate specification of glaciers in the Swiss Alps. Until recently, the reference file containing primary ground characteristics was only available at a grid spacing of 1° resolution, so its use in high resolution *RCMs* is inadequate. Modern techniques used in the exploitation of high-resolution geographical data bases combined with existing satellite imagery now enable the resolution of surface characteristics with much improved definition, hence leading to greater confidence on the spatial distribution of the simulated fields computed by the land-surface scheme in *RCMs*.



*M. Beniston and M.M. Verstraete (eds.),*
*Remote Sensing and Climate Modeling: Synergies and Limitations,* 247-271.

# 1. INTRODUCTION

Currently, even sophisticated regional climate models (*RCMs*) are limited in the spatial detail they can resolve. The significance of the simulated spatial details depends to some extent on the resolution of the specified surface fields the model needs to infer surface parameters. There are fixed values that are known quite accurately, such as the planetary data (the earth's radius and its angular velocity of rotation, the acceleration of gravity, *etc.*), but the spatially-varying data used to specify boundary conditions such as surface topography, surface roughness, surface albedo, soil moisture, heat capacity, *etc.*, need to be used with caution.

A significant amount of the spatial variability of climate in the model is generated at the surface. The procedure used for the treatment of subgrid-scale boundary layer and surface processes are collected in the model's physical parameterisation module. The boundary-layer and land-surface schemes that use these parameterisations need, in addition to the resolved variables, a number of parameter fields such as the local proportion of vegetation and bare ground that are used to derived surface boundary conditions. The boundary turbulent fluxes of sensible heat and evapotranspiration are important for the maintenance of the energy and water balance at the surface. Solar radiation is the main source of energy available for the climate system and the reflection at the surface depends on the surface albedo. Outgoing infrared terrestrial radiation is a function of the ground surface temperature which in turn is related to the amount of energy absorbed by the ground. All these fluxes thus depend on the surface characteristics and its state that need to be accurately determined since they have a profound effect on the lower atmosphere, on the atmospheric circulation, and ultimately on climate.

Land-cover features are in close interaction with climate at the surface. There is both a significant latitudinal and altitudinal dependence of land-cover types. Mountain glaciers are found in highland regions at all latitudes, and represent only a very small fraction of the global cryosphere; thus mountain glaciers tend to influence climate at the regional scale. Their surface characteristic generate particular local weather and climatic conditions (Van Den Broeke, 1997) that also affect surrounding regions. These conditions are induced by the surface temperature of a glacier which is below freezing, by the reflection of a large amount of the incoming solar energy, and by the fact that they retain most of the snowfall during much of the year (Chap. 4; Paterson 1995). In Switzerland, glaciers covering close to 1500 $km^2$, represents about 3% of the total area of the country.

As a first approach, prescribing fixed lower boundary conditions for glaciers is valid because the time scale for change in extent of these ice masses are much longer than the current integration time of the model.

This paper describe the results obtained from numerical simulations performed with the Canadian Regional Climate Model (*CRCM*) to assess the sensitivity of the simulated surface variables and turbulent fluxes when glaciers are included as part of the surface boundary conditions in the Swiss Alps. Section 2.1 describes the *CRCM*, section 2.2 section briefly reviews the concepts that form the basis for the parameterisation of the surface processes over land in *CRCM* and its relation with the surface characteristics it employs, section 2.3 discusses experimental setup and the procedure used to define the lower boundary conditions in the experiments, section 3 analyses the results, and concluding remarks to this work are included in section 4.

## 2. DESCRIPTION OF *CRCM*

The Canadian regional climate model used in this study is described in Caya and Laprise (1999). The numerical formulation is based on a semi-Lagrangian semi-implicit non-hydrostatic dynamical kernel (Laprise *et al.*, 1997), and on the physical parameterisations of the second-generation Canadian General Circulation Model (*GCMII*; McFarlane *et al.*, 1992). The model physics includes the unresolved transfer processes, cloud formation, the generation of precipitation and latent heat release, as well as surface energy balance and hydrology. Even though the physics of the *CRCM* as been "tuned" for *GCMII*, it is used here as such although a minor change to the cloud onset function has recently been made (Laprise *et al.*, 1998). *CRCM* being a limited-area model, it is necessary to prescribe its boundary conditions. The lateral and uppermost nesting consists of driving the *CRCM* with a time series of observed or model-generated atmospheric fields, namely pressure, temperature, water vapour and horizontal wind components, at the external boundaries. At the surface a set of geophysical and other land/ocean fields are prescribed onto *CRCM* surface grid (SSTs, orography height, soil and vegetation characteristics, for example). In this work, the terrain height of the Alps is of prime importance because its complexity determines lower level circulations that affect the values of wind velocity. The overall nesting procedure used here is designed to be one-way, *i.e.*, *CRCM* does not feed-back into the driving data.

### 2.1 Surface and boundary-layer parameterisations

This section briefly reviews the concepts that form the basis for the parameterisation of the boundary layer and surface processes over land in *CRCM* and its relation with the surface characteristics it employs. The surface parameters are read as input from a land-cover and soil reference dataset. The processes described here and relevant for our study are the surface winds, turbulent vertical flux at the ground, aspects of solar reflection and emitted infrared radiation at the ground, and the surface energy and moisture balance.

The procedures used for the treatment of subgrid-scale vertical transfer processes as well as the land surface processes used in *CRCM* are those used in *GCMII*, and most of the material found in this section is described in Boer *et al.* (1984) and in McFarlane *et al.* (1992). The source/sink terms of momentum, heat energy, and water vapour at the surface represent effects of turbulent processes that operate at scales which are smaller than those resolved in the *CRCM*, even with the finest possible resolution of the model. The inclusion of their effects on the resolved scales is nevertheless still necessary. They enter the prognostic equations of the primary atmospheric variables in a parameterised form.

The Reynolds stress due to the vertical exchange of horizontal momentum $(\vec{\tau}_s)$ enters the prognostic conservation equations as a consequence of vertical momentum flux divergence. This stress has a significant influence on the simulated winds. The first momentum level is often several tens of meters above the modelled surface so that lowest level windspeed may be of limited use. The *GCMII* physics includes a diagnostic computation of the anemometer-level windspeed according to similarity theory, $|\vec{V}_{anem}| = (u_*/k)\ \ln(z/z_0)$, where $|\vec{V}_{anem}|$ is the anemometer wind speed, $u_*^2 = C_M|\vec{V}_a|^2$, $u_*$ is the friction velocity, $C_M$ is a momentum transfer coefficient, $|\vec{V}_a| = (u_a^2 + v_a^2)^{1/2}$ is the lowest model-level wind speed, $z_o$ is the roughness height for momentum, $z$ is the height above the surface (the anemometer is typically at 10 m), and $k$ is the von Karman constant. The turbulent vertical fluxes and the flux of solar and infrared radiation affect the lowest model level turbulent diffusion of heat and moisture as well as the surface energy balance and hydrology. The treatment of surface processes employs a single layer for heat and a "bucket" model for the soil moisture regime. The surface "ground" prognostic variables are the temperature, $T_g$, the soil moisture, $W$, and the snow mass, $S_M$. The surface energy balance equation at the surface can be written as:

$$G = (1 - \alpha)S + F_L - H_s - LE_s - M \tag{1}$$

where $G$ is the heat flux into the ground, $S$ and $F_L$ are respectively the incoming solar and net infrared radiative fluxes at the ground, $\alpha$ is the surface albedo, $H_s$ is the sensible heat flux, $E_s$ is the evapotranspiration, $L$ is the latent heat which value depends of the state of the water at the surface, and $M$ represents the energy change associated with the melting of frozen soil moisture and snow. The ground surface temperature over land is computed using the force-restore method and consequently the heat flux into the ground is expressed as in McFarlane *et al.* (1992):

$$G = C_*[\partial T_g/\partial t + \omega(T_g - T_o)] / 2 \tag{2}$$

where $\omega$ is the diurnal frequency, and $T_o$ is taken to be a 24-h moving average of $T_g$. The soil effective heat capacity, $C_*$, is given by the product of the soil heat capacity, $C_g$, and the exponential damping depth of the diurnal temperature wave, $d_g = (2\lambda_g/\omega C_g)^{1/2}$, $\lambda_g$ being the soil thermal conductivity.

The model makes use of bulk aerodynamic formulæ for the vertical transfer of heat, moisture and momentum within the constant flux layer. The parameterised fluxes assume that the transfer process is proportional to the local gradient of wind velocity, temperature, and humidity respectively between the surface and the atmosphere multiplied by the wind speed. The general expression for the surface fluxes are the same as used in *GCMII* (see Boer *et al.*, 1984 and McFarlane *et al.*, 1992):

$$F_\psi = \rho C_{F_\psi} \left|\vec{V}_a\right| (\psi_g - \psi_a) \tag{3}$$

where $\psi$ is the transported quantity, $F_\psi = [(\tau_x, \tau_y)_s, H_s, E_s]$, $\rho$ is the air density, $\tau_x$ and $\tau_y$ are the Renolds stress components at the surface, $C_{F_\psi}$ = $(C_M, C_H c_p, C_E\beta)$, where $C_M$ is the momentum transfer coefficient, $C_H$ and $C_E$ are the transfer coefficients for sensible heat and for moisture respectively, $c_p$ is the specific heat of air, $\psi_a = [(u_a, v_a), T_a, r_a]$, where $(u_a, v_a)$ are the lowest model level air velocity components, $T_a$ and $r_a$ are boundary layer temperature and mixing ratio, $\psi_g = [(u_g, v_g), T_g, r_g]$, where $(u_g, v_g) = (0, 0)$, $r_g = r_{sat}$ is the saturation mixing ratio at the earth's surface, and $\beta$ is the evapotranspiration factor. The transfer coefficients at the surface are functions of the atmospheric stability accounted for a Richardson number dependence, and on the roughness height in the general form $(C_M, C_H) = [C_{DM}F_M(\mathrm{Ri}_B, z_o/z_L), C_{DH}F_H(\mathrm{Ri}_B, z_o/z_L)]$, where $\mathrm{Ri}_B$ is the bulk Richardson number, $z_L$ is the lowest prognostic level for momentum, and $(C_{DM}, C_{DH})$ are neutral drag coefficients. In this version of the model, the surface moisture flux coefficient is taken to be equal to the ones for heat and momentum ($C_E = C_H = C_M$).

Concerning the surface moisture in the model, soil wetness is expressed by a non-dimensional variable ($w = W/W_C$), where $W$ is the total soil moisture per unit area including the liquid and solid water phases, and $W_C$ is the water-holding capacity of the soil. The budget equation follows the standard one, *i.e.*, the time evolution of the soil moisture is equal to the liquid precipitation rate ($P_L$), plus the melting of snow rate ($M_S$), minus the evaporation rate ($E_L + E_F$; evaporation and sublimation), minus the runoff ($R$). The snow mass budget equation is simply equal to the solid precipitation ($P_S$) minus snow evaporation ($E_S$) minus the melting rate of snow. The evaporation rates must add up to the total evapotranspiration rate, *i.e.*, $E_s = E_L + E_F + E_S$. The definition of $E_S$ involves an evapotranspiration factor $\beta = \beta(\beta_g, \delta_S)$. This is an efficiency factor that depends on ground wetness, which is a function of the snow-free evapotranspiration factor $\beta_g$=min $[W/sW_C, 1]$, where $s$ is a slope factor, and $\delta_S$ the fractional snow cover.

Generally, the surface albedo, the transfer coefficients, the roughness height, the evapotranspiration factor, the snow masking depth, the soil heat capacity, the soil thermal conductivity and the soil water-holding capacity are functions of the land-cover types and soil characteristics.

The land surface scheme in *GCMII* uses land-cover characteristics coming from the Wilson and Henderson-Sellers (1985) reference surface data at a resolution of 1° lat x 1° lon. *GCMII* currently makes use of a set of 22 land-cover categories (apart from open water and sea-ice surfaces) in conjunction with a subset of 9 types of soil data. Table 1 lists the land-cover types and their associated parameter values. Soil types are classified according to their possible combinations of colour (dark, medium, light), and texture (fine, intermediate, coarse). At each model grid square, the most frequently occurring primary (*LC*1) and secondary (*LC*2) land-cover and the most frequently occurring soil type as well as their relative proportion are assigned, $\delta_{LC}$ and $\delta_{GND}$ respectively, where $\delta_{LC} + \delta_{GND} = 1$. Once the primary and secondary land-cover are assigned to a grid square, a lookup procedure is used to obtain parameter values needed for the land-surface scheme, namely the soil depth, the evapotranspiration factor, and the snow masking depth. These values are then combined linearly with weights of 2/3 and 1/3 respectively to produce the effective soil depth, $D_s$, the effective evapotranspiration factor, $s$, and the effective snow masking depth, $S_{mask}$, at each model grid square. The soil type is used to determine the bare ground colour and texture at each grid square. Then, the resulting soil depth is used in conjunction with the soil texture (determining its porosity) to define the effective water-holding capacity of the soil column, $W_C$. The heat capacity, $C_g$, and thermal conductivity, $\lambda_g$, of the soil are dependent on the soil moisture and on the soil mineral content while the contribution of the air is not taken into account. The proportion of the soil mineral content is derived from the soil texture. The effective heat capacity of the soil, $C_*$, taking into account the contribution due to the snow at the surface, is computed as the weighted average of the snow-free and the snow-covered surface.

The model currently makes use of $C_{DM}$. The relationship between drag coefficient, measurement height, and surface roughness under statically neutral conditions in the surface layer is given by $C_{DM} = k^2 \left[\ln z/z_0\right]^{-2}$. The neutral drag coefficient over land, used for either the momentum, heat and moisture transfer are taken from Cressman (1960). The latter is basically a form drag, very coarse in resolution (2.5° x 2.5°), which closely resembles the shape of the large-scale orography.

*Table 1.* Land-cover used in *CRCM*. Parameters assigned to each type are the soil depth, $D_s$, the slope factor, $s$, and the snow masking depth, $S_{mask}$

| Type | Description | $D_s$ (m) | $s$ | $S_{mask}$ (m) |
|---|---|---|---|---|
| 1 | Glacier | * | 0.25 | 0.01 |
| 2 | Inland lake | | | |
| 3 | Unused | | | |
| 4 | Evergreen needle leaf tree | 1.5 | 0.25 | 12.0 |
| 5 | Evergreen broad leaf tree | 1.5 | 0.25 | 8.0 |
| 6 | Deciduous needle leaf tree | 1.5 | 0.25 | 8.0 |
| 7 | Deciduous broad leaf tree | 2.0 | 0.25 | 8.0 |
| 8 | Tropical broad leaf tree | 2.0 | 0.25 | 8.0 |
| 9 | Drought deciduous tree | 1.0 | 0.75 | 4.0 |
| 10 | Evergreen broadleaf shrub | 1.0 | 0.50 | 2.0 |
| 11 | Deciduous shrub | 1.0 | 0.50 | 1.0 |
| 12 | Thorn shrub | 1.0 | 0.75 | 1.0 |
| 13 | Short grass and forbs | 1.0 | 0.75 | 0.1 |
| 14 | Long grass | 1.0 | 0.50 | 0.15 |
| 15 | Arable | 1.0 | 0.50 | 0.1 |
| 16 | Rice | 1.0 | 0.50 | 0.1 |
| 17 | Maize | 1.0 | 0.50 | 0.1 |
| 18 | Cotton | 1.0 | 0.50 | 0.1 |
| 19 | Sugar | 1.0 | 0.50 | 0.1 |
| 20 | Irrigated crop | 0.75 | 0.25 | 0.1 |
| 21 | Urban | 0.5 | 0.75 | 4.0 |
| 22 | Tundra | 0.75 | 0.50 | 0.15 |
| 23 | Swamp | 1.0 | 0.25 | 0.01 |
| 24 | Desert | 0.5 | 0.75 | 0.01 |

* for Glacier $W_C$ is fixed at 200 kg/m$^2$

A climatological surface albedo is also assigned to each grid square. It is specified through two spectral intervals, namely the visible band albedo, $\alpha_v^{LC}$ (0.3 - 0.68 μm), and the near infrared band albedo, $\alpha_{nir}^{LC}$ (0.68 - 4 μm), according to a weighted average of the values of the land-cover categories from the Wilson and Henderson-Sellers (1985) dataset. During the course of a simulation, the albedo may be increased due to snow that accumulates over the surface and then reduced as the snow pack ages. Primary and secondary land-cover have their own snow masking depth and the surface is considered as fully covered with snow when snow mass exceeds the effective snow masking depth. Soil types have their own albedo values in the two spectral bands, respectively $\alpha_v^{GND}$, and $\alpha_{nir}^{GND} = 2\alpha_v^{GND}$. During the course of a simulation, the values for dry conditions may also be reduced for wet soil conditions, up to 7 % when $W = W_C$. The resulting spectrally-averaged albedo used in the surface energy balance equation, $\alpha$, of a partial snow cover surface over land is taken to be the linear combination of the snow-free and the snow-covered albedo values. The weights are determined from the fractional coverage of bare ground, the land-cover and that of the snow mass that has cumulated over the grid square.

This summarizing description of the modelled land-surface process is aimed at emphasizing the close link which exists between the ground-surface prognostic variables of temperature, snow, and moisture, the surface flux of momentum, sensible heat, and moisture as a function of the land-cover and soil types, and the wind speed just above the surface. Modifying the land-cover type may have profound effect on the surface fluxes, on the modelled albedo, then on the surface energy and moisture balances, on the surface winds, and consequently on the atmospheric circulation. Thus, the prescription of land-cover and soil types may well be crucial for regional climate simulations at high resolution.

*GCMII* operates with a resolution typically coarser than 1°, so that representative parameter values are specified by averaging the 1° x 1° data. However, *CRCM* operates with resolution typically much finer than 1° so that representative parameter values are coarser than the mesh size and the problem of "tiling" often arises. In a small country such as Switzerland, only one land-cover and soil type are defined (two at most) so practically no spatial variability are found in the surface parameters. In order to overcome this problem, the reference surface data for use in *RCMs* must be at a resolution higher than 1°. A few high-resolution datasets do exist, but take into account land-cover and soil categories different from those used in the model; pre-processing of these data is necessary.

## 2.2 Experimental setup

In order to assess the role of land-cover types in the Swiss Alps, we simulate conditions prevailing on the particular winter of 1990 with the *CRCM* with and without alpine glaciers. This winter is characterised by its strong storms over the Alps. To do so, the *CRCM* is run in a cascade "self-nesting" mode which consists of downscaling first *NCEP-NCAR* reanalysis (Kalnay *et al.*, 1996) at 60 km with 20 vertical levels, and an archival period of 6 h during a three-month period from January 1 to March 15, 1990. These results are then used to nest a simulation within the same model but now at 5 km resolution and 30 vertical levels, and an archival period of 1 h during a three-day period from February 26 to March 1, 1990. The latter finally serves to nest a simulation at 1 km with 46 vertical levels over a one-day period, *i.e.*, February 27, 1990. Because the 5- and 1-km simulations are expensive to run in terms of computer resources, and require a large amount of memory on our local server, it was necessary to run them on a much reduced-size domain and for shorter periods compared to the 60-km simulation. The technique involves the use of several subdomains as shown in Fig. 1. The downscaling of *NCEP-NCAR* data is carried out on domain *A*. The intermediate nesting on domain *B* provides a better spatial scale transition, since it allows the atmospheric circulation to adapt to the complex terrain. Over domain *C*, the model is run at 1 km resolution with a very realistic topographic representation of the Alps. Here, numerous details are now accurately captured, the Rhône Valley is resolved as well as many of its tributary valleys, and also the appropriate areas where glaciers are located. At 60-km resolution the land-sea mask and the height of the orography are, as for the land-cover and soil types, taken from the 1° x 1° reference file. The SSTs and sea-ice coverage are taken from the 1°x 1° resolution dataset of GISST (Rayner *et al.*, 1996). At 5- and 1-km, the height of the orography is properly resolved on the model grid.

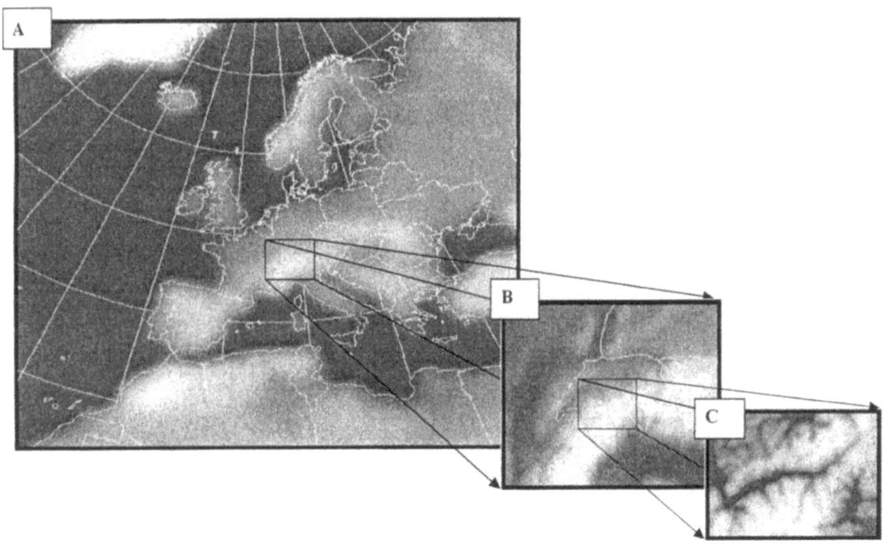


*Figure 1.* Domains of integration. The outer domain (A) is used to downscale *NCEP-NCAR* reanalysis at 60 km. Intermediate domain (B) is used for the 5-km integration, and the inner domain (C), is used for the 1-km integration

Two experiments are performed at 1 km resolution: first, a "control" experiment with uniform land-cover and soil type over southern Switzerland, and secondly, a "perturbed" experiment in which glaciers are realistically resolved on the grid.

The land-cover and soil types prescribed in the control experiment are taken from the 1° dataset of Wilson and Henderson-Sellers. The primary land-cover consists of type Nb. **13** (Table 1; short grass and forbs) uniform over the domain, the secondary land-cover is of type Nb. **4** (evergreen needleleaf tree) north of the Alps, and Nb. **24** (desert) in the Alps. The soil type is uniformly distributed throughout the domain characterised by a medium colour and an intermediate texture. The parameter values used in the control simulation and defined from the land-cover and soil types are depicted in Table 2. The soil texture defines a porosity of 48 %. The "dry" soil albedo may be reduced by 7 % for completely wet conditions. Throughout the domain the background land-cover visible and near infrared albedo are nearly constants.

*Table 2.* Parameter values used in the *CRCM* at 1-km resolutions over southern Switzerland. *LC*1-2 stands for primary and secondary land cover type (see Table 1); other symbols are defined in the text

| | **Experiments** | | | |
|---|---|---|---|---|
| | "control" | | "perturbed" | |
| | | | Glacier | |
| **Parameter** | north of Alps | Alps | $25\% < \delta_{glac} < 75\%$ | $\delta_{glac} > 75\%$ |
| *LC*1 | 13 | 13 | 1 | 1 |
| *LC*2 | 4 | 24 | 24 | 1 |
| $\alpha_v^{LC}$ | 0.11 | 0.11 | 0.66 - .43 (1-$\delta_{glac}$) | 0.66 |
| $\alpha_{nir}^{LC}$ | 0.23 | 0.23 | 0.22 + .23 (1-$\delta_{glac}$) | 0.22 |
| $\alpha_v^{GND}$ "dry" | 0.23 | 0.23 | - | - |
| $\alpha_v^{GND}$ "dry" | 0.45 | 0.45 | - | - |
| $D_s$ (cm) | 116.5 | 83.5 | - | - |
| $W_C$ (kg m$^{-2}$) | 559.2 | 400.8 | 200.0 | 200.0 |
| $s$ | 0.58 | 0.75 | 0.42 | 0.25 |
| $S_{mask}$ (cm) | 402.7 | 7.0 | 1.0 | 1.0 |
| $C_{DM}$ | 0.0044 | 0.0045 | ($C_{DMcontrol}$ + 0.0023)/2 | 0.0023 |
| $\delta_{GND}$ | 0.10 | 0.13 | 1 - $\delta_{glac}$ | 0.0 |

In the perturbed experiment, the Swiss glaciers have been superimposed over the control land-cover and soil types fields. The original glacier dataset has been produced by the Swiss Federal Office of Statistics (OFS, 1999) at 100 m resolution compiled over the period 1979-1985. This data has then been aggregated on the *CRCM* 1-km grid. This field represent the fractional coverage of glaciers ($\delta_{glac}$) over Switzerland and the local values are used to infer the primary and the secondary land-cover types, the fractional area of bare soil, the albedo within the two spectral bands, and the neutral drag coefficient (or the roughness height) at each *CRCM* grid square. It is assumed that the soil type is the same as the one defined in the control experiment. Even though the "dry" soil albedo is the same, the resulting albedo of the bare soil fraction may well be different from the control experiment due to a different soil moisture evolution. The procedure devised to build the file containing the "perturbed" fixed lower boundary conditions is the following: the aggregated glacier coverage field is used as an input file and where the fractional coverage of the glaciers exceeds 75 % locally, it is assumed that the primary and the secondary land-cover type are both prescribed as "glacier" (Table 1; Nb. **1**). The effective heat capacity of the glacier ice is fixed at 148'492.42 J $m^{-2}$ $K^{-1}$ ($d_g \approx 13$ cm) but is allowed to vary according to the snow mass. The glacier drag value, ($C_{DM} = 0.0023$) has been chosen to be smoother than the one used in the control experiment, but is representative of an heterogeneous ice surface. Its associated roughness height is in the range of values typical over glaciers (Smeets *et al.*, 1999). Where the fractional coverage of the glacier lies between 25 and 75 % locally, it is assumed that the primary land-cover type is fixed as "glacier" (Nb. **1**), the secondary land-cover type is fixed as "desert" (Nb. **24**), and this is considered as a transition zone between glaciers and bare soil. The visible band albedo decreases and the infrared band albedo increase linearly from their original "glacier" values as the fractional coverage of bare soil increases to values that match the local "dry" soil values. The fractional area of the grid covered by bare soil is fixed as the remaining part of that is not covered by glaciers. The neutral drag coefficient varies as the local average of the glacier drag value and that defined in the control experiment. The effective heat capacity of the glacier ice is fixed as the one described in the previous case. Finally, where the fractional coverage of the glaciers lies below 25 %, all the parameter values are kept the same as defined in the control experiment.

During both the control and the perturbed simulations, the archival period is at hourly intervals. Although the lower boundary conditions of the land-cover are changed, the initial values of surface temperature, $T_{go}$, soil wetness $W_o$, and snow mass, $S_{Mo}$, are kept the same to avoid systematic bias in the analysis.

## 3. RESULTS

In this section, only the results of the 1-km model resolution for this particular winter condition are discussed. Results of a multi-scale and multi-seasonal analysis of the impact of the land-cover upon the regional surface conditions in Switzerland will be addressed in a forthcoming paper.

In the following, the "perturbed" simulation refers to the simulation in which glaciers are presents in the Bernese Alps and in southern Switzerland, while the "control" refers to the simulation without glacier ice.

Since there are fluxes and variables whose values are averaged between archival periods in the model (*ex.* $G$, $H_s$, $E_s$, $V_{anem}$), time series for the February 27, 1990 period will be illustrated from 0100 UTC to 2300 UTC inclusively. The effects of changing the land-cover are analysed in terms of the differences in mean field values and surface parameters and variables over glacier ice in the Alps between the control and the perturbed experiments.

## 3.1 Energy budget at the surface

Differences in surface albedo between the perturbed and the control simulations are remarkable over glaciers ($LC1 = 1$). There is also a significant effect where the proportion of glacier declines in the transition zones in the periphery of the glaciers. This is the particularly the case in the Bernese Alps. Over that region (486 km$^2$ approx.), the spectrally averaged albedo, $\alpha$, has risen from 0.38 in the control to 0.76, and from 0.38 in the control to 0.74 in the perturbed simulation average over glaciers (164 km$^2$ approx.) and over the transition zone (322 km$^2$ approx.), respectively, during the period. This contributes to a decrease in the net solar radiation by more than 300 W m$^{-2}$ locally over the glacier in the perturbed simulation as shown in Fig. 2.

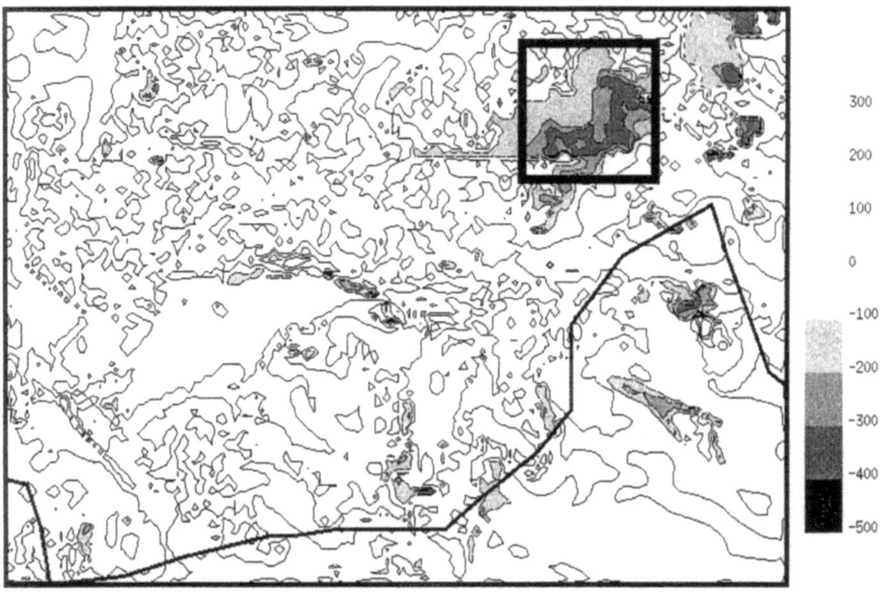


*Figure 2.* Net solar radiation differences at 1200 UTC on February 27, 1990 in southern Switzerland simulated with *CRCM* at 1 km resolution ("perturbed"-"control"). The solid line is the Swiss-Italian border, contour intervals is 100 Wm$^{-2}$, negative values are shaded, and the box indicates the location where the analyses over Bernese Alps are performed in the text

Since the upper atmospheric conditions are similar in both simulations, it may be assumed that the incident solar radiation, $S$, is similar in both simulations ($\Delta S < 15$ %). Thus, the net solar radiation differences are mostly due to the surface albedo effects. It can be seen in Fig. 2 that remarkable effects are located in the Bernese Alps ice field north of the Rhône valley. Since the main interest is in evaluating the effects of the land-cover change on the surface conditions, most of the analysis found in this text is carried out mainly over this region that includes glacier areas as well as transition zones in the periphery of the glaciers. In Fig. 2, the region under study is identified on the grid, that is glacier where $LC1 = LC2 = 1$ ($\delta_{glac} > 75$ %), and the transition zone where $LC1 = 1$, $LC2 = 24$ (25 % $< \delta_{glac} <$ 75 %).

The differences in radiation budget components between the perturbed and the control simulations in the Bernese Alps are shown in Fig. 3a. The net all-wave radiation differences at the surface ($\Delta Q^*$; $Q^* = (1-\alpha)\ S + F_L$) is dominated by the effects of the reflection of the solar radiation, averaged over both glaciers and transition zone, where the net solar radiation is reduced by more than 175 W m$^{-2}$ compared to the control simulation. The net infrared longwave differences are slightly negative (-2 and -1 W m$^{-2}$) over glaciers and transition zone, respectively, and have a negligible effect on the net radiation budget differences during February 27, 1990. The differences in energy budget components over the Bernese Alps are shown in Fig. 3b. While the differences in the heat flux entering the ground is slightly negative on the daily average, $\Delta G = -0.3$ and -0.7 W m$^{-2}$ for glaciers and transition zone respectively, the behaviour of the sensible and latent heat fluxes is somewhat different. The sensible heat flux magnitude is significantly reduced in the perturbed simulation. The direction of the fluxes is preserved (positive from the surface to the atmosphere) but over glaciers, it decreases by 250 W m$^{-2}$, whereas it decreases only by 65 W m$^{-2}$ in the transition zone compared to the control simulation on the daily average. The downward latent heat flux (negative from the atmosphere to the surface) is decreased (negatively) by more than 200 W m$^{-2}$ over glaciers on the daily average but only by 15 W m$^{-2}$ in the transition zone compared to the control simulation on the daily average. There is also a significant diurnal influence in the sensible and latent heat fluxes in the transition zone. The downward sensible heat flux is reduced more during the day than during the night, the downward latent heat flux is reduced during the day but increases during the night. The changes in both sensible and latent heat fluxes are balanced within a few W m$^{-2}$ by the decrease in the net radiative flux at the surface; the result is a slight decrease of energy available for the underlying surface. Note that the energy change associated with the freezing of soil moisture and snow, $M$, is zero since there is no liquid water available in the soil and surface temperature is maintained below freezing. Table 3 summarises the effects and shows the energy budget components (using hourly means), spatially averaged over the glaciers and transition zone in the Bernese Alps, and averaged for the day of February 27, 1990.

*Table 3.* Daily averages of radiation and energy balance components in $Wm^{-2}$, area averaged over the Berneese Alps in the control simulation and averaged over transition zone and glacier respectively in the perturbed simulation. $F_L$(atm) is the downward atmospheric infrared flux and $F_L$(sfc) (= $\sigma T_g^4$) is the upward terrestrial infrared flux where $F_L = F_L(\text{atm}) - F_L(\text{sfc})$

| Energy component | "Control" area averaged over Bernese Alps | "Perturbed" Bernese Alps | |
|---|---|---|---|
| | | $25\% < \delta_{glac} < 75\%$ | $\delta_{glac} > 75\%$ |
| $S$ | 141.3 | 163.2 | 154.9 |
| $\alpha S$ | 53.3 | 120.5 | 118.2 |
| $F_L(atm)$ | 199.7 | 196.7 | 198.0 |
| $F_L(sfc)$ | 271.6 | 272.9 | 268.9 |
| $H_s$ | 459.8 | 324.2 | 279.7 |
| $LE_s$ | -448.8 | -366.4 | -314.9 |
| $G$ | -3.1 | -3.9 | -3.2 |
| $M$ | 0.0 | 0.0 | 0.0 |

a)

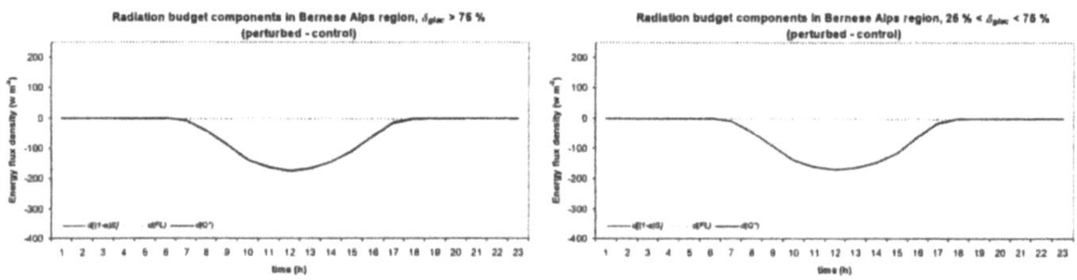


b)

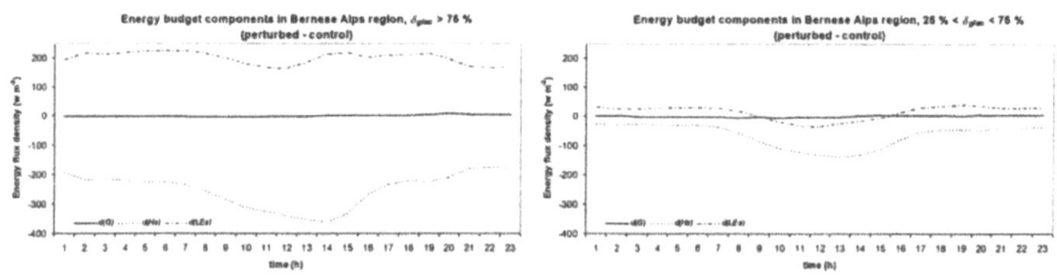


c)

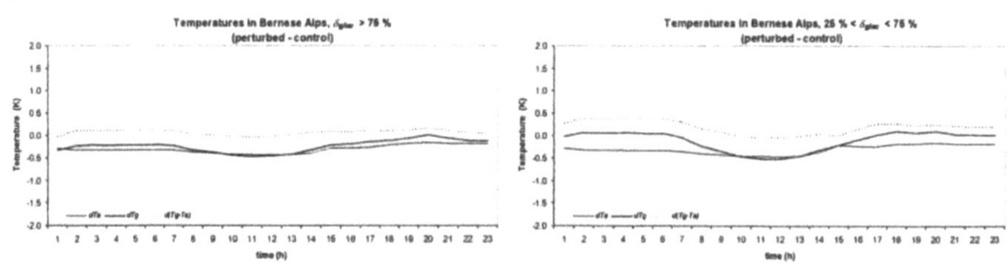

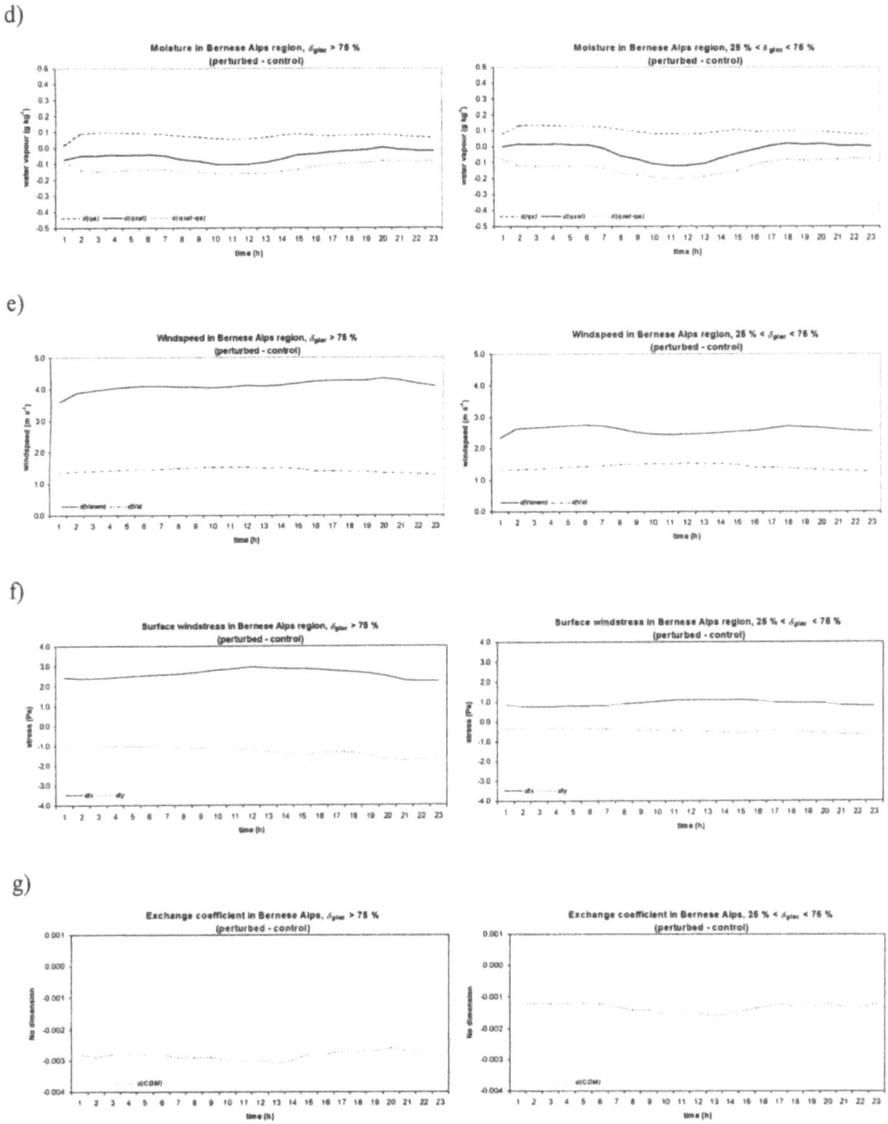


*Figure 3.* Differences in surface fluxes, variables and coefficients between the "perturbed" and the "control" simulations area averaged over the Bernese Alps glaciers and transition zone regions from 0100 to 2300 UTC on February 27, 1990

## 3.2 Ground surface and boundary-layer air temperatures

During this particular winter day, there is a slight increase in the vertical temperature differences on the daily average over the surface for most areas covered by glaciers in the Alps between the perturbed and the control simulations. In the Bernese Alps for example, while there is a slight decrease in the ground surface temperature on average ($\Delta T_g$ = -0.24°C and -0.12°C, respectively, for glaciers and transition zone), but there is a similar decrease in the boundary layer air temperature ($\Delta T_a$ = -0.31°C). The stronger temperature drop tends to be associated with locations where there is less snow accumulating over glaciers and over the transition zone as well as land cover compared to the control simulation. Elsewhere in the region surrounding the glaciers, it may happen that surface temperature increases slightly. Generally, where there is "bare" glacier or where snow mass is below 10 g kg$^{-1}$, surface temperature tends to decrease. Figure 3c shows the differences in air and ground temperatures between the control and the perturbed simulations in the Bernese Alps over the glacier and transition zone during February 27, 1990. In both simulations, the ground surface temperature stays below freezing. It is colder over the glacier on average but it is warmer during night and colder during the day in the transition zone, compared to the control simulation. The surface air temperature is colder on the average and the resulting difference between ground and surface air temperature increases slightly on average during this day.

## 3.3 Moisture regime at the surface

Figures 3d shows the soil saturation specific humidity (noting that $q_{sat} \approx r_{sat}$), the air specific humidity and their difference ($q_a \approx r_a$), between the perturbed and the control simulations in the Bernese Alps glaciers and transition zone during February 27, 1990. During that day, the air near the surface exhibits a moisture excess over glaciers of 0.07 g kg$^{-1}$ and 0.1 g kg$^{-1}$ over the transition zone on average compared to the control simulation. The saturation specific humidity is always lower over glaciers ($\Delta q_{sat}$ = -0.05 g kg$^{-1}$), but over the transition zone it is generally lower during the day and higher during the night, the daily average being slightly negative, $\Delta q_{sat}$ = -0.03 g kg$^{-1}$. Their differences are usually negative on a daily average ($\Delta(q_{sat}-q_a)$ = -0.12 g kg$^{-1}$), which imply that the vertical saturation deficit has increased in magnitude. Compared to the control integration in the Bernese Alps, the evaporation (deposition) has decreased over glaciers ($\Delta E_s$ = 6.1 mm d$^{-1}$) but only slightly over the transition zone ($\Delta E_s$ = 0.5 mm d$^{-1}$). The precipitation (solid) decreased over glaciers ($\Delta P_s$ = -3.7 mm d$^{-1}$) and over the transition zone ($\Delta P_s$ = -1.6 mm d$^{-1}$) compared to the control simulation. The difference averaged over the day in the Bernese Alps is a slight ($P_s$ - $E_s$) deficit compared to the control simulation over glaciers and transition zone. Snow accumulation on the ground has thus decreased in the Bernese Alps compared to the control simulation, about $\Delta M_s$ = -5.6 kg m$^{-2}$ (-1.4 cm) over glaciers, and $\Delta M_s$ = -1.3 kg m$^{-2}$ (-0.4 cm) in the transition zone on average during the day. Both simulations produced the same value of the soil moisture: the normalised moisture variable, $w$, shows that there is no liquid phase present throughout the day and the solid phase remains at 0.9 of their respective capacity. Since the soil is not saturated with water on average during the day, it is presumed that the runoff may be neglected.

## 3.4 Effects on the winds and windstress at the surface

Figure 3e shows the differences in anemometer-level and lowest model-level windspeeds between the perturbed and the control simulations in the Bernese Alps on February 27, 1990. Over glaciers and the transition zone, the anemometer-level windspeed increased more than the lowest model-level windspeed, $\Delta V_{anem} = 4.1$ m s$^{-1}$ over glaciers, and $\Delta V_{anem} = 2.6$ m s$^{-1}$ in the transition zone compared to $\Delta V_a = 1.4$ m s$^{-1}$ for both on average during this day. This increase of the lowest model-level windspeed indicates that the vertical gradient of the wind is also increasing in the surface layer. Figure 3f shows the differences in the surface wind stress. The average wind blowing in the southeast direction increased its speed in the perturbed simulation but the stress components are effectively decreasing by 2.6 and 1.3 Pa over glaciers, and by 0.9 and 0.4 Pa over the transition zone respectively in both $x$ and $y$ directions.

## 3.5 Bulk Richardson number and the exchange coefficient at the surface

Figure 3g shows the differences in the momentum exchange coefficient between the perturbed and the control simulations over glaciers and over the transition zone in the Bernese Alps on February 27, 1990. The bulk Richardson number is negative but close to zero in both simulations, but in the perturbed simulation, its value increases slightly during the day and decreases slightly during the night, resulting in a modest increase of its value over glaciers ($\Delta Ri_B = 0.002$), the daily average being a small decrease over the transition zone ($\Delta Ri_B = -0.0006$). The daily average exchange coefficient also decreased over glaciers from 0.006 in the control simulation on average to 0.003 in the perturbed simulation, and from 0.005 to 0.001 over the transition zone, the drop being generally more pronounced during the day, *i.e.*, from 0700 to 1900 UTC.

## 3.6 Discussion

The position on the model grid where the largest effects on most surface variables and fluxes are generated is where glaciers cover the entire grid square, $LC1 = LC2 = 1$, *i.e.*, where $\delta_{glac} > 75$ %. However, significant effects also appeared in the transition zone between glaciers and their surroundings, *i.e.*, where $LC1 = 1$, and $LC2 = 24$, where 25 % < $\delta_{glac}$ < 75 %. The Bernese Alps are characteristics of such an area in Switzerland for this particular version of the model.

One of the most important feature of glaciers is their high albedo. The reflection of a large portion of the incident solar radiation is of primary importance in their overall low energy status (Chap. 3; Oke 1987). The albedo effect is important since the reflection of solar energy is mainly due to ice and snow components, without any contribution of either vegetated surfaces or bare soil. In the model, fresh snow albedo has values of 0.90 and 0.70 in the visible and near-infrared bands respectively overlying an already highly reflective glacier. This is the reverse case for soil as previously explained where the variation of the surface albedo as a function of the fractional area of bare soil has led to different expressions according to the two wavelength bands. Where the gridpoint is partially covered with glaciers, *i.e.*, where $LC1$=1 and $LC2$=24, the importance of the albedo effect decreases with increasing bare soil fraction; this is the case in the transition zone in the periphery of the Bernese Alps and in southern Switzerland near the French and Italian borders, where the albedo decreases to their "original" values rapidly with distance away from the glacier. Over glaciers, there is less net all-wave radiation available at the surface compared to the control simulation. A smaller effect of the upward infrared energy is due to the colder surface temperature in the perturbed simulation over glaciers but this minor contribution does not modulate the net all-wave radiation budget at the surface. In summary, as shown in Table 3, the radiation budget is significantly reduced following a significant increase in the reflected solar radiation. The surface is thus losing less sensible heat, and there is also a decrease in the latent heat flux directed towards the surface, meaning in this case that the deposition rate has decreased. Note here that strong values of sensible and latent heat fluxes are mainly due to high wind velocities ($V_a > 28$ m s$^{-1}$; 11 on Beaufort scale). The result is thus a slight net energy loss ($\Delta G < 0$) at the surface, leading to a slight decrease in the surface temperature. In the Bernese Alps, part of the area is considered as snow-covered ($\delta_S = 1$), and the thermal characteristics of the surface are those of snow in both simulations on that particular day. The surface temperature decrease in the Bernese Alps, $\Delta T_g$=-0.24°C and -0.12°C respectively for glaciers and transition zone is consistent with the comparatively small loss of net energy at the surface in the perturbed experiment. Note that there is no significant tendency for the surface ground temperature to increase or decrease over glaciers during this particular day; this is due to snow accumulation in both simulations that directly influences the thermal characteristics of the surface. Even if the surface temperature is slightly decreasing over glaciers, the saturation deficit between earth's surface and the air is decreasing in the perturbed experiment. The evaporation flux directed towards the surface ($\beta = 1$) decreases as does the precipitation flux on average. Snow accumulation on the ground is similar over both glaciers and the transition zone during the day. The anemometer-level windspeed is also increasing over the Bernese Alps and particularly over the glaciers. Even if the vertical gradient of the surface wind is increasing over glaciers in the perturbed simulation, both components of the surface stresses are effectively decreasing.

It can be seen in Eq. 3 that the intensity of the surface fluxes of momentum, heat and moisture are proportional to the surface vertical gradient of winds, temperature and mixing ratio (or specific humidity), to the lowest model-level windspeed, and to their exchange coefficients. The surface vertical gradient of wind is increasing, consequently this would lead to potentially enhance the wind stress and the exchange of momentum at the surface. The surface wind stresses are actually decreasing in both the $x$ and $y$ directions, however. The exchange coefficient for heat and moisture are also increasing in magnitude. However, the results indicate a net decrease in the upward heat flux and a net decrease in the downward moisture flux at the surface. The non-linear nature of the formulation of the transfer leads to flux values that cannot be predicted solely on the basis of the vertical gradient of the quantity to be transfered at the surface. The knowledge of the behaviour of the stability function that determines the transfer coefficients is also required.

In order to quantify the changes to the surface turbulent fluxes in more detail, Eq. 3 is partitioned into two parts: a ventilation factor ($W_\psi = \rho C_{F_\psi} |\vec{V_a}|$), multiplied by a vertical difference factor ($\delta\psi = \psi_g - \psi_a$). A preliminary analysis has shown that the behaviour of the surface fluxes is well represented by the product of the daily and spatially averaged factors. Consequently, the relative change of the fluxes may be written as:

$$\frac{\overline{\Delta F_\psi}}{\overline{F_\psi}} = \frac{\overline{\Delta W_\psi}}{\overline{W_\psi}} + \frac{\overline{\Delta \delta\psi}}{\overline{\delta\psi}} + \frac{\overline{\Delta W_\psi \Delta \delta\psi}}{\overline{W_\psi \delta\psi}} + R_F \qquad (4)$$

where overbars represent daily and spatially averaged quantities, $\Delta$ represents the difference of averaged values between the perturbed and the control simulations, and $R_F$ is the residual. The relative change of the momentum, heat and moisture fluxes is thus represented by the sum of the relative change of the ventilation factor, plus the relative change of the vertical difference factor, plus the relative change of the correlation between the variation of both factors combined, and a residual. In the diagnostic equation 4, the residual includes the errors due to using fluxes as the product of averaged ventilation and vertical differences factors on the basis of hourly averaged quantities. A summary of the relative effects of the daily average components of the fluxes over glaciers and over the transition zone in the Bernese Alps is shown in Table 4.

*Table 4.* Relative changes in % of the daily average momentum, heat and moisture flux components averaged over the Bernese Alps glaciers (a) and over the transition zone (b). The overbar $\left(\overline{\ }\right)$ represents spatially and daily average and the Δ operator indicates the difference between average values in the perturbed run minus the averages in the control run over glaciers and transition zone respectively

a)

| **Over glaciers** / **Flux** | $\frac{\overline{\Delta F_\psi}}{\overline{F_\psi}}$ | $\frac{\overline{\Delta W_\psi}}{\overline{W_\psi}}$ | $\frac{\overline{\Delta\delta\psi}}{\overline{\delta\psi}}$ | $\frac{\overline{\Delta W_\psi \Delta\delta\psi}}{\overline{W_\psi \delta\psi}}$ | $R'$ |
|---|---|---|---|---|---|
| **momentum** | -45.0 | -48.9 | 4.9 | -1.8 | 0.8 |
| **Sensible heat** | -47.3 | -48.9 | 1.9 | -1.0 | 0.7 |
| **evapotranspiration** | -39.0 | -48.9 | 16.3 | -8.0 | 1.6 |

b)

| **Over transition zone** / **Flux** | $\frac{\overline{\Delta F_\psi}}{\overline{F_\psi}}$ | $\frac{\overline{\Delta W_\psi}}{\overline{W_\psi}}$ | $\frac{\overline{\Delta\delta\psi}}{\overline{\delta\psi}}$ | $\frac{\overline{\Delta W_\psi \Delta\delta\psi}}{\overline{W_\psi \delta\psi}}$ | $R'$ |
|---|---|---|---|---|---|
| **momentum** | -16.8 | -21.3 | 4.2 | -0.7 | 1.0 |
| **Sensible heat** | -16.7 | -21.3 | 6.7 | -1.3 | 0.8 |
| **evapotranspiration** | -4.0 | -21.3 | 21.6 | -4.6 | 0.3 |

Over glaciers, the reduction of the momentum flux magnitude (-45.0 %), of the upward sensible flux (-47.3 %) as well as of the downward evapotranspiration (-39.0 %) at the surface is primarily due to the reduction of the intensity of the ventilation factor, -48.9 %. As seen before, the vertical difference of the windspeed at the surface tends to increase the momentum flux magnitude and act as an antagonist to the change in the ventilation factor but its magnitude is not important enough to compensate for the negative change in the ventilation factor. The momentum correlation factor and the residual play a minor role in the surface momentum flux. The vertical difference of the temperature also tends to increase the sensible flux but its magnitude is too weak to compensate for the negative change in the ventilation factor. The correlation factor and the residual also play a minor role in the surface sensible heat flux. Finally, the relative magnitude of the vertical difference of the saturation deficit between the surface and the lower atmosphere plays a certain role in reducing the downward flux of water vapour but it is not enough to compensate for the negative change in the ventilation factor which is, this time, helped by the correlation factor. The residual does not play a significant role here either.

Over the transition zone, there is a weaker reduction of the momentum flux magnitude (-16.8 %), of the upward sensible flux (-16.7 %) as well as of the downward evapotranspiration (-4.0 %). Regarding the momentum and the sensible heat fluxes, the changes are primarily due to the reduction of the intensity of the ventilation factor, -21.3 %. The vertical difference factors play a secondary role as antagonists but their effects are not sufficient to compensate for the relative change in the ventilation factor, the correlation factor and the residual playing a negligible role. In the case of the evaporation, the effect of the ventilation factor (-21.3 %) is compensated entirely by the effect of the vertical difference factor (+21.6 %), the total effect being determined by the correlation between the vertical difference and the ventilation factors (-4.6 %). This means basically that on average during this day over the transition zone, the space and time variation of the windspeed and the variation of transfer coefficient are varying together to produce the final value of the flux.

In order to analyse in more details the reasons of the relative magnitude of the ventilation effect on the surface turbulent fluxes, the same decomposition can be apply to partition the ventilation coefficient into two parts, an exchange coefficient factor ($C_\psi = \rho C_{F_\psi}$) multiplied by a wind factor ($|\vec{V}_a|$). The magnitude of relative decrease of the ventilation coefficient over glaciers and over the transition zone are respectively -48.9 % and -21.3 %, is in part due to the decrease of the exchange coefficient factor ($\overline{\Delta C_{F_\psi}}/\overline{C_{F_\psi}}$ = -51 %, and -25 % over glaciers and over transition zone respectively). The change in the wind factor act as an antagonist to the change in the exchange coefficient factor ($\overline{\Delta|\vec{V}_a|}/\overline{|\vec{V}_a|}$ = 4 % over both glaciers and transition zone respectively), the correlation between the changes in wind and exchange coefficient, and the residual having negligible effects. Even though the changes of the bulk Richardson number diagnosed on the daily average ($\overline{\Delta RI_B}/\overline{RI_B}$ = -5 % and 2 % over glaciers and over transition zone respectively) would imply less stable conditions above glaciers and more stable conditions above the transition zone on the average compared to the control simulation, and thus potential increase and increase of the exchange coefficient over glaciers and transition zone, the use of a smaller neutral drag coefficients in the perturbed simulation, $C_{DM}$ = 0.0023 over glacier instead of 0.0044 over land, and a mean value over the transition zone ($C_{DM}$ = 0.0033; see Table 2) relative to the control simulation, dominates the effects of the change in the surface exchange coefficient and consequently on the ventilation coefficient.

Finally, the relative change of the anemometer-level windspeed, $\overline{\Delta|\vec{V}_{anem}|}/\overline{|\vec{V}_{anem}|}$ = 15 % and 10 % over glaciers and over transition zone respectively, has a positive contribution coming from the increase of the first model-level windspeed, 4 % on the average, a important positive contributions coming from the increase of the scaling of the exchange coefficient, [ $C_{C_M} = C_M^{1/2}/k\ln(z/z_o)$ ], where $\overline{\Delta C_{C_{DM}}}/\overline{C}_{C_{DM}}$ = -30 % and -13 % % over glaciers and over transition zone respectively, consequently the correlation between the variation of the scaling of the exchange coefficient and the change of the first model-level windspeed, as well as the contribution coming from the residual must contribute all together to 41 % and 19 % over glaciers and over transition zone respectively. This strong effect may be explained in terms of the mutual influence between the departure of the mean winds and the departure of the square root of the exchange coefficient; as the frictional drag decreases on the average (in terms of either $z_o$ or $C_{DM}$), the windspeed above the surface must increases on average by reducing the momentum exchange between the surface and the overlying air and *vice versa*.

## 4. CONCLUSION

This study is intended to evaluate quantitatively the effects of land-cover change on surface fluxes and prognostic variables using the *CRCM* interfaced with *GCMII* physics, the latter using a simple land-surface scheme and boundary-layer parameterisation, under the atmospheric conditions particular of a stormy winter day over western Europe. Although the surface conditions were substantially modified in the perturbed simulation through the specification of alpine glaciers in southern Switzerland, it is realised that these have also an influence in surrounding areas. In this experiment, changing the land-cover type from a typical vegetated surface to glaciers over areas in the Alps affect primarily the surface albedo which in turn modified the reflected solar radiation flux, the radiation budget at the surface, and subsequently the partitioning of the sensible heat flux and latent heat flux. Secondly, even though the vertical gradient of windspeed, temperature and moisture at the surface are modified, the negative change in the momentum, heat and moisture turbulent surface fluxes are mainly modulated by the ventilation coefficient, the latter being dominated by the change in the neutral drag coefficient. On average during this particular day over the Bernese Alps, the increase in anemometer-level windspeed is a result of the increased in the lowest model-level windspeed, and the decrease of the roughness height (or of $C_{DM}$), the decrease in the surface ground temperature, seen mainly during the day, is due to the decrease of the net available energy at the surface, and the increase in the surface saturation specific humidity.

This study thus helps to understand through a specific case-study that there are no simple relationships between the land-cover definition and the conditions above and emphasises the needs for the specification of land-use in high-resolution climate models. Although a sophisticated regional climate model is not absolutely ncccssary for this kind of study, the analysis of the simulated results emphasises the fact that conditions were also affected in the periphery of the perturbed areas. Snow accumulating on glaciers appear to be an important factor in determining the thermal behaviour of the surface. The analysis given above in the text should then bc used with caution and should not be generalised to all glaciers, for all atmospheric conditions and for all seasons.

There are observational and modelling indications that the weather and climate are modulated by land-surface characteristics. Kalthoff *et al.* (1999) have analysed temporally and spatially, on the basis of station observations in the upper Rhine valley, the behaviour of the surface energy budget and concluded that the orography, precipitation and land-use were the main influences. Harrison (1975) studied the elevation component of soil temperature variation in Britain and concluded that the change in soil type (particularly through changes in thermal conductivity) between lowlands and uplands is a major factor affecting the seasonal distribution of temperature. Soil texture also seems to play a role in the spring moisture regime since the upland soils experienced much slower rates of drying in soil surface horizons. Ecosystems, although represented crudely in climate models, influence weather and climate over periods of seconds to years through exchanges of energy, moisture and momentum between the land surface and the atmosphere [see Pielke *et al.*, (1998) for an overview]. Garratt (1993) provided a comprehensive summary of *GCM* atmospheric boundary-layer surface schemes, and the main results from sensitivity studies have shown that regional and global climate depend on albedo, surface moisture, surface roughness, and vegetation. This suggests that there is a need to account for soil and vegetation effects in such models. Segal *et al.* (1988) have shown that under favourable environmental conditions, vegetated areas adjacent to dry bare soil regions may provide substantial gradients of sensible heating which result in the onset of thermally induced mesoscale circulation. Seth and Giorgi (1996) have studied the organised mesoscale circulations induced by vegetation using a *RCM* and drawn similar conclusions for spatial scales less than 300 km. Pielke *et al.* (1997) demonstrated that the use of high-resolution land-cover in their *RCM* (RAMS) had a substantial influence on the overlying atmosphere. In addition, there is evidence that local land use practices influence regional climate in adjacent areas Stohlgren *et al.* (1998).

Therefore, there are clear indications that properly-resolved land-cover data (vegetation and soil types, their relative proportions over an area, and their radiative, thermal and aerodynamic properties) including glacier areas, should be taken into account in high resolution regional climate models.

The advance in the development of geographical data bases and satellite imagery now allows the definition of a high resolution land-cover (in terms of the types displayed in Table 1 for instance), and soil characteristics of the surface with much more detail than 1° resolution over Switzerland. Work is currently under way to investigate the effects of resolving the complete land-cover types over Switzerland at high resolution in order to assess the seasonal effects of glaciers and other types of surface on Swiss climate, where the underlying working hypothesis is that variability in the surface climate can be generated with greater details in surface conditions.

## 5. ACKNOWLEDGEMENTS

This research was supported by the Swiss National Science Foundation under Grant 2100-049525.96. The authors would like to thank the "regional climate modelling" team at the University of Québec at Montréal (UQAM), Canada, for their constructive comments and technical assistance.

## 6. REFERENCES

Boer, G. J., N. A. McFarlane, R. Laprise, J. D. Henderson, and J.-P. Blanchet, 1984: The Canadian climate centre spectral atmospheric general circulation model. *Atmos.-Ocean*, **22,** 397-429.

Caya, D., and R. Laprise, 1999: A semi-implicit semi Lagrangian regional climate model: the Canadian RCM. *Mon. Wea. Rev.*, **127,** 341-362.

Cressman, G. P., 1960: Improved terrain effects in barotropic forecasts. *Mon. Wea. Rev.*, **88,** 327-342.

Garratt, J. R., 1993: Sensitivity of climate simulations to land-surface and atmospheric boundary-layer treatments - a review. *J. Climate*, **6,** 419-449.

Harrison, S. J., 1975: The elevation component of soil temperature variation. *Weather,* **30,** 397-409.

Kalnay, E., and Coauthors, 1996: The NCEP/NCAR 40-years Reanalysis project. *Bull. Amer. Meteor. Soc.*, **77,** 437-471.

Kalthoff, N., F. Fiedler, M. Kohler, O. Kolle, H. Mayer, and A. Wenzel, 1999: Analysis of energy balance components as a function of orography and land use and comparison of results with the distribution of variables influencing local climate. *Theor. Appl. Climatol.*, **62,** 65-84.

Laprise, R, D. Caya, M. Giguere, G. Bergeron, H. Côté, J.-P. Blanchet, G. J. Boer, and N. A. McFarlane, 1998: Climate and climate change in Western Canada as simulated by the Canadian regional climate model. *Atmos.-Ocean,* **36,** 119-167.

Laprise, R, D. Caya, G. Bergeron, and M. Giguère, 1997: The formulation of the André Robert $MC^2$ (mesoscale compressible community) model. *Atmos.-Ocean,* **35,** 195-220.

McFarlane, N. A., G. J. Boer, J.-P. Blanchet, and M. Lazare, 1992: The Canadian climate centre second generation general circulation model and its equilibrium climate. *J. Climate,* **5**, 1013-1044.

OFS, Office Fédéral de la Statistique: GEOSTAT, sect. H, Utilisation du sol, Berne (Suisse), ed. 02, 1999.

Oke, T. R., 1987: *Boundary layer climate.* 2nd ed., Methuen, 435 pp.

Paterson, W. S. B., 1995: *The physics of glaciers.* 3rd ed., Pergamon, 480 pp.

Pielke Sr., R. A., T. J. Lee, J. H. Copeland, J. L. Eastman, C. L. Ziegler, and C. A. Finley, 1997: Use of USGS-provided data to improve weather and climate simulation. *Ecological Applications*, **7,** 3-21.

Pielke Sr., R. A., R. Avissar, M. Raupach, A. J. Dolman, X. Zeng, and A. S. Denning, 1998: Interactions between the atmosphere and terrestrial ecosystems: influence on weather and climate. *Global Change Biology,* **4,** 461-475.

Rayner, N. A., E. B. Horton, D.E. Parker, C. K. Folland, and R. B. Hackett, 1996: *Version 2.2 of the global sea-ice and sea surface temperature data set, 1903-1994.* Climate Research Technical Note 74, Hadley Centre Met. Office, Bracknell, 21 pp.

Segal, M., R. Avissar, M. C. McCumber, R. A. Pielke, 1988: Evaluation of vegetation effects on the generation and modification of the mesoscale circulations. *J. Atmos. Sci.,* **45,** 2268-2292.

Seth, A., and F. Giorgi, 1996: Three-dimensional model study of organized mesoscale circulations induced by vegetation. *J. Geophys. Res.*, **101,** 7371-7391.

Smeets, C. J. P. P., P. G. Duynkerke, H. F. Vugts, 1999: Observed wind profiles and turbulence fluxes over an ice surface with changing surface roughness. *Boundary-Layer Meteorol.,* **92,** 101-123.

Stohlgren, T. J., T. N. Chase, R. A. Pielke, T. G. F. Kittel, and J. S. Baron, 1998: Evidence that local land use practices influence regional climate, vegetation, and stream flow patterns in adjacent natural areas. *Global Change Biology*, **4,** 495-504.

Van Den Broeke, M., 1997: Structure and diurnal variation of the atmospheric boundary layer over a mid-latitude glacier in summer. *Boundary-Layer Meteorol.,* **83,** 183-205.

Wilson, M. F., and A. Henderson-Sellers, 1985: A global archive of land cover and soils data for use in general circulation climate models. *J. Climatol.*, **5,** 119-143.

# Using Satellite Data Assimilation to Infer Global Soil Moisture Status and Vegetation Feedback to Climate

WOLFGANG KNORR[1] and JAN-PETER SCHULZ[2]
[1]*Max Planck Institute for Biogeochemistry, Jena, Germany*
[2]*Danish Meteorological Institute, Copenhagen, Denmark*

Abstract: The importance of land surface and vegetation characteristics for climate has long been hypothesisized and is reflected by increasingly sophisticated land surface schemes used in climate models. However, accurate parameterisation of land surface processes is still hampered by the complexity of the processes, and by data availability at the global scale required for general circulation models. It is, therefore, desirable to utilise additional data sources for land surface models, of which satellite data appear to be the most promising in terms of availability and spatial and temporal coverage. Here, monthly satellite-derived fields of the fraction of Absorbed Photosynthetically Active Radiation (fAPAR) are assimilated into a land surface and vegetation model, the Biosphere Energy-Transfer Hydrology scheme (BETHY). Assimilation offers the advantage that uncertainties of both the satellite-derived fAPAR and model parameters can be accounted for. Since fAPAR can also be predicted by the model, this information is not discarded as in other approaches where satellite data are used as forcing. During assimilation, a number of model parameters are adjusted until a cost function reaches its minimum. This cost function is defined by the squared deviation between monthly model-simulated and satellite-derived fAPAR as well as between initial and adjusted model parameters, both normalised by their assumed error variances. One of the adjusted parameters, the maximum plant-available soil moisture, is used in a subsequent sensitivity study with the ECHAM-4 climate model. The results show that changes in this parameter as a result of satellite data assimilation can lead to significant changes in simulated soil moisture and 2m air temperature over large parts of the tropics, where soil water storage is usually underestimated in climate and vegetation models. A comparison of BETHY simulations with soil water measurements from Amazonia supports this finding, and also shows that using fAPAR as forcing would have lead to inconsistencies between the carbon balance, predicting a strong decrease in fAPAR at negative carbon gains, and the value of fAPAR prescribed from the satellite data. The study aims at demonstrating the potential of assimilating satellite data into land surface models, as well as the significance of vegetation for the land surface



*M. Beniston and M.M. Verstraete (eds.),*
*Remote Sensing and Climate Modeling: Synergies and Limitations,* 273-306.

climate. It is further intended to indicate a methodology for the assimilation of satellite data into general circulation models that include an interactive, i.e. climate-responsive, vegetation component.

# 1. INTRODUCTION

It is now generally accepted that land surface processes can have a significant impact on the global climate, [e.g. *Budyko*, 1956; *Budyko*, 1974; *Budyko and Sokolov*, 1978; *Garratt*, 1993; *Geiger*, 1965; *Geiger et al.*, 1995; *Mintz*, 1984]. Various different land surface schemes for use in climate and numerical weather prediction models have been developed, ranging from simple bucket schemes [*Budyko*, 1956; *Manabe*, 1969] up to fairly complex models including an explicit representation of vegetation [e.g. *Dickinson et al.*, 1986; *Sellers et al.*, 1986]. These models have been tested in intercomparison studies, revealing a wide range in model performances [*Henderson-Sellers et al.*, 1996; *Schulz et al.*, 1998]. Numerical experiments have been used to identify the most important factors of land surface-atmosphere interactions, which have turned out to include evapotranspiration [*Shukla and Mintz*, 1982], water-holding capacity [*Milly and Dunne*, 1994], albedo [*Charney et al.*, 1977] and surface roughness [*Sud et al.*, 1988]. All of those land surface parameters can change dramatically during deforestation and other forms of land conversion, with important consequences for regional and global climate [*Chase et al.*, 2000; *Lean and Warrilow*, 1989; *Polcher and Laval*, 1994; *Shukla et al.*, 1990]. To a large extent, these surface fluxes and parameters are controlled by the vegetation cover, which in turn is largely determined by the climate [*Box*, 1981; *Holdridge*, 1947]. Those links then create biogeographic feedbacks between terrestrial vegetation and the atmosphere, which have been found to alter climate sensitivities to imposed changes in surface cover [*Gutman*, 1984], with indications of creating new climate-vegetation equilibria [*Charney et al.*, 1975; *Claussen et al.*, 1997; *Ganopolski et al.*, 1998]. Adequate representation of the plant-controlled surface hydrology has also been found to be important for the quality of numerical weather prediction [*Viterbo and Beljaars*, 1995] and for the simulated climate in a number of nested limited-area models [*Christensen et al.*, 1997].

Much of the close link between vegetation activity and climate-relevant land surface processes can be attributed to the large water requirement of land plants. Because vegetation growth is limited by water availability over the largest part of the terrestrial surface, plants tend to maximise water use by controlling transpiration through their stomata, small pores on the leaf surface [*Jones*, 1983], as well as through efficient rooting strategies [*Kleidon and Heimann*, 1998]. As a consequence, more than half of land evapotranspiration happens as plant-controlled stomatal transpiration [*Budyko*, 1974; *Budyko and Sokolov*, 1978]. This explains why a number of recent land surface schemes have attempted to include physiological processes, such as carbon uptake, in conjunction with energy and water exchanges [e.g. *Sellers et al.*, 1996].

On a global scale, the state and evolution of land surfaces can best be characterised using data from space-borne remote sensing platforms that have now been available for approximately 20 years. The presence of vegetation can be identified rather easily with optical sensors because its reflectance shows a strong contrast between the visible and the near-infrared optical domains [*Verstraete*, 1994]. Provided the link between vegetation activity, the terrestrial water balance and climate is captured in an appropriate modelling framework, remote sensing of vegetation can in principle be used to infer information about other climate relevant parameters, such as soil moisture status. One such modelling framework that uses the technique of data assimilation will be introduced in the present study.

When combined with models, remote sensing data can be used for either initialisation, forcing, evaluation, or assimilation. Which of these is used in a particular case will depend on the requirements and the intended use of the model as well as the type of information extracted from the remote sensing data. Initialisation and forcing always require appropriate algorithms to derive parameters from the remote sensing data that can be used in the models concerned. In the case of evaluation, an analogous strategy is to compare parameters derived from remote sensing with state variables of the model that represent the same physical or biophysical quantitity. Since the signal recorded by the satellite sensor is usually the result of complex radiative interaction between the physical system under investigation and other factors, such as the state of the atmosphere or the surface background, this strategy requires the solution of an inverse problem [see e.g. *Verstraete and Pinty*, 1996; *Verstraete and Pinty*, 2000]. An alternative strategy that can be used for evaluation is therefore to add a description of the measurement process to the model: in this case, it is possible to predict the signal that should have been observed by the sensor under the given viewing conditions in a forward manner, and compare it to the actual measurement. – Both techniques can also be used for assimilation of satellite data, which in effect is an extension of the method of model evaluation through formalisation of subsequent model adjustments (see below for further discussion of this method).

The question whether forcing and initialisation, or evaluation and assimilation of remote sensing data is the preferable method will usually depend on the use of the derived parameter within the model. This parameter may be considered either a cause for the process modelled, or an effect of it, and consequently appear as either input for initialisation or forcing, or as output used for evaluation or assimilation. For example, since atmospheric circulation models have atmospheric parameters as an output, assimilation is the preferred use of atmospheric remote sensing data, such as in numerical weather prediction [e.g. *Anderson et al.*, 1994]. Remotely sensed thermal and microwave radiation, both related to the energy and moisture balance of land surfaces, have also been used for assimilation in the context of regional hydrological studies [*Blyth*, 1993; *McNider et al.*, 1994; *Ottle and Vijal-Madjar*, 1994], or numerical weather prediction [*van den Hurk et al.*, 1997]. The opposite situation is found for a number of vegetation models designed to study the global carbon cycle: The design of those models has been focussed on the process of carbon uptake and cycling at a given vegetation distribution, and has consequently used a remotely sensed measure of vegetation productivity, fAPAR, as a model input [*Potter et al.*, 1993; *Prince*, 1991; *Ruimy et al.*, 1996]; fAPAR (fraction of vegetation-absorbed photosynthetically active radiation) is a quantitity that can be derived rather well from optical remote sensing. If, however, vegetation cover and productivity are predicted as a result of climate and soils input data, as in this study, fAPAR may appear as a model output and can thus be used for either evaluation or assimilation.

The main advantage of using fAPAR, or any other parameter, for forcing is that there is no need to model the relevant processes by which this quantity is modified. If, on the other hand, this parameter is used for assimilation, it must at least be possible to represent those processes in some simplified form, such that the adjusted model, following assimilation, is able to reproduce them correctly. For example, it is not necessary to include the extent to which human activity modifies vegetation activity by irrigation and agriculture, if fAPAR is used to force a vegetation model. If fAPAR is assimilated, however, the impact of agriculture and other factors not modelled explicitly may be accounted for by modifying the length of the growing season, the water balance, and the vegetated fraction of a model grid cell (see below).

The advantage of data assimilation is generally that it allows to account for varying degrees of uncertainty in the satellite data; a special case of which are data gaps. This is important for optical satellite remote sensing, because in some very cloudy regions (e.g. tropical rainforests) and at high latitude during the winter (when the sun is far away from zenith), no reliable observations of surface conditions may exist for one to several months per year. This situation has led to the development of data where additional information has been used to fill in gaps occurring in the satellite data [*Los et al.*, 1994]. Here, assimilation offers the important advantage that remotely sensed information may be combined with other, independent information in systematic and accountable way. In the case of fAPAR, this independent information consists of climate and soils data that, which, in an appropriate model, can be used to derive vegetation cover on a global scale [e.g. *Box*, 1981; *Haxeltine and Prentice*, 1996]. This is particularly important when process descriptions carry large uncertainties, as it is the case for vegetation modelling [*Knorr*, 2000]: assimilation allows to assign uncertainties to both remote sensing derived and model internal parameters, and, through combining both sources of information, to reduce the overall modelling uncertainty. In contrast to this method, forcing with satellite data does not allow for adjusting the model itself, with the consequence that other state variables of the model may be inconsistent with the one that is prescribed from the satellite data. An example where forcing the model with high fAPAR at very low soil moisture content results in strongly negative carbon gain of the vegetation is given in Section 5.

Despite the advantages described above, no global-scale assimilation of fAPAR or other vegetation related remotely sensed parameters has yet been attempted, neither for vegetation nor for climate models containing a suitable vegetation description within their surface schemes. The reason may be that vegetation models are still a relatively recent development, and that only few surface schemes are able to predict vegetation cover and activity [e.g. IBIS, *Foley et al.*, 1996]. However, some evaluation of vegetation models against satellite data has been performed, either against derived quantities like fAPAR [*Knorr*, 1997], or by combining the vegetation model with a radiative transfer model of atmosphere, vegetation and soil [*Knorr et al.*, 1995].

In this study, global monthly fields of fAPAR derived from satellite data are compared to monthly fields of fAPAR as predicted by the Biosphere Energy-Transfer Hydrology (BETHY) scheme [*Knorr*, 1997; *Knorr*, 2000], which is a model of surface processes, photosynthesis and land-biosphere carbon balance forced "off-line" by monthly climate and fixed soil input data. Both fields are intended to represent one average year under mean climatic conditions. Differences between them are then used to adjust the model in a suitable assimilation procedure. Although the model uses fields of actual vegetation types as input, the amount of leaf area for each month is derived entirely from the water and carbon balance.

Assimilation and forcing of the BETHY scheme with satellite derived fAPAR is first tested against observations of soil moisture from a rainforest site in Amazonia. This site represents conditions of natural vegetation under severe water limitation for a significant part of the year. One of the model parameters that has a particularly large impact on the water balance and is adjusted during assimilation is the maximum plant-available soil moisture content. Global fields of this model parameter, adjusted following the assimilation procedure, are then used in a sensitivity analysis with the ECHAM-4 climate model [*Roeckner et al.*, 1996], with the purpose of assessing the impact of including information contained in the satellite data on the simulated climate. The results should then indicate what benefit could be gained from including an active biosphere in a climate model, i.e. a combined surface and vegetation model that predicts vegetation amount based on the simulated climate itself.

A further aim of this study is to outline and demonstrate a methodology that could lead to some valuable improvements in the fields of climate simulation and vegetation modelling. Its main advantage is that it ensures consistency between derived parameters, model dynamics and observations within their specific degrees of accuracy, and an adequate representation of the coupling between atmospheric circulation and the terrestrial carbon and water cycles.

## 2. SHORT DESCRIPTION OF THE BETHY SCHEME

The Biosphere Energy-Transfer and Hydrology scheme (BETHY) is essentially an off-line simulator of soil-vegetation-atmosphere interaction, where "off-line" stands for description of surface processes without the need for coupling with an atmospheric general circulation model (GCM). It consists of four parts: energy and water balance, photosynthesis, carbon balance, and phenology [*Knorr*, 1997; *Knorr*, 2000].

### 2.1 Energy and water balance

The energy and water balance part, which is the most comprehensive, considers three pools of water, which are soil water, intercepted water residing on the vegetation, and snow. Precipitation can fall as rain directly on the ground, or hit the vegetation, filling up the interception reservoir up to 0.1 mm times leaf area index (LAI, see subsection on phenology) after which throughfall occurs, or it falls as snow and contributes to the snow reservoir. (There is no intercepted snow, so that snow is always assumed to lie beneath the vegetation.) Rainfall on the ground, throughfall and snow melt all fill up the ground water pool up to a value of $w_{max}$; if this value is exceeded, runoff occurs. This is essentially a bucket scheme, in which the total soil water content can vary between a minimum at the wilting point and a maximum where runoff and vertical drainage start to occur. The modelled soil water content, $w \leq w_{max}$, is equal to the actual soil water content minus the amount stored in the soil at wilting point; this quantity is sometimes called "plant-available" soil moisture content. To ensure conistency with the

ECHAM-4 climate model, the maximum plant-available soil moisture content, $w_{max}$, is taken everywhere as 65% of the saturated soil water content, a value derived from the same external input data as for the ECHAM-4 model (see below). The water balance is updated at a daily time step.

Apart from runoff, water leaves the surface as evapotranspiration, where it becomes part of the energy balance. Potential soil evaporation and snow evaporation are assumed equal to the equilibrium rate [*Jarvis and McNaughton*, 1986], until either the snow pool is depleted, or soil moisture becomes limiting. The limitation of soil evaporation is assumed to depend on the time since the last rainfall event, following *Ritchie* [1972]. Evaporation of intercepted water and transpiration, the latter by far the largest flux for most cases, are both computed with the Penman-Monteith formula [*Monteith*, 1965]. Energy balance and radiation are all computed hourly.

The Penman-Monteith formula requires the specification of the combined conductance of all stomata, or leaf pores, of the canopy, denoted $G$. This value is set to infinity for evaporation of intercepted water, and to $G_c = G_{c,0} / (1 + b_e \Delta e)$ for transpiration, where $G_{c,0}$ is stomatal conductance at a standard leaf-internal $CO_2$ concentration, $C_{i,0}$, assumed when water is not limiting, $\Delta e$ the vapour pressure deficit above the canopy, and $b_e$ a factor controlling stomatal closure in response to soil water. $G_{c,0}$ is computed from the diffusion equation across the stomata from the photosynthesis rate without water limitation, $A_{c,0}$ (see next subsection):

$$G_{c,0} = \frac{1.6 A_{c,0}}{C_a - C_{i,0}} \frac{R T_K}{p} \tag{1}$$

where $C_a$ is the atmospheric $CO_2$ concentration, $C_{i,0}$ is assumed 65% of $C_a$ for plants with C3, and 37% for plants with C4 photosynthesis, $T_K$ is the air temperature in Kelvin, $p$ air pressure, and $R$ the universal gas constant. $b_e$ is set to a value that limits the transpiration rate at 13:00 h (assumed to represent the situation of highest atmospheric demand) to a root supply rate of $S = c_w \, w / w_{max}$, with $c_w$ set to 1 mm/hour [*Federer*, 1982].

The other conductance term needed for the calculation of transpiration and evaporation of intercepted water is the aerodynamic conductance between the canopy and the free air. Its value varies between approximately 0.200 m/s for forests and 0.025 m/s for grasslands, and is computed from *Brutsaert* [1982]:

$$G_a = \frac{k^2 u}{\left[ \ln\left( \frac{z-d}{z_0} \right) \right]^2} \tag{2}$$

with $k = 0.41$ (von Karman constant), wind speed $u$, reference height $z$, roughness length $z_0$, and zero plane displacement $d$. For vegetated surfaces, the following parameters have been fitted to data [*Kelliher et al.*, 1993], assuming a globally fixed wind speed of $u = 3$ m/s: $z_0 = 0.1\ h_c$, $z = h_c + 2$ m, and $d = 0.56\ h_c$, taking vegetation dependent values for the canopy height, $h_c$. Its values, chosen to deliver good agreement with typical values for forests and grasslands, are 30 cm for short grass, 1 m for shrubs, 2 m for long grass, 15 m for temperate, boreal and tropical deciduous trees, and 30 m for tropical evergreen trees; they are determined according to the vegetation map used as input (see below). Previous sensitivity tests with the BETHY scheme have shown that both carbon uptake and transpiration are rather insensitive against a variation of u between half and double the globally fixed value taken here.

The radiative balance considers incoming solar radiation at the surface, derived from incoming surface PAR (photosynthetically active radiation) with a conversion factor depending on cloudiness and solar angle [*Pinker and Laszlo*, 1992], outgoing longwave radiation based on a surface emissivity of 0.97, and sky radiation depending on air temperature, air vapour pressure, and cloudiness [*Brutsaert*, 1982, p. 137]. Daily averages of incoming PAR are interpolated linearly between monthly values from input data, which are then used to compute hourly PAR and broadband solar radiation, following a method by *Weiss and Norman* [1985]. Cloudiness is kept constant over a day and is inferred from the ratio of incoming to potential PAR. Net radiation is computed separately for vegetation and bare soil to computed separate evaporation rates from vegetation and soils. The broadband albedo of vegetation albedo is set to 0.15 globally, while soil albedo is prescribed from input data, taking 0.15, 0.20 and 0.35 for dry dark, medium and bright soils, respectively, and 0.07, 0.10 and 0.18 for the same soils when they are wet [*Wilson and Henderson-Sellers*, 1985]. A soil heat flux is also included, assumed to be a fixed fraction (0.036) of the overall net radiation [*Verma et al.*, 1986].

## 2.2 Photosynthesis

Two photosynthesis models are used, depending on the photosynthetic pathway of the vegetation: a model for C3 plants by *Farquhar et al.* [1980], and an adaptation of this model for C4 plants developed by *Collatz et al.* [1992]. In both cases, the net canopy leaf assimilation rate, $A_c$, and the canopy-integrated rate of leaf or "dark" respiration, $R_{d,c}$, are calculated, with dark respiration assumed to be proportional to the maximum carboxylation rate (with a factor of 0.011 or 0.042 at 25°C, for C3 or C4 plants, respectively), a limiting rate set by the amount of primary $CO_2$ fixating enzymes present in the leaf. This maximum carboxylation rate and other parameters determining vegetation productivity are prescribed by fixed values for each vegetation type [*Beerling and Quick*, 1995]. PAR absorption by leaves is calculated with the two-stream approximation [*Sellers*, 1985] for three layers, using direct and diffuse PAR computed as in *Weiss and Norman* [1985]. The time step of this model part is one hour.

## 2.3 Carbon balance

The photosynthesis calculation passes daily integrals of $A_c$ and $R_{d,c}$ to the carbon balance to compute gross and net primary productivity (GPP and NPP, respectively) on a daily time step. While GPP describes the amount of carbon taken up through photosynthesis, NPP stands only for the part converted into plant material and not lost as respiration. There are two types of plant respiration, maintenance respiration, $R_M$, and growth respiration, $R_G$, such that NPP = GPP − ($R_M$ + $R_G$). Canopy-integrated leaf respiration, R_d,c, is assumed to account for 40% of $R_M$, and $R_G$ for 25% of NPP [*Ryan*, 1991].

## 2.4 Phenology

In the context of this study, where the focus is on the water and energy balance, the significance of NPP lies in the fact that it limits leaf growth and thus controls, among other factors, the leaf area index (LAI). This is accounted for by the phenology part, in which the LAI is computed as the minimum of a temperature limited value, $\Lambda_T$, a water limited value, $\Lambda_W$, and a growth limited one, $\Lambda_G$, i.e.

$$\Lambda = \min \{\Lambda_T; \Lambda_W; \Lambda_G\} \tag{3}$$

The value of $\Lambda$ is updated every 10 days.

Temperature limitation of LAI, which has its reason in a frost avoidance strategy of the vegetation, is prescribed as in *Dickinson et al.* [1986]:

$$\Lambda_T = \begin{cases} 0 & \text{if } T_{0.5} < T_\phi \\ \Lambda_{\max}\left[1 - \left(\dfrac{T_{\phi,\max} - T_{0.5}}{T_{\phi,\max} - T_\phi}\right)\right] & \text{if } T_\phi < T_{0.5} < T_{\phi,\max} \\ \Lambda_{\max} & \text{if } T_{0.5} > T_\phi \end{cases} \tag{4}$$

with the temperature at 0.5 m soil depth, $T_{0.5}$, and standard values of $\Lambda_{max} = 5$, $T_\phi = 5°C$ and $T_{\phi,max}=15°C$, with the exception of agriculture, where $T_{\phi,max} = 12°C$. $T_\phi$ describes the daily mean temperature at which leaves are formed in spring and shed in autumn. Water limitation is modelled such that $\Lambda_W$ is set to the value that maximises NPP as long as the LAI increases. At decreasing LAI, $\Lambda_W$ is set to the LAI value of the preceding time step, and as soon as NPP becomes negative, $\Lambda_W$ is set such that NPP equals zero, thus decreasing the LAI just enough to avoid carbon losses. Eventually, $\Lambda_G$ accounts for 50% of NPP invested into leaf growth, but cannot be lower than $\Lambda_G=0.5$ to allow initial leaf growth.

The actual LAI, $\Lambda$, computed with this scheme is distributed in a partially "clumpy" fashion, depending on the expected maximum annual leaf area index. The grid area fraction covered by vegetation, $f_c$, is computed from

$$f_c = 0.9 \min\{\mathrm{MI}; 1\} \min\{\Lambda_T (T_w) / \Lambda_c; 1\} \tag{5}$$

where $T_w$ is the mean temperature of the warmest month, MI is the annual moisture index (annual precipitation devided by annual potential evapotranspiration) and $\Lambda_c = 3$ a critical threshold value below which vegetation starts becoming patchy rather than evenly distributed. Especially in semi-arid areas, clumping can have a significant effect on the energy balance, leading to a general increase in NPP [*Knorr*, 1997].

## 2.5 Spatial and temporal resolution

The BETHY scheme can be run on any spatial grid, with several vegetation types present at each grid point. Here, the resolution chosen is 111 km by 111 km (on an equal area grid with 1° by 1° resolution at the equator) and up to three vegetation types per grid cell. The model is run for 11,069 land grid points excluding Antarctica.

While the BETHY scheme is driven by monthly climate data, the strongly non-linear behaviour of the water balance makes it necessary to compute this process internally on a daily time step, distributing monthly precipitation according to the stochastic model of *Geng et al.* [1986]. This model requires daily probabilities of wet days which are, together with temperature and incoming PAR, interpolated linearly from monthly to daily mean values.

To account for diurnal variations, radiation, energy balance and photosynthesis are calculated every 10 days on an hourly time step, such that those diurnal processes are recomputed every time the LAI is updated. Computation of the diurnal temperature cycle assumes maximum temperature as mean plus half the diurnal amplitude, reached at 14:00 hours, with a sinusoidal time course between dawn and dusk and a linear decrease during the night [*Rosenberg*, 1974, p. 91]. Specific humidity is assumed to be close to saturated humidity at dawn, and to grow during the day by an amount depending on the ratio of actual to potential evapotranspiration [*Müller*, 1982; *Rosenberg*, 1974].

On the output side of the BETHY scheme, those internal calculations at higher temporal resolution are averaged for each grid point over a month and over vegetation types to provide carbon and water fluxes as well as LAI and fAPAR at the same time resolution as the input data. To ensure that results are independent of initial conditions, the model is run for six years with mean-climate input data, of which the first three are considered as spinup and the last three are again averaged for each month to compute a mean response of vegetation and land surfaces to the climatic forcing. A detailed description of the BETHY scheme can be found in [*Knorr*, 1997].

## 2.6 Model input data

The BETHY vegetation and surface modelling scheme derives its climatic input data largely from maps originally based on ground station meteorological observations. This insures a more realistic simulation of surface climate than it would be possible within a GCM. Input data all have a monthly time step and are converted to the 1° equal-area grid used here.

Monthly climatological means of precipitation, near-surface diurnal mean and amplitude are taken from the climatology of Cramer and Leemans [Cramer, pers. comm.][*Leemans and Cramer*, 1991], the number of wet days per month from Friend (pers. comm.) based on [*Müller*, 1982], and incident PAR from the International Satellite Cloud Climatology Project (ISCCP) [*Pinker and Laszlo*, 1992].

Other input data are the global maps of soil brightness and land-cover by [*Wilson and Henderson-Sellers*, 1985], from which relative abundances of up to three vegetation types out of a list of 23 are derived. The 23 vegetation types are also assigned a list of values [*Knorr*, 1997] specifying photosynthetic capacity [*Beerling and Quick*, 1995] and pathway, phenological type, height, and leaf area per dry mass [*Schulze et al.*, 1994]. For reasons of compatibility with the ECHAM-4 climate model (cf. above), the total soil water-holding capacity, $w_s$, is taken from published data [*Patterson*, 1990], and converted to $w_{max}$ by taking 65% of that value. This allows translating adjusted fields of $w_{max}$, – after assimilation – again into saturated soil-water content, to be used in a sensitivity test with ECHAM-4.

# 3. SATELLITE DATA PROCESSING

The satellite data used are from the NOAA Global Vegetation Index (GVI) archive [*Gutman et al.*, 1995] for 1989 and 1990, and with a resolution of 1/7° latitude by longitude. Subsequent data processing by CESBIO [*Berthelot et al.*, 1994] includes conversion to reflectances, and atmospheric correction using the SMAC code [*Rahman and Dedieu*, 1994].

Although most of the cloud contamination in the original daily AVHRR fields have already been removed in the GVI archive by selecting maximum contrast between channel 1 and 2 raw counts, further cloud screening and monthly compositing can lead to a still significant reduction in the residual cloud amount [*Gutman et al.*, 1996]. The usual approach is to select those dates where the normalised difference vegetation index (NDVI) derived from channel 1 and 2 reflectances is at a maximum [*Holben*, 1986]. However, this maximum compositing technique tends to introduce signifant angular bias of artificially high "greenness" [*Meyer et al.*, 1995], and temporal biases at the start and end of a growing season [*Gutman et al.*, 1996]. It has also been found that the NDVI is rather sensitive to changes in soil background colour and atmospheric composition; if those influences are considered noise, then it can be shown that more modern indices achieve a better signal-to-noise ratio [*Goel and Qin*, 1994; *Leprieur et al.*, 1994].

In this study, channels 1 and 2 of AVHRR are combined to compute the Global Environment Monitoring Index [GEMI, *Pinty and Verstraete*, 1992], designed to be robust against changes in soil brightness [*Leprieur et al.*, 1994] and atmospheric perturbations [*Flasse and Verstraete*, 1994]. To avoid temporal and angular bias, cloud-screened weekly values from the GVI data set are averaged, with no maximum compositing performed. Cloud identifcation is based on negative deviations of weekly values from a filtered GEMI time series, based on GEMI's property of assuming low to negative values over cloud scenes [*Flasse and Verstraete*, 1994]. In addition, data points with so called "hot-spot" viewing conditions are also eliminated, as well as those where view or sun zenith angles are larger than 60°. Hot-spot conditions are assumed whenever a function $G$ (defined in *Verstraete et al.* [1990]) assumes values less than 0.25.

The data fields are then averaged over 1° latitude by longitude, after which areas of high residual atmospheric contamination are identified by comparing time-interpolated maximum composite GEMI with the monthly average according to the standard procedure. If the ratio of monthly maximum to monthly average is above 1.1, the 1° pixel is discarded. The annual average data coverage achieved with this method is 74.8% of global land areas.

As the last step of satellite data processing, values of GEMI are translated into fAPAR using the following approximation:

$$\text{fAPAR} = \begin{cases} 0 & \text{if GEMI} \leq 0.380 \\ 1 & \text{if GEMI} \geq 0.720 \\ -0.122 - 1.061\ \text{GEMI} + 3.639\ \text{GEMI}^2 & \text{else} \end{cases} \quad (6)$$

The relationship is based on a regression between GEMI and fAPAR computed with a radiative transfer model [*Gobron et al.*, 1997], and accounts for increasing residual atmospheric contamination over humid, densely vegetated areas. The error in fAPAR is estimated to lie between 0.05 and 0.10.

## 4. THE METHOD OF DATA ASSIMILATION

As already outlined in the introduction, the strategy persued in this study is to derive a parameter related to land surface properities, fAPAR, from the remote sensing data (see Section 3) and then to assimilate this quantity into the BETHY scheme by minimizing the deviation between the satellite fAPAR and the value computed within the photosynthesis part (see Section 2) through modificiation of certain parameters within the BETHY scheme. In a susequent simulation, those optimised parameter fields are then used within an atmospheric GCM.

### 4.1 Temporal and spatial scales

This strategy raises a number of issues related to the treatment of the atmospheric effect on the remote sensing signal, and to the time scales involved. Since those issues are also relevant for a wider range of applications where remote sensing data are assimilated within atmospheric and land surface models, they are first discussed in a more general context [cf. *Verstraete and Pinty,* 2000].

As the signal measured on board the satellite, i.e. the top-of-atmosphere (TOA) radiance, depends on both the state of the surface and of the atmosphere, an atmospheric model with an adequate surface representation should carry all the necessary information to simulate the observed satellite signal as a function of its internal variables. In an appropriate assimilation scheme, it would then be possible to modify both atmospheric and surface state variables of this model while assimilating TOA radiances directly. The reason why such a scheme has not been implemented, however, lies in the differences between temporal and spatial scales at which atmospheric and surface properties change: while surface properties are very variable in space but only change on a weekly to monthly time scale, atmospheric variables tend to be spatially coherent over hundreds of kilometers but can change within hours. This discrepancy in scales creates scaling problems, but has made weather forecasting possible without taking surface properties into account.

In numerical weather prediction, the impact of surface properties is essentially treated as a perturbation that introduce some degree of uncertainty into the derivation of atmospheric parameters from the satellite data [*Le Dimet and Talagrand*, 1986]. In fact, TOA radiances are not normally assimilated directly; instead, certain atmospheric parameters are derived from the satellite data and then assimilated. For example, the 3-D variational analysis scheme implemented in February 1996 by the European Centre for Medium Range Weather Forecast [ECMWF, *Anderson et al.*, 1994] uses, among other data, thermal infrared and microwave soundings from NOAA's TIROS Operational Vertical Sounder [TOVS, *Smith et al.*, 1979] to assimilate wind, temperature and humididty.

The opposite applies to the assimilation scheme presented in this study: the focus is on the monthly response of vegetation and the surface energy and water balance to climatic forcing, independent of short-term fluctuations of the atmospheric state. During the derivation of fAPAR, those atmospheric fluctuations are partly removed by cloud screening algorithm and monthly averaging, while the impact of the average atmospheric state on the satellite signal is removed by atmospheric correction. The impact of the remaining uncertainties about the state of the atmosphere is minimised by choosing a vegetation index relatively robust against those fluctuations, and are accounted for by assigning an uncertainty for the derived value of fAPAR (see below).

For longer simulations than medium-range weather forecasts, however, both surface and atmospheric variables matter [*Verstraete*, 1989]. A suitable strategy for this kind of situation is outlined by Fig. 1: atmospheric variables are assimilated into a coupled atmosphere-land surface model at a short time step – daily or shorter – while holding surface variables constant. At longer intervals, e.g. weekly or monthly, surface variables are then assimilated, using the atmospheric state predicted by the model to account for the atmospheric effect on the TOA radiances measured by the satellite.

The modelling framework presented here represents a first step in this direction. Assimilation of fAPAR is carried out "off-line" because it is compuationally less expensive than assimilation within a GCM, and because coupled vegetation-atmosphere GCMs are still a very new development that have not yet been tested extensively. Another reason is that the spatial resolution of 1° is closer to the 15 km resolution of the satellite data than the resolution of most GCMs, so that the potential of satellite-derived fAPAR for assimilation into surface schemes can be explored in more detail. As far as time scales are concerned, monthly averages of fAPAR derived from satellite data are compared to monthly averages computed by the BETHY scheme. The assumptions are that both represent mean climatological conditions, and that, to a first approximation, differences between the simulated and satellite-derived fAPAR arise solely from parameterisations of the BETHY scheme, and not from deficiencies of the climatic input data, or a mismatch of the periods covered by satellite and climate data.

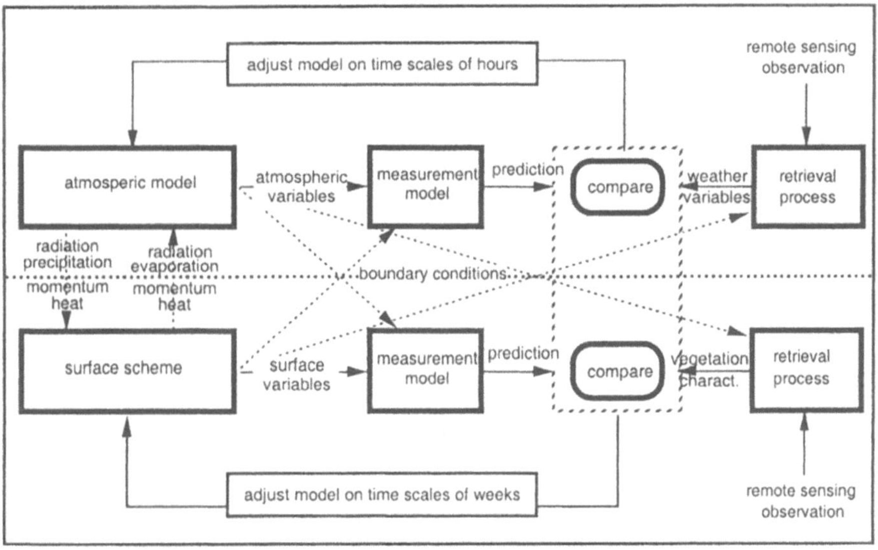


*Figure 1.* Illustration of the proposed assimilation schemes for surface and atmospheric variables

## 4.2 The assimilation scheme

The purpose of the present study is thus twofold: to introduce an assimilation scheme for land surface models that uses a quantity related to vegetation productivity; and to assess the impact of derived surface properties on the simulated climate. If there is a significant impact, this can be taken as an indication that assimilation of surface variables in addition to atmospheric variables could be used to improve GCM simulations. Since in the modelling framework used here, this requires the inclusion of an active biosphere within the surface model, the degree of the impact can also be taken as an indication of the importance of vegetation feedback to climate.

This preliminary setup preceding a full assimilation of atmospheric and surface variables within a GCM follows the lower part of Fig. 1. fAPAR is chosen as a critical parameter that controls primary productivity and transpiration, one that at the same time can be derived rather accurately from the contrast between the red and near-infrared reflectances measured from space [*Asrar et al.*, 1994; *Goel and Qin*, 1994].

The assimilation is performed by a parameter re-estimation of the BETHY scheme, in which three key parameters are modified (see below). Those parameters have earlier been identified to as those to which the carbon and moisture balance are most sensitive [*Knorr*, 1997; *Knorr*, 2000].

This fAPAR assimilation scheme consists of four steps:

1. Global monthly fields of fAPAR are derived from NOAA-11/AVHRR data at 1 degree resolution after rigorous screening of cloudy and unfavorable angular conditions.
2. The BETHY scheme is driven with mean climate data to compute both land-surface carbon and water exchanges and fAPAR for 12 months at 1 degree spatial resolution.
3. Satellite derived and model predicted values of fAPAR are compared.
4. Model parameters $x_1...x_n$ are modified until a cost function, $J$, reaches its minimum:

$$J(x_1 \ldots x_n) = \sum_{m=1}^{12} \frac{\left[f_m - g_m(x_1 \ldots x_n)\right]^2}{\sigma_f^2} + \sum_{j=1}^{n} \frac{(x_i - y_i)^2}{\sigma_{y,i}^2} \qquad (7)$$

This minimisation is carried out separately for each grid point.

The cost function, $J$, accounts for the difference between monthly satellite derived fAPAR, $f_m$, and model computed fAPAR, $g_m$, scaled by the assumed error variance of fAPAR, $\sigma^2_f$. An additional term is used to increase J when the model parameters, $x_i$, deviate from the standard values, $y_i$, assumed to represent *a priori* knowledge, with error variance of $\sigma^2_{y,i}$ for the values $y_i$.

The three model parameters chosen as $x_1...x_3$ are: $x_1 \equiv w_{max}$, representing water limitation ($w_{max}$ is the bucket size, i.e. the maximum amount of plant-available soil moisture), $x_2 \equiv T_\phi$, representing temperature limitation ($T_\phi$ is the leaf onset and shedding temperature, cf. Eq. 4), and $x_3 \equiv f_c$, representing other, residual limitations that typically have longer times scales than decades, such as human land use or nutrient availability ($f_c$ is the vegetated cover fraction, cf. Eq. 5). $J$ is then minimised by modifying $x_1$ to $x_3$ at each grid cell, re-running the model each time for the six-year simulation period, with the first three years neglected as spinup (see above). The parameters, their assumed error variances and their allowed value ranges are listed in Table 1. The constrained minimisation technique used is the downhill simplex method

[*Press et al.*, 1992], which, compared to gradient methods, is extremely robust against abrupt changes in $J(x_1 \ldots x_n)$.

*Table 1.* Measured quantities and model parameters used in the assimilation scheme, together with assumed uncertainties ($\sigma$), allowed ranges, and their significance in the model context

| General symbol | in model | quantity/ parameter | $\sigma$ | min. | max. | significance |
|---|---|---|---|---|---|---|
| $f_m$ | fAPAR | satellite-derived monthly fAPAR | 0.2 | 0 | 1 | radiometric measure indicative of vegetation productivity |
| $x_1$ | $w_{max}$ | bucket size | $4w_{max}$ | 0 | $\infty$ | water limitation |
| $x_2$ | $T_\phi$ | leaf onset/ shedding temp. | 2°C | $-\infty$ | $T_{\phi,,max}$-2°C | temperature limitation |
| $x_3$ | $f_c$ | vegetation fractional cover | 0.5 | 0 | 1 | other limitations |

# 5. RESULTS

## 5.1 Achieved consistency with observed fAPAR

Before assessing the results of the model simulation in detail, it is important to insure that the assimilation scheme has been successful at improving the consistency of modelled fAPAR with the remote sensing data. A $\chi^2$ statistics is used here as a criterion for the goodness-of-fit, taking into account the assumed error in the satellite-derived fAPAR, $\sigma_f = 0.2$:

$$\chi^2 = \frac{1}{n}\sum_{m=1}^{n_m}\frac{(f_m - g_m)^2}{\sigma_f^2} \tag{8}$$

where $f_m$ is satellite-derived fAPAR, $g_m$ the mean annual cycle of modelled fAPAR, and $m$ the $n_m$ months with valid satellite data. This is shown in Fig. 2 for the simulation with BETHY before and after applying the fAPAR assimilation scheme. Simulations with $\chi^2 > 1$ can be considered to deviate significantly from observations, and there appear to be two large regions for which this occurs: the boreal forest belt of Canada, Scandinavia and northern Russia, and the semi-arid tropics, especially the Brazilian, African and Australian savannas and bushlands. Some heavily populated regions also appear, especially in North Africa and the Middle East, Western Europe, Bangladesh, and China.

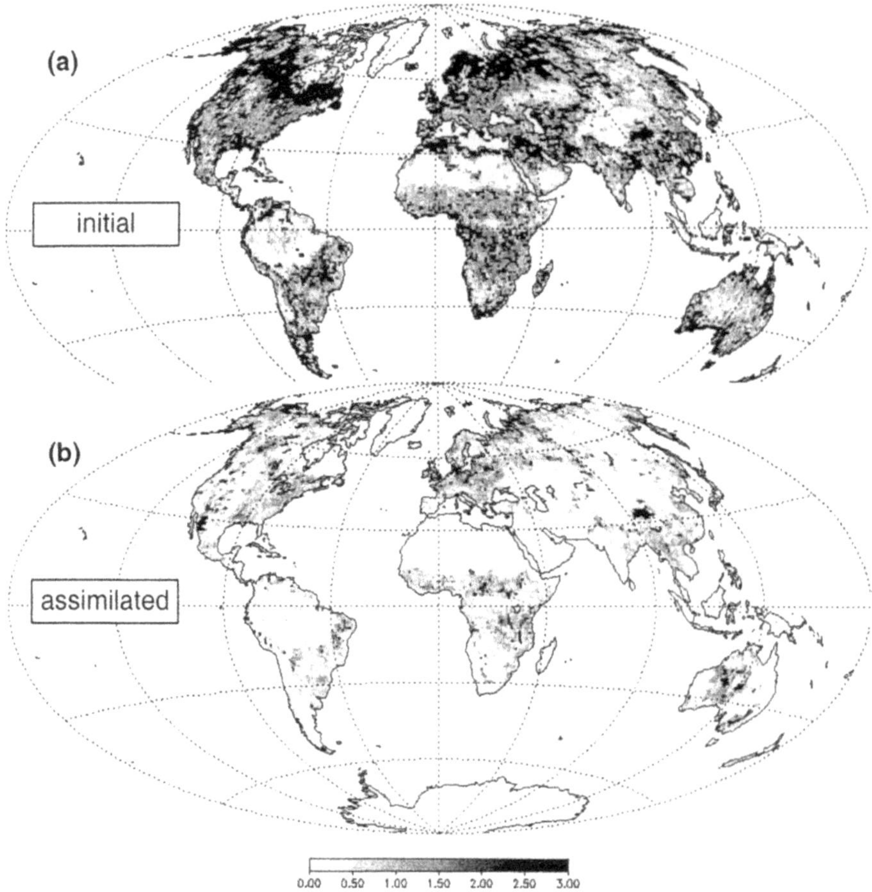


*Figure 2.* $\chi^2$-statistics of the fit between average monthly simulated and satellite observed fAPAR with the BETHY model. Panel (a) was generated using initial model parameters, panel (b) after assimilation of the satellite fAPAR fields

After assimilation, consistency clearly improves and is now mostly better than $\chi^2 = 1$. It turns out that in the boreal areas, vegetation cover had been largely overestimated, probably because nitrogen limitation of forest density has not been accounted for explicitly. The model also predicts spring-time greening too late in Western Europe, which leads to high values there. In these two cases, it is mainly adjustment of the parameters $f_c$ and $T_\phi$ which leads to improved consistency with measured fAPAR.

In most of the semi-arid tropics, however, vegetation cover is most sensitive to the maximum plant-available soil moisture content, $w_{max}$. Here, vegetation cover has been underestimated in the initial model run, i.e. before assimilation, with the exception of some of the Brazilian savannas, where the initial value of $w_{max}$ is very high (see next subsection).

## 5.2 Impact on simulated soil moisture

Since the water cycle is the main focus here, further analysis will concentrate on changes in the parameter $w_{max}$ and the simulated plant-available soil water content. Changes in $w_{max}$, the maximum plant-available water holding capacity of the soil, are documented in Fig. 3, showing the initial (a) and adjusted fields (b), and the difference between the two (c).

Thc initial ficlds, i.c. thosc assumcd bcfore assimilation and derived from the analysis of Patterson as used in the ECHAM-4 climate model, largely reflect changes in soil porosity, with relatively small changes in the assumed rooting depth [*Patterson*, 1990]. However, the maximum amount of soil water that can be used by vegetation, and thus becomes available as latent heat flux to the atmosphere, is also determined by the depth of the rooting system [*Kleidon and Heimann*, 1998]. This impact of vegetation can be seen in the adjusted $w_{max}$ field of Fig. 3b: whereas southern Africa and parts of Australia have rather low water storage capacities in the initial fields, the densitity of the vegetation there leads to rather high values after assimilation. The situation for South America turns out to be a little more complex: high $w_{max}$ is predicted by the assimilation scheme for the Northeast, lower than initial $w_{max}$ for the Cerrado southwest of that area, and again very high values at the edge of the Amazon rainforest (cf. Fig. 3c). For this latter region, the spatial pattern exhibited by the initial fields, with a large area of high soil water holding capacity extending across the rather sharp rainforest/savanna boundary, is very different from the adjusted ones. This pattern found in the initial fields is probably unrealistic, because for the same climate, evergreen rainforest trees require more soil water storage than drought-deciduous savanna vegetation (see next subsection).

The simulated plant-available soil water content in the tropics is shown in Fig. 4 for the northern-hemisphere (March, top) and southern-hemisphere dry seasons (September, bottom). An important result for the northern tropics is that after assimilation, there is more soil moisture available in the Sahel region during March, where the zone of partially wet soils extends further north into the Sahara. This is also reflected by an increase of $w_{max}$ in that region shown in Fig. 3c. For the southern tropics, which include most of the Amazon basin, it is important to note that an increase in $w_{max}$ has lead to more available soil moisture during the dry season in most of southern Africa, the Brazilian northeast, parts of the Amazon rainforest, and northern Australia.

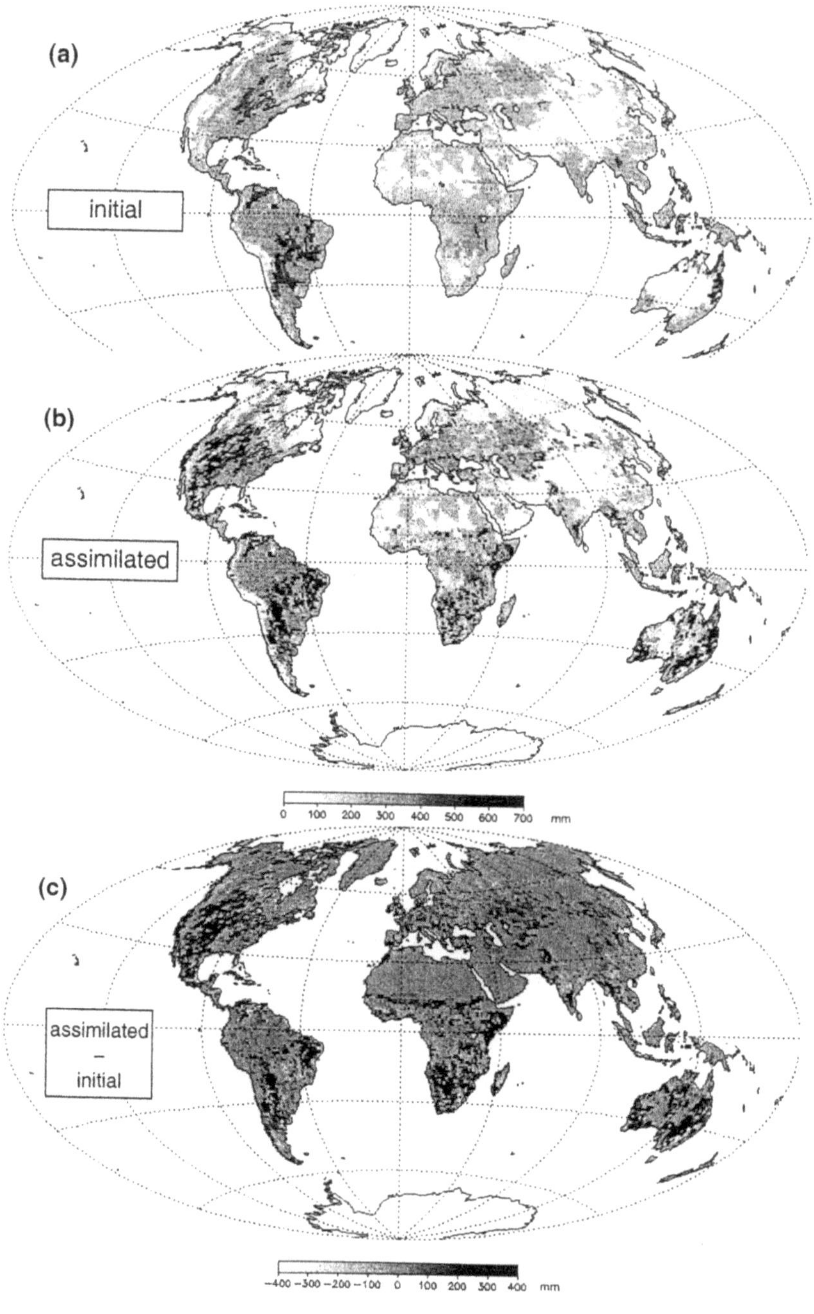


*Figure 3.* Maximum plant-available soil moisture, in mm, used as initial fields in the assimilation scheme (a), and as derived after maximising consistency with satellite derived fAPAR (b). Panel (c) shows the difference between he two fields as the effect of satellite assimilation

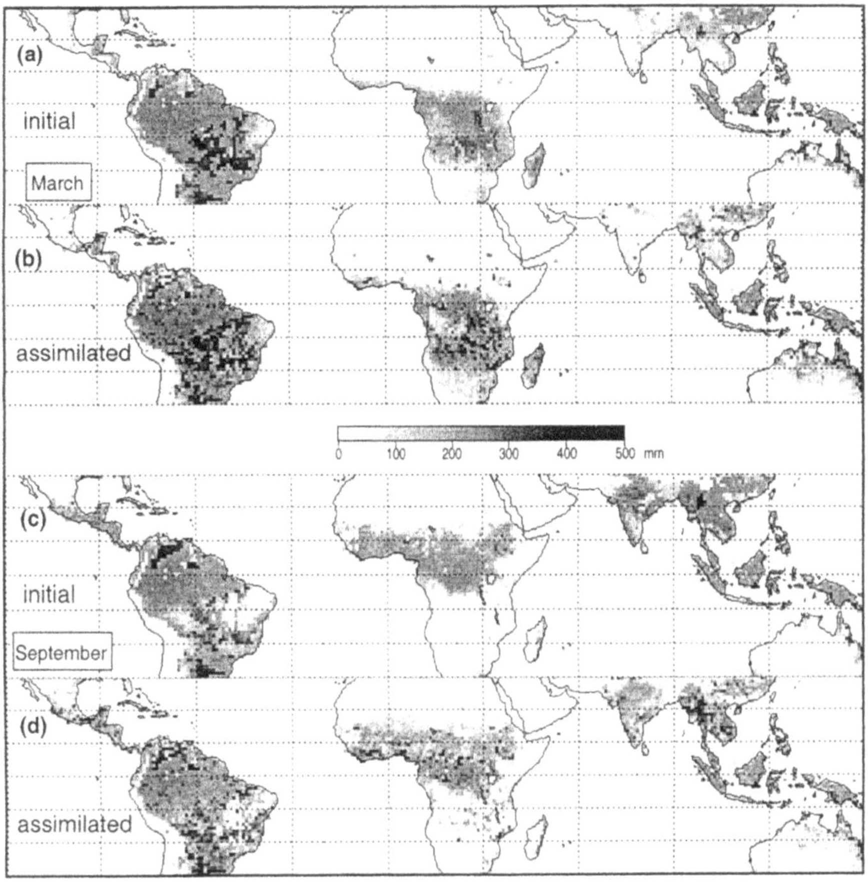


*Figure 4.* Simulated plant available soil moisture for the tropics with the BETHY land surface and vegetation model, for March (dry season in the north) and September (dry season in the south). (a) and (c) have been generated using the initial fields of maximum plant-availabe soil moisture, (b) and (d) with optimised fields of this parameter as derived through assimilation

## 5.3 Comparison with soil moisture measurements from Amazonia

So far, the analysis has shown that requiring concistency with observed vegetation cover has had an impact on the simulated soil water content in large parts of the tropics, with a tendency towards greater maximum soil water storage. In fact, it has been discovered that often trees in tropical rainforests can survive prolonged dry periods by developing very deep roots [*Nepstad et al.*, 1994], with important consequences for the soil water balance. It has also been argued that, at least in water limited environments, plants should optimise the rooting strategy for water use, which leads to much deeper roots and higher soil water capacity than traditionally assumed in climate or vegetation models [*Kleidon and Heimann*, 1998].

To investigate whether the model simulations shown before are consistent with observations, the BETHY model is run on the basis of daily precipitation data for 1992 reported by Nepstad and co-workers – with temperature data for 1992 unchanged – for their Paragominas site (3°S, 48°W). Initial rooting depth was set to 2 m, with $w_{max} = 220$ mm computed from soil texture information as in a previous application of the model [*Knorr*, 1997].

Simulated and satellite-derived fAPAR for this site are show in Fig. 5a. The model run with a-priori parameters shows a pronounced decrease in fAPAR during the dry season as a result of leaf shedding forced by declining soil water, which disagrees both with the satellite data, and with published reports for this site. If, however, simulated fAPAR after assimilation is compared to satellite-derived values, the agreement improves considerably, with a lower variability for the model calculated than for the satellite derived values. In fact, satellite-derived values of fAPAR tend to be higher during the dry season, and lower during the wet season. This is most likely a result of residual contamination of the satellite signal by high water vapour concentration during the wet season. This particular result demonstrates how data assimilation can also help to filter out noise in input data by accounting for both model and data uncertainties.

More important for the surface energy balance than fAPAR is the actual evapotranspiration. As Fig. 5b shows, high evapotranspiration rates persist through the dry season after assimilation, although somewhat reduced due to declining soil water reserves, while they approach zero with the a-priori setup of the model. This large difference in the energy and water balance between the two model versions can be attributed entirely to persisting soil water reserves during the dry season, as Fig. 5c shows: while in the a-priori model version with 2 m root depth, the soil dries out almost completely, observations [*Nepstad et al.*, 1994] reveal much higher reserves in that part of the soil. The difference, however, is not attributable to less evapotranspiration, but to higher soil water storage; this is revealed by Fig. 5d: the slope of the simulated and observed soil water curves at the beginning of the dry season (May to June) are all similar, but in the assimilated case, the model infers very large water storage in accordance with measurements down to a depth of 8 m, a depth where there were still active roots. (It should be noted that this particular site was found to be free of soil water intrusion).

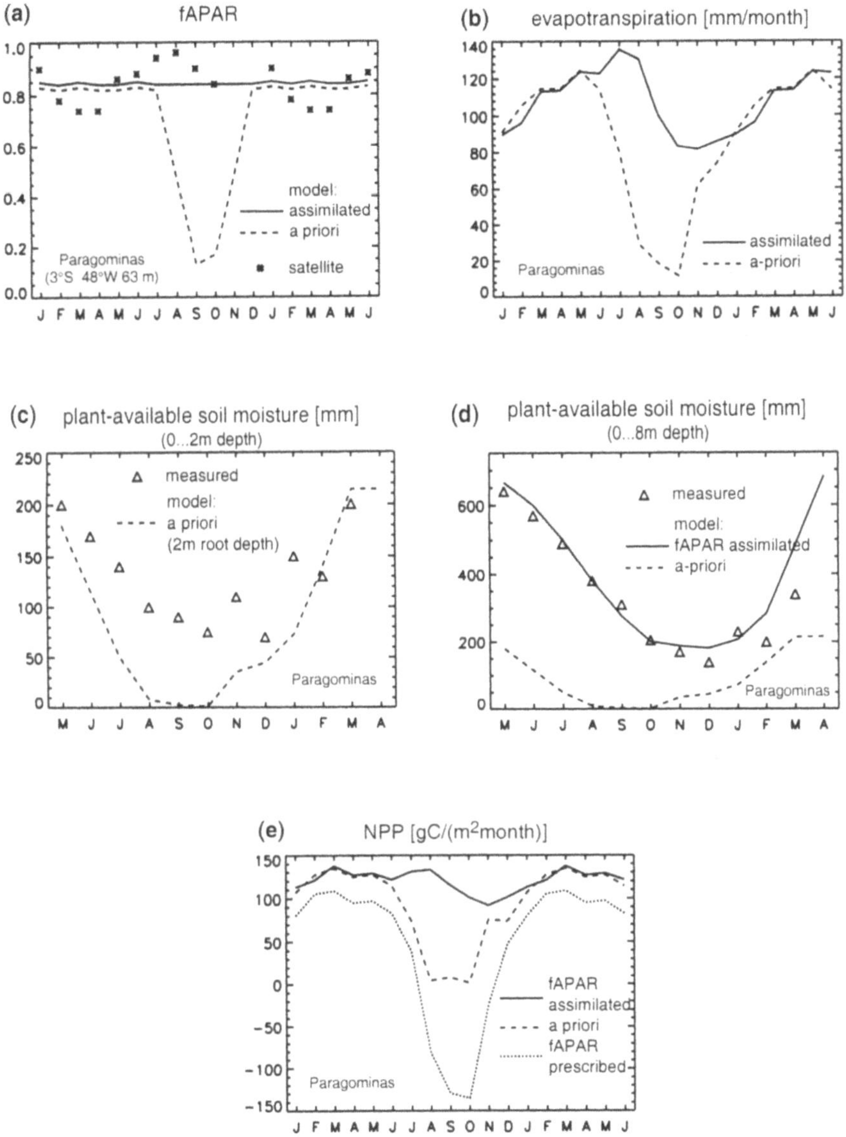


*Figure 5.* Satellite derived and simulated fAPAR (a), simulated evapotranspiration (b), and NPP (e), and simulated plant-available soil moisture vs. measurements by Nepstad et al. (1994) for the top 2 m (c) and 8 m (d), all for the Paragominas site in the eastern Amazonian rainforest (3°S, 48°W). The BETHY model was run in three modes with observed daily rainfall data: without assimilation (*a priori*), with assimilation of the satellite derived fAPAR values (*assimilated*, all except c), and with satellite-derived fAPAR used as forcing (*prescribed*, only e)

To better understand the mechanism through which changes in $w_{max}$ can lead to an improved consistency with observed fAPAR, and to contrast this method with simply prescribing the LAI according to satellite observations, Fig. 5e shows the simulated net primary productivity (NPP) also for the case of prescribed fAPAR: when the LAI is set the observed constant value of 5, NPP declines rapidly during the dry season and reaches large negative values caused by persisting plant respiration at closed stomata. By contrast, in the normal a priori run, LAI declines to stabilise NPP around zero, while in the assimilated case, NPP persists rather unchanged. This latter type of carbon balance has been found for a rainforest site with a similar climate [*Grace et al.*, 1995], who have carried out eddy correlation measurements of $CO_2$ fluxes in Rondonia, Brazil. This indicates that, because of the carbon costs of maintaining extensive foliage, the presence of green vegetation can be used to deduce a source of transpired soil water. In a climate model that contains an interactive vegetation component, this information could then be used to check its consistency with rather easily observable satellite information.

## 5.4 Impact of inferred soil water capacity on simulated climate

Rather than using a model with a fully coupled interactive vegetation component, of which only a few are currently being developed, the inferred field of $w_{max}$ is used in a sensitivity study to test the impact of the additional satellite information on the climate simulated by the ECHAM-4 general circulation model (GCM). ECHAM-4 [*Roeckner et al.*, 1996] is the fourth generation of the ECHAM GCMs, a series of spectral climate models developed at the Max-Planck-Institut für Meteorologie, Hamburg. Its land surface scheme takes into account vegetational effects on the energy and moisture cycles, such as the interception of precipitation or the stomatal control of evapotranspiration, including a parameterisation of soil moisture stress in dry regions.

The land surface characteristics of ECHAM-4 are described by a set of global annual mean land surface parameters [*Claussen et al.*, 1994], including quantities like surface background albedo, leaf area index and fractional vegetation cover. This data set has been constructed by allocating parameter values from different sources to major ecosystem complexes [*Olson et al.*, 1983]. The global distribution of the total soil water-holding capacity was derived from a high-resolution data set [*Patterson*, 1990].

In order to assess the impact of the newly derived distribution of soil water reservoirs, two global experiments are conducted. One simulation is performed with the standard version of the ECHAM-4 GCM and serves as a control simulation, while in the other run the total soil water-holding capacity is replaced by $w_s = w_{max} / 0.65$, in accordance with the convention used in ECHAM-4 to convert total water-holding capacity to the plant-available amount (see above). Both model versions are integrated for a 10-year period after at least five years of spin-up to exclude any remaining impact of soil moisture initialization. A T42 spatial resolution was chosen which is equivalent to about 2.8° × 2.8° on a latitude-longitude grid. Both simulations use an annual cycle of monthly mean climatological sea surface temperatures. As Fig. 3c shows, there is no change in soil water holding capacity for unvegetated regions, so that only the effect of the addional satellite-based information is assessed.

Before comparing simulations with the modified $w_s$ to the standard control run, Fig. 6 is used to compare the simulated control climate of ECHAM-4 with the climate map of *Legates and Willmott* [1990a]. Comparing to Fig. 4, it turns out that there is some agreement between the tropical semi-arid areas experiencing increased dry-season soil water content after assimilation, and those where ECHAM-4 overestimates the 2m temperature. Although much of this difference can probably be attributed to model dynamics and radiation parameterisation, soil moisture might also play a role here.

As Fig. 7 shows, the satellite-inferred changes from the standard $w_{max}$ derived by the BETHY model are able to compensate some of this discrepancy when used in ECHAM, at least for the southern tropics. Increased soil water storage leads to increased evapotranspiration, cooling the air near the surface. In southern Africa, where the a priori soil water capacity was rather low (cf. Fig. 3), and for parts of South America, these changes can amount to as much as 3°C. However, for March (not shown), there is only very little change, in the South because soil water reservoirs are filled in both cases, and for the northern tropics, because the change in soil water storage is rather small. Consequently, the simulated temperature in summer and early fall is reduced, which is shown in Fig. 8a; compared to the climatology by *Legates and Willmott* [1990a, b], it is actually reduced into the right direction.

It is likely that other factors than soil moisture contribute to the described differences between near-surface temperature simulated by ECHAM-4 and the climatology by Legates and Willmott. Too little precipitation leading to too much surface drying should not be the reason, at least in the case of southern Africa, as Fig. 8b shows. For example, the Arabian Desert is also simulated too warm, and there is certainly no vegetation not accounted for in the climate model. However, it could also be expected that the assimilation procedure underestimates $w_{max}$, because interannual changes in precipitation are not taken into account.

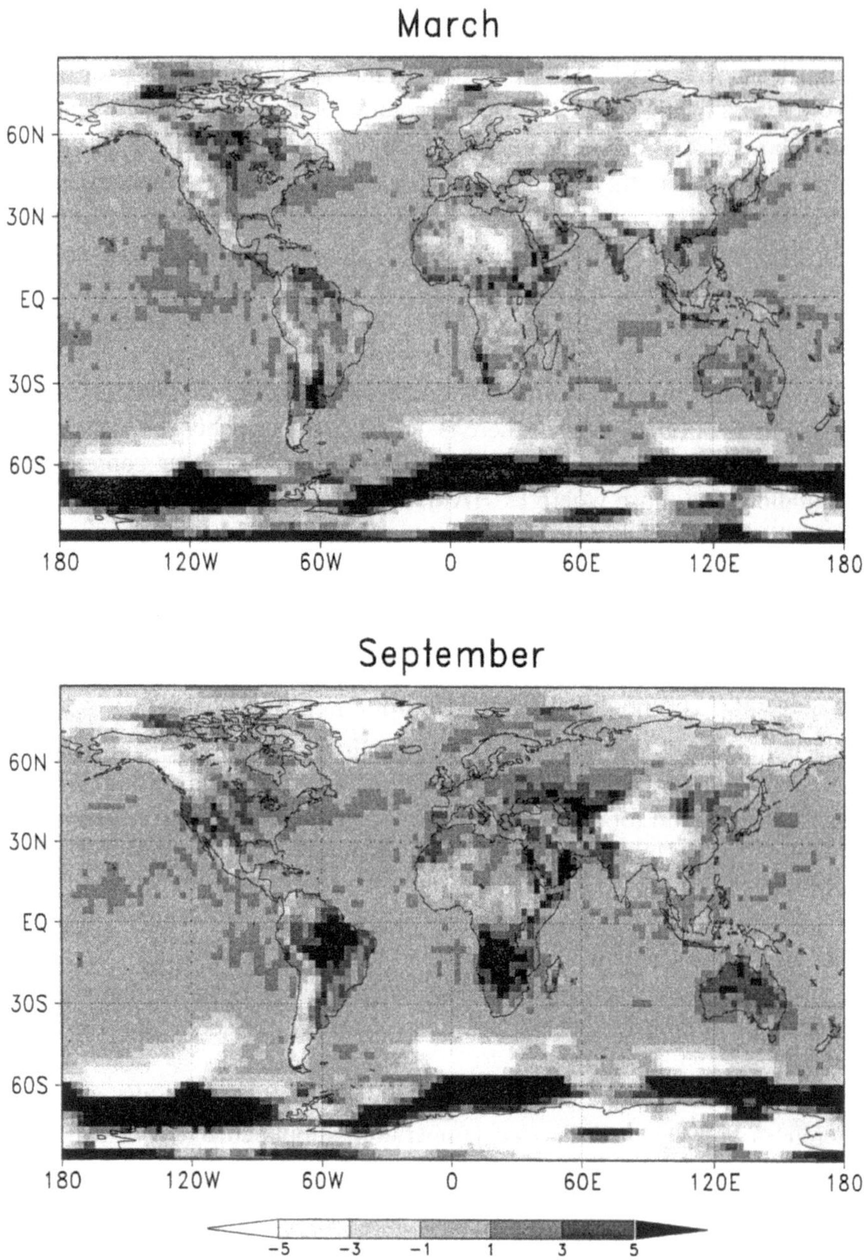


*Figure 6.* Difference between the 10-year mean 2m air temperatures in °C simulated with the standard ECHAM-4 GCM, used as control simulation, and the climatology of Legates and Willmott [1990a], for March (dry season in the northern tropics) and September (dry season for southern tropics)

Higher year-to-year variability would force the vegetation to develop even deeper roots than estimated here with a mean climate. This is suggested by the single-point simulations shown in Fig. 5, using precipitation data from the relatively dry year of 1992. Some estimates of maximum rooting depth by vegetation type [*Canadell et al.*, 1996] also show rather large values, suggesting that some factors determining rooting strategy may be missing in this analysis. All taken together, soil moisture storage does appear to have a significant impact on climate, and the inclusion of vegetation leads to further possibilities of validating the results of climate model simulations against global satellite data.

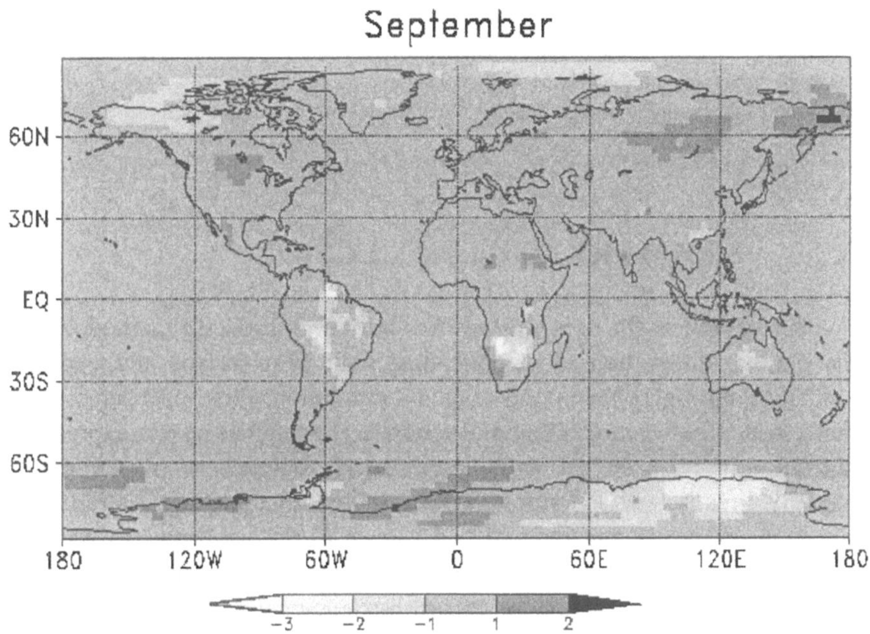


*Figure 7.* Difference of the 10-year mean simulated 2m air temperature, in °C, for September between the ECHAM-4 simulation with modified soil water capacity based on the fAPAR assimilation with the BETHY land surface and vegetation model minus the ECHAM-4 control simulation using standard soil water capacity. Changes in southern Arctic due to a high variability of sea ice cover are not significant

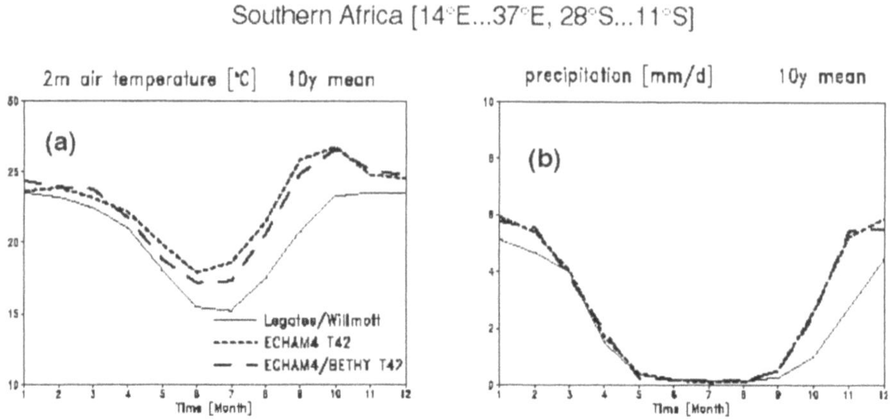


*Figure 8.* Seasonal cycle of 2m air temperature (a) and precipitation (b) for a portion of southern Africa, from the ECHAM-4 control simulation and the simulation with modified soil water capacity, both compared to the climatologies of Legates and Willmott [1990a, b]

# 6. SUMMARY AND CONCLUSIONS

The purpose of the present study has been to introduce a methodology for the optimal integration of satellite data with land surface and vegetation modelling through a technique of data assimilation. First, it is demonstrated that a match between satellite and modelled vegetation cover, expressed as fAPAR, can be achieved without the usual approach of prescribing this quantity directly to a land surface or vegetation scheme. Assimilation has the advantage of preserving internal consistency of the land model's water and carbon balances with the vegetation cover. As a more detailed one-point simulation shows, violation of this consistency requirement can result in greatly underestimated land surface evapotranspiration rates; if the carbon balance is also computed, plant productivity can even become contradictory to the observed presence of vegetation. Another advantage is that gaps in the satellite data can easily be taken into account – something that makes traditional methods of satellite data use complicated – and that model predictions of fAPAR can effectively be used to filter out residual noise in the satellite data.

In a further analysis, focussing on the effect of the data assimilation on simulated soil moisture, it is shown that consistency with satellite derived fAPAR requires rather large values of maximum soil water storage in much of the tropics. Even though the initial, a priori map of soil water storage capacity taken from the ECHAM-4 general circulation model shows rather high values for maximum plant-available soil water – around 250 to 400 mm in the tropics, against only 160 mm globally for the previous ECHAM-3 –, optimised values are even higher and often exceed 500 mm. This has a significant impact on the simulated soil moisture content during the dry season, and for the simulated land climate in the ECHAM-4 GCM. While those changes in simulated 2 m land temperatures are relatively small when compared to differences between GCM and observed temperatures, the changes still compensate for some of the tendency of ECHAM-4 to simulate dry-season temperatures that are too high.

It is certainly true that much of the detail of the results presented here will depend on the qualitity of the precipitation data used to drive the vegetation and land surface model. The data used here are a widely accepted standard for global vegetation models and have been used extensively. However, coverage of meteorological stations can be sparse in parts of the tropics, and some of the very high values of soil moisture storage might be sensitive to possible data problems. Nevertheless, the general tendency of large tropical water reservoirs has also been found from other investigations, based on rooting depth [*Canadell et al.*, 1996; *Nepstad et al.*, 1994], optimised vegetation growth [*Kleidon and Heimann*, 1998], or the atmospheric moisture balance over the Amazon basin [*Zeng*, 1999]. This result demonstrates the importance of vegetation for the global land surface water budget, and stresses the importance of including vegetation effects in GCMs. If vegetation is represented as an interactive part of the land surface such that its distribution and amount are predicted, satellite derived greeness can also be used as check on the realism of the simulated climate and surface parameterisation. The present study is thus intended as a first step in that direction.

There is a further issue related to data availability in land surface and large-scale hydrological modelling. Separate attempts to use microwave remote sensing data to estimate soil moisture have so far only achieved limited results, while similar assimilation techniques have been developed for satellite derived surface temperatures (see above). The advantage of the technique presented here is that, since soil moisture, skin temperature and vegetation cover are all predicted by the surface model, these parameters can in principle be used for integrating diverse types of remotely sensed information into a common data assimilation scheme.

## 7. ACKNOWLEDGMENTS

The idea for developing the methodology presented here is the result of numerous discussions the first author has had with members of the remote sensing and vegetation modelling groups at the EU Joint Research Centre, at

Ispra, namely Bernard Pinty, Michel Verstraete, and Philippe Martin. Support by Wolfgang Cramer and Andrew Friend, who have provided climate data fields, is also greatfully acknowledged.

# 8. REFERENCES

Anderson, E., J. Pailleux, J.-N. Thépaut, J.R. Eyre, A.P. McNally, G.A. Kelly, and P. Courtier, Use of cloud-cleared radiances in three/four-dimensional variational data assimilation, *Quaterly Journal of the Royal Meteorological Society*, *120*, 627-653, 1994.

Asrar, G., M. R.B., and B.J. Choudhury, Spatial heterogeneity in vegetation canopies and remote sensing of absorbed photosynthetically active radiation: a modeling study, *Remote Sensing of Environment*, *41*, 627-653, 1994.

Beerling, D.J., and W.P. Quick, A new technique for estimating rates of carboxylation and electron transport in leaves of C3 plants for use in dynamic global vegetation models, *Global Change Biology*, *1*, 289-294, 1995.

Berthelot, B., G. Dedieu, F. Cabot, and S. Adam, Estimation of surface reflectance and vegetation index using NOAA/AVHRR: Methods and results at global scale, in *6th international symposium on physical measurements and signatures in remote sensing*, ISPRS, Val d'Isère, France, 1994.

Blyth, K., The use of microwave remote sensing to improve spatial parameterization of hydrological models, *Journal of Hydrology*, *152*, 103-129, 1993.

Box, E.O., *Macroclimate and plant forms: An introduction to predictive modeling in phytogeography*, 174 pp., Junk, Den Haag, 1981.

Brutsaert, W., *Evaporation into the atmosphere*, 299 pp., Reidel, Dordrecht, The Netherlands, 1982.

Budyko, M.I., *Heat balance of the Earth's surface*, 255 pp., 1956.

Budyko, M.I., *Climate and Life*, 508 pp., Academic Press, New York, 1974.

Budyko, M.I., and A.A. Sokolov, *Water Balance of the Earth*, 663 pp., UNESCO Press, 1978.

Canadell, J., R.B. Jackson, J.R. Ehleringer, H.A. Mooney, D.E. Sala, and E.D. Schulze, Maximum rooting depth of vegetation types at the global scale, *Oecologia*, *108*, 583-595, 1996.

Charney, J.G., W.J. Quirk, S.-H. Chow, and J. Kornfield, A comparative study of the effects of albedo change on drought in semi-arid regions, *Journal of Atmospheric Sciences*, *34*, 1366-1385, 1977.

Charney, J.G., P.H. Stone, and W.J. Quirk, Drought in the Sahara: A biogeophysical feedback mechanism, *Science*, *187*, 434-435, 1975.

Chase, T.N., R.A. Pielke, T.G.F. Kittel, R.R. Nemani, and S.W. Running, Simulated impacts of historical land cover changes on global climate in northern winter, *Climate Dynamics*, *16*, 93-105, 2000.

Christensen, J.H., B. Machenhauer, R.G. Jones, C. Schär, P.M. Ruti, M. Castro, and G. Visconti, Validation of present-day regional climate simulations over Europe - LAM simulations with observed boundary conditions, *Climate Dynamics*, *13*, 489-506, 1997.

Claussen, M., U. Lohmann, E. Roeckner, and U. Schulzweida, A global data set of land-surface parameters, Max-Planck-Institut für Meteorologie, Hamburg, Germany, 1994.

Claussen, M., U. Lohmann, E. Roeckner, and U. Schulzweida, The greening of the Sahara during the mid-holocene - results of an interactive atmosphere-biome model, *Global Ecology and Biogeography Letters*, *6*, 369-377, 1997.

Collatz, G.J., M. Ribas-Carbo, and J.A. Berry, Coupled photosynthesis-stomatal conductance model for leaves of C4 plants, *Australian Journal of Plant Physiology*, *19*, 519-538, 1992.

Dickinson, R.E., A. Henderson-Sellers, P.J. Kennedy, and M.F. Wilson, Biosphere-Atmosphere Transfer Scheme (BATS) for the NCAR Community Climate Model, National Center for Atmospheric Research, Boulder, CO, 1986.

Farquhar, G.D., S. von Caemmerer, and J.A. Berry, A biochemical model of photosynthesis in leaves of C3 species, *Planta*, *149*, 78-90, 1980.

Federer, C.A., Transpirational supply and demand: plant, soil, and atmospheric effects evaluated by simulation, *Water Resources Research*, *18*, 355-362, 1982.

Flasse, S., and M.M. Verstraete, Monitoring the environment with vegetation indices: comparison of NDVI and GEMI using AVHRR data over Africa, in *Vegetation, Modelling and Climate Change Effects*, edited by F. Veroustraete, and R. Ceulemans, pp. 107-135, Academic Publishing, The Hague, The Netherlands, 1994.

Foley, J.A., I.C. Prentice, N. Ramankutty, S. Levis, D. Pollard, S. Sitch, and A. Haxeltine, An integrated biosphere model of land surface processes, terrestrial carbon balance, and vegetation dynamics, *Global Biogeochemical Cycles*, *10* (4), 603-628, 1996.

Ganopolski, A., C. Kubatzki, M. Claussen, V. Brovkin , and V. Petoukhov, The influence of vegetation-atmosphere-ocean interaction on climate during the mid-holocene, *Science*, *280*, 1916-1919, 1998.

Garratt, J.R., Sensitivity of climate simulations to land-surface and atmospheric boundary-layer treatments - a review, *Journal of Climate*, *6*, 419-449, 1993.

Geiger, R., *The Climate near the Ground*, Harvard University Press, Cambridge MA, 1965.

Geiger, R., R.H. Aron, and P. Todhunter, *The Climate near the Ground*, 528 pp., Vieweg-Verlag, Braunschweig, Germany, 1995.

Geng, S., F. Penning de Vries, and I. Supit, A simple method for generating daily rainfall data, *Agricultural and Forestry Meteorology*, *36*, 363-376, 1986.

Gobron, N., B. Pinty, M.M. Verstraete, and Y. Govaerts, A semi-discrete model for the scattering of light by vegetation, *Journal of Geophysical Research*, *102*, 9431-9446, 1997.

Goel, N.S., and W. Qin, Influences of canopy architecture on relationships between various vegetation indices and LAI and FPAR: A computer simulation, *Remote Sens. Rev.*, *10*, 309-347, 1994.

Grace, J., J. Lloyd, J. McIntyre, A. Miranda, P. Meir, H. Miranda, C. Nobre, J. Moncrieff, J. Massheder, Y. Mahli, I. Wright, and J. Gash, Carbon dioxide uptake by an undisturbed tropical rain forest in South-West Amazonia, *Science*, *270*, 778-780, 1995.

Gutman, G., Numerical experimants on land surface alterations with a zonal model allosin for interaction between geobotanic state and climate, *Journal of Atmosperic Sciences*, *41*, 2679-2685, 1984.

Gutman, G., D. Tarpley, A. Ignatov, and S. Olson, The enhanced NOAA global land dataset from the Advanced Very High Resolution Radiometer, *Bulletin of the American Meteorological Society*, *76*, 1141-1156, 1995.

Gutman, G., D. Tarpley, A. Ignatov, and S. Olson, The relative merit of cloud/clear identification in the NOAA/NASA Pathfinder AVHRR 10-day composites, *International Journal of Remote Sensing*, *17*, 3295-3304, 1996.

Haxeltine, A., and I.C. Prentice, BIOME3: an equilibrium terrestrial biosphere model based on ecophysiological constraints, resource availability, and competition among plant functional types, *Global Biogeochemical Cycles*, *10* (4), 693-709, 1996.

Henderson-Sellers, A., K. McGuffie, and A.J. Pitman, The Project for intercomparison of land-surface parameterization schemes (PILPS): 1992 to 1995, *Climate Dynamics*, *12*, 849-859, 1996.

Holben, B.N., Characteristics of maximum-value composite images from temporal AVHRR data, *International Journal of Remote Sensing*, *7*, 1417-1434, 1986.

Holdridge, L.R., Determination of world formations from simple climatic data, *Science*, *105*, 193-215, 1947.

Jarvis, P.G., and K.G. McNaughton, Stomatal control of transpiration: scaling up from leaf to region, *Advances in Ecological Research*, *15*, 1-49, 1986.

Jones, H.G., *Plants and Microclimate*, 323 pp., Cambridge University Press, Cambridge, U.K., 1983.

Kelliher, F.M., R. Leuning, and E.-D. Schulze, Evaporation and canopy characteristics of coniferous forests and grasslands., *Oecologia*, *95*, 152-163, 1993.

Kleidon, A., and M. Heimann, Optimised rooting depth and its impacts on the simulated climate of an atmospheric general circulation model, *Geophysical Research Letters*, *25*, 345-348, 1998.

Knorr, W., Satellitengestützte Fernerkundung und Modellierung des globalen $CO_2$-Austauschs der Landvegetation: Eine Synthese, Max-Planck-Institut für Meteorologie, Hamburg, Germany, 1997 (available in English through http://www.bgc-jena.mpg.de/~wolfgang.knorr).

Knorr, W., Annual and interannual $CO_2$ exchanges of the terrestrial biosphere: process-based simulations and uncertainties, *Global Ecology and Biogeography*, *9*, 225-252, 2000.

Knorr, W., N. Gobron, P. Martin, B. Pinty, M.M. Verstraete, and G. Dedieu, Constraining a climate driven vegetation model with satellite data, in *International Colloquium on Photosynthesis and Remote Sensing*, edited by G. Guyot, pp. 269-279, EARSEeL, Montpellier, France, 1995.

Le Dimet, F.X., and O. Talagrand, Variational algorithms for analysis and assimilation of meteorological observations: theoretical aspects, *Tellus A*, *38*, 91-110, 1986.

Lean, J., and D.A. Warrilow, Simulation of the regional climatic impact of Amazon deforestation, *Nature*, *342*, 411-413, 1989.

Leemans, R., and W. Cramer, The IIASA climate database for mean monthly values of temperature, precipitation and cloudiness on a terrestrial grid, Institute of Applied Systems Analysis, Laxenburg, Austria, 1991.

Legates, D.R., and C.J. Willmott, Mean seasonal and spatial variability in global surface air temperature, *Theoretical and Applied Climatology*, *41*, 11-21, 1990a.

Legates, D.R., and C.J. Willmott, Mean seasonal and spatial variability in gauge-corrected, global precipitation, *International Journal of Climatology*, 10, 111-127, 1990b.

Leprieur, C., M.M. Verstraete, and B. Pinty, Evaluation of the performance of various vegetation indices to retrieve vegetation cover from AVHRR data, *Remote Sens. Rev.*, *10*, 265-284, 1994.

Los, S.O., C.O. Justice, and C.J. Tucker, A global 1° by 1° NDVI data set for climate studies derived from the GIMMS continental NDVI data, *International Journal of Remote Sensing*, *15*, 3493-3518, 1994.

Manabe, S., Climate and the ocean circulation. I. The atmospheric circulation and the hydrology of the earth's surface, *Monthly Weather Review*, *97*, 739-774, 1969.

McNider, R.T., A.J. Song, D.M. Casey, P.J. Wetzel, W.L. Crosson, and R.M. Rabin, Towards a dynamic-thermodynamic assimilation of satellite surface temperature in numerical atmospheric models, *Monthly Weather Review*, *122*, 2784-2803, 1994.

Meyer, D., M.M. Verstraete, and B. Pinty, The effect of surface anisotropy and viewing geometry on the estimation of NDVI from AVHRR, *Remote Sensing Reviews*, *12*, 3-27, 1995.

Milly, P.C.D., and K.A. Dunne, Sensitivity of the global water cycle to the water-holding capacity of land, *Journal of Climate*, *7*, 506-526, 1994.

Mintz, Y., The sensitivity of numerically simulated climates to land-surface boundary conditions, in *The Global Climate*, edited by J.T. Houghton, pp. 79-105, Cambridge University Press, Cambridge, U.K., 1984.

Monteith, J.L., Evaporation and environment, *Symposium of the Society for Experimental Biology*, *19*, 205-234, 1965.

Müller, M.J., *Selected climatic data for a global set of standard stations for vegetation science*, Junk, Den Haag, The Netherlands, 1982.

Nepstad, D.C., C.R. de Carvalho, E.A. Davidson, P.H. Jipp, P.A. Lefebvre, G.H. Negeiros, E.D. da Silva, T.A. Stone, S.E. Trumbore, and S. Vieira, The role of deep roots in the hydrological and carbon cycles of Amazonian forests and pastures, *Nature*, *372*, 666-669, 1994.

Olson, J.S., J.A. Watts, and L.J. Allison, Carbon in live vegetation of major world ecosystems, Oak Ridge National Laboratory, Oak Ridge, 1983.

Ottle, C., and D. Vijal-Madjar, Assimilation of soil moisture inferred from infrared remote sensing in a hydrological model over the HAPEX-MOBILHY region, *Journal of Hydrology*, *158*, 241-264, 1994.

Patterson, K.A., Global distributions of total and total-available soilwater-holding capacities, M.S. thesis, 119 pp., University of Delaware, Newark DE, 1990.

Pinker, R.T., and I. Laszlo, Global distribution of photosynthetically active radiaton as observed from satellites, *Journal Climate*, *5*, 56-65, 1992.

Pinty, B., and M.M. Verstraete, GEMI: A non-linear index to monitor global vegetation from satellites, *Vegetatio*, *101*, 1335-1372, 1992.

Polcher, J., and K. Laval, The impact of African and Amazonian deforestation on tropical climate, *Journal of Hydrology*, *155*, 389-405, 1994.

Potter, S.C., J.T. Randerson, C.B. Field, P.A. Matson, P.M. Vitousek, H.A. Mooney, and S.A. Klooster, Terrestrial ecosystem production: a process model based on global satellite and surface data, *Global Biogeochemical Cycles*, *7*, 811-841, 1993.

Press, W.H., S.A. Teukolsky, W.T. Vetterling, and B.P. Flannery, , in *Numerical Recipes in Fortran*, pp. 402-406, Cambridge University Press, Cambridge, U.K., 1992.

Prince, S.D., A model of regional primary productivity for use with coarse resolution satellite data, *International Journal of Remote Sensing*, *12*, 1313-1330, 1991.

Rahman, H., and G. Dedieu, SMAC: A simplified method for the atmospheric correction of satellite measurements in the solar spectrum, *International Journal of Remote Sensing*, *15*, 123-143, 1994.

Ritchie, J.T., Model for predicting evaporation from a row crop with incomplete cover, *Water Resourcers Research*, *8*, 1204-1213, 1972.

Roeckner, E., K. Arpe, L. Bengtsson, M. Christoph, M. Claussen, L. Dümenil, M. Esch, M. Giorgetta, U. Schlese, and U. Schulzweida, The atmospheric general circulation model ECHAM4: Model description and simulation of present-day climate, Max-Planck-Institut für Meteorologie, Hamburg, Germany, 1996.

Rosenberg, N.J., *Microclimate: The Biological Environment*, 315 pp., Wiley, New York, 1974.

Ruimy, A., G. Dedieu, and B. Saugier, TURC: A diagnostic model of continental gross primary productivity and net primary productivity, *Global Biogeochemical Cycles*, *10*, 269-285, 1996.

Ryan, M.G., Effects of climate change on plant respiration, *Ecological Applications*, *1*, 157-167, 1991.

Schulz, J.-P., Dümenil, L., J. Polcher, C.A. Schlosser, and Y. Xue, Land surface energy and moisture fluxes: Comparing three models, *Journal of Applied Meteorology*, *37*, 288-307, 1998.

Schulze, E.-D., F.M. Kelliher, C. Körner, J. Lloyd, and R. Leuning, Relationships among maximum stomatal conductance, ecosystem surface conductance, carbon assimilation rate, and plant nitrogen nutrition: a global ecology scaling exercise, *Ann. Rev. Ecol. Syst.*, *25*, 629-660, 1994.

Sellers, P.J., Canopy reflectance, photosynthesis and transpiration, *International Journal of Remote Sensing*, *6*, 1335-1372, 1985.

Sellers, P.J., Bounoua, L., G.J. Collatz, D.A. Randall, D.A. Dazlich, S.O. Los, J.A. Berry, I. Fung, C.J. Tucker, C.B. Field, and T.G. Jensen, Comparison of radiative and physiological effects of doubled atmospheric $CO_2$ on climate, *Science*, *271*, 1402-1406, 1996.

Sellers, P.J., Y. Mintz, Y.C. Sud, and A. Dalcher, A Simple Biosphere Model (SiB) for use within general circulation models, *Journal of Atmospheric Science*, *43*, 505-53, 1986.

Shukla, J., and Y. Mintz, Influence of land-surface evapotranspiration on the earth's climate, *Science*, *215*, 1498-1501, 1982.

Shukla, J., C. Nobre, and P.J. Sellers, Amazon deforestation and climate change, *Science*, *247*, 1322-1325, 1990.

Smith, W.L., H.M. Woolf, C.M. Hayden, D.Q. Wark, and L.M. McMillin, The TIROS-N Operational Vertical Sounder, *Bulletin of the American Meteorological Society*, *60*, 1177-1187, 1979.

Sud, Y.C., J. Shukla, and Y. Mintz, Influence of land surface roughness on atmospheric circulation and precipitation: A sensitivity study with a general circulation model, *Journal of Applied Meteorology*, *27*, 1036-1054, 1988.

van den Hurk, B.J., W. Bastiaanssen, H. Pelgrum, and E. Meijgaard, A new methodology for assimilation of initial soil moisture fields in weather prediction models using Meteosat and NOAA data, *Journal of Applied Meteorology*, *36*, 1271-1283, 1997.

Verma, S.B., D.D. Baldocchi, D.E. Anderson, D.R. Matt, and R.J. Clement, Eddy fluxes of $CO_2$, water vapor and sensible heat over a deciduous forest, *Boundary Layer Meteorology*, *36*, 71-91, 1986.

Verstraete, M.M., Land surface processes in climate models: status and prospects, in *Climate and Geo-Sciences*, edited by A. Berger, S. Schneider, and J.C. Duplessy, pp. 321-340, Kluwer, Dordrecht, The Netherlands, 1989.

Verstraete, M.M., Retrieving canopy properties from remote sensing measurements, in *Imaging Spectrometry - a tool for Environmental Observations*, edited by J. Hill, and J. Mégier, pp. 109-123, ECSC, EEC, EAEC, Brussels and Luxemburg, 1994.

Verstraete, M.M., and B. Pinty, Designing optimal spectral vegetation indices for remote sensing applications, *IEEE Transactions in Geoscience and RemoteSensing*, *34*, 1254-1265, 1996.

Verstraete, M.M., and B. Pinty, Environmental information extraction from satellite remote sending data, in *Inverse Methods in Global Biogeochemical Cycles*, edited by P. Kasibhatla et al., pp. 125-137, American Geophysical Union, Washington D.C., 2000.

Verstraete, M.M., B. Pinty, and R.E. Dickinson, A physical model of the bidirectional reflectance of vegetation canopies. 1. Theory, *Journal of Geophysical Research*, *95*, 11775-11765, 1990.

Viterbo, P., and A.C.M. Beljaars, An improved land surface parameterization scheme in the ECMWF model and its validation, *Journal of Climate*, *8*, 2716-2748, 1995.

Weiss, A., and J.A. Norman, Partitioning solar radiation into direct and diffuse, visible and near-infrared components, *Agricultural and Forestry Meteorology*, *34*, 205-213, 1985.

Wilson, M.F., and A. Henderson-Sellers, A global archive of land cover and soils data for use in general circulation models, *Journal of Climate*, *5*, 119-143, 1985.

Zeng, N., Seasonal cycle and interannual variability in the Amazon hydrologic cycle, *Journal of Geophysical Research*, *104* (D8), 9097-9106, 1999.

# The Use of Remotely-sensed Data for the Estimation of Energy Balance Components in a Mountainous Catchment Area

P.A. BRIVIO, R. COLOMBO and M. MERONI
*Remote Sensing Dept., IRRS - CNR, Milan, Italy*

Abstract: The knowledge of the spatial distribution of biophysical parameters related to the surface energy balance, such as surface albedo and surface temperature, is of great interest for various applications, such as the modelling of atmospheric behaviour and the monitoring of water resources.

Satellite-based remotely-sensed data may provide an important contribution in the estimation of energy fluxes, at the surface-atmosphere interface, through the determination of biophysical parameters in a distributed way.

In this study the determination of actual evapotranspiration is estimated as a key input to the hydrological balance at catchment scale. The experiment was conducted using high resolution satellite data of Landsat Thematic Mapper in an high mountainous catchment (Valmasino) of the Italian Alps. The watershed surface covers an area of 188 km$^2$ and elevation ranges from 250 m to 3650 m, including different land cover types from prairie to forest.

Remotely-sensed images were integrated with ground based meteorological measurements and with a Digital Elevation Model in a GIS environment to solve latent heat flux as residual term of the one-dimensional surface energy balance equation.

Daily values of evapotranspiration, estimated from spatially distributed instantaneous latent heat fluxes, are compared with daily rate of actual evapotranspiration computed according to the Priestley-Taylor and Penmann-Monteith methods.

## 1. INTRODUCTION

Although it is widely recognised that many aspects of land-atmosphere interactions greatly affect the forecasting capabilities of both general and regional circulation models, comprehension of the landscape patchiness of



*M. Beniston and M.M. Verstraete (eds.),*
*Remote Sensing and Climate Modeling: Synergies and Limitations*, 307-327.

water and heat exchange processes at the land surface is limited by the inadequate amount of data generally available in rugged mountainous terrain. The land surface characteristics in Alpine regions present complex interaction and high variability on short intervals of space and time, giving rise to difficult problems in the modelling of related processes.

According to the International Convention on the Protection of the Alps (CIPRA, 1991), the Alpine environment is one of the most sensitive European terrestrial ecosystems. Aggressive development of tourism, long-term effects arising from changes in climate and loss of a protective forest layer could lead to an increase in the frequency and severity of natural disasters. Under such conditions the Convention demands immediate comprehensive counter-measures.

The processes which require more thorogh invesitgation within Alpine environments, at different spatial and temporal scales, have been identified as a result of the experience gained over the last few decades in modelling the evolution of the geosphere/biosphere system and its links to climate (Rott and Rast, 1999).

Monitoring of complex areas by remotely-sensed imagery and ground data may provide an important contribution to develop and test new techniques for land surface parameterisations, for the evaluation of soil-vegetation-atmosphere-transfer (SVAT) schemes, and for biogeochemical (BGC) remote sensing (RS) driven model input (Olioso et al., 1999; Running et al., 1999). Many biophysical and biochemical parameters can be realistically derived from remotely-sensed data and employed as input in distributed SVAT and BGC models for determining the energy balance components and estimating the actual evapotranspiration.

Evapotranspiration is generally estimated by conventional ground-based methods, such as the Bowen ratio coupled with a net radiometer (Spittelehouse and Blac, 1980) or eddy covariance technique measurements (Baldocchi et al., 1988). However, although these techniques do provide accurate measurements over an homogeneous area surrounding the meteorological station, the results are not directly applicable to large areas or to natural heterogeneous surfaces. Satellite RS can offer the way for deriving biophysical and biochemical parameters without the loss of accuracy associated with spatial interpolation techniques among point measurements.

Several researchers have applied RS data for estimating evapotranspiration by using different modelling techniques (Moran et al., 1989; Goodin, 1995; Pegrum and Bastiaanssen, 1996; Kustas et al., 1989; Anderson et al., 1996; Hall et al., 1991), with successful results over semiarid rangeland basins (Chehbouni et al., 1997; Hurtado et al., 1997; Kustas et al.1994), and over nearly full agricultural canopy covers (Hurtado et al., 1994). However, few experiments have been made in highly forested complex terrain (Kaneko and Hino, 1996; Vidal et al., 1994; Wigmosta et al. 1994) and in heterogeneous Alpine environments (Kaimal and Finnigan, 1994; Menzel and Lang, 1998).

In this study, one such experiment conducted in a mountainous forested environment is presented, and related problems are analysed. Instantaneous latent heat fluxes were mapped for all vegetated surfaces of the Valmasino catchment in the Italian Alps by integrating RS data and ground measurements in a single-layer resistance model.

## 2. STUDY AREA AND MATERIALS

In order to investigate the land-atmosphere relationships and to model the exchange processes over mountainous vegetated surfaces the Valmasino watershed was selected.

### 2.1 Watershed characteristics

#### 2.1.1 Geographical and morphological characteristics

The Valmasino watershed (Fig.1) is located in the Central Alps at the border between Italy and Switzerland and it is situated between latitudes 46°09' N and 46°17' N and longitudes 9°34' E and 9°45' E.

The upstream drainage area measures 188 $km^2$ with important glaciers located in the upper part. Its physiographical characteristics are formed by four valleys with the main Masino river flowing from north to south and four lower order valley elongated E-W. The Valmasino watershed represents a typical Alpine morphology characterised by great relief energy, high variability of steep slopes, variable aspects and cast shadows. Elevation ranges between 265 m a.s.l. (Ardenno village) and 3678 m a.s.l. at the summit of the Monte Disgrazia peak. As shown in Table 1, more than 90% of the entire catchment surface is located above 1000 meters and more than 70% of surfaces have a slope greater than 25°.

*Table 1.* Morphometrical characteristics of the Valmasino watershed

| Altitude (m a.s.l.) | Area (%) | Slope (%) | Area (%) | Aspect (°) | Area (%) |
|---|---|---|---|---|---|
| 265-1000 | 8 | 0-25 | 5 | 315-45 | 17 |
| 1000-2000 | 36 | 25-50 | 24 | 45-135 | 24 |
| 2000-3000 | 53 | 50-80 | 31 | 135-225 | 35 |
| > 3000 | 3 | > 80 | 40 | 225-315 | 24 |

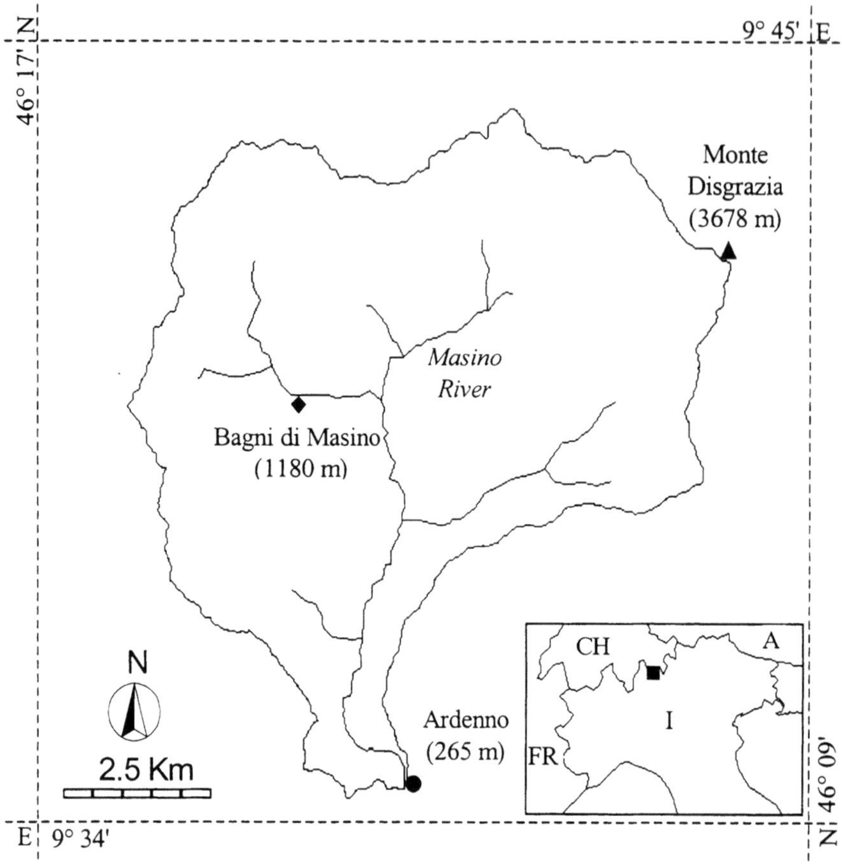


*Figure 1.* The Valmasino catchment experiment site located in the Central Italian Alps

### 2.1.2 Climatological and hydrological characteristics

The meteorological station of Sondrio located at 298 m a.s.l. collected the most important time series data set, beginning in 1958 (ERSAL, 1992). The mean annual air temperature is 11 °C. The isothermal trend describes the role of the slope exposure, with the eastern slopes colder and the related agriculture activities taking place at lower elevations. The vertical thermal gradient is strongly affected by inversion phenomena. Annual potential evapotranspiration is 696 mm. Mean annual rainfall is 948 mm with the rainy season in summer (maximum monthly rainfall depth: 106 mm, in August) and a dry season during winter, with a long period of scarce rainfall (minimum monthly rainfall depth: 39 mm, in February). Extreme events can be considered to occur when more than 100 mm of rain falls in 24 hours. The hydrological regime of the Masino river is typically Alpine, showing maximum water flow during the late spring-early summer and minimum flow during the winter season.

### 2.1.3 Land cover and vegetation characteristics

The Valmasino catchment is sparesely urbanised and the dominant land use is devoted to pasture. The alluvial plain is dedicated to agricultural activities, mainly maize and staple meadows. Slopes and high areas are covered by spontaneous vegetation. Forested areas, including deciduous and evergreen forests, cover about 30%, in the remnant areas rock outcrops and debris are dominant.

Vegetation in the study area can be schematised in three zones changing with altitude (Giacomini, 1960). The lower part of the catchment includes the mountainous zone, characterised by fully covered evergreen, deciduous and mixed forests with main tree species represented by *Fagus sylvatica*, *Picea abies* and *Larix decidua*. The sub-Alpine zone is characterised by open woodland and shrubs, with the main trees species of *Pinus mugo*, *Pinus cembra* and *Larix decidua*. Brushes and tundra mainly constitute the vegetation cover of the Alpine zone (above 2000 m), with dominance of *Rhododendron*.

## 2.2 Data used

A digital elevation model (DEM) of the catchment with 30 m x 30 m grid size was derived from digitised vector contour lines from a 1:25,000 scale map, with an equidistance of 50 m. The drainage network and catchment boundary were digitised directly from the topographic map.

Meteorological measurements were collected from the permanent meteorological station located at Bagni di Masino at an elevation of 1180 m. The station is sited on a meadow with a height of 15 cm.

Two Landsat-5 Thematic Mapper (TM) scenes, acquired in the late spring season (30-05-1996 Path 194, Row 28) and in the late winter season (07-03-1997 Path 193, Row 28), were used in this study.

Field surveys were carried out to identify training and testing areas to be used in satellite image supervised classification. Vegetation was characterised during surveys by means of vegetation height and vegetation fractional cover measurements.

# 3. METHODOLOGY

Latent heat flux for the Valmasino watershed was estimated as a residual term of the one-dimensional surface energy balance equation following the procedure schematised in Figure 2.

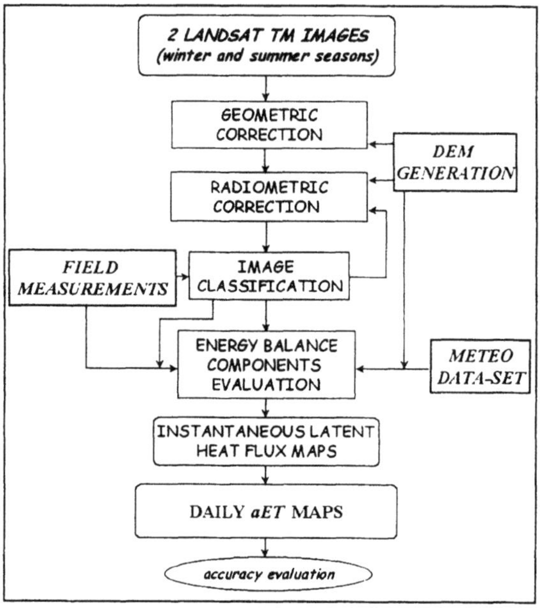


*Figure 2.* Bock diagram illustrating the overall methodology

The methodological approach is specifically based on the direct use of remote sensing derived information for watershed characterisation and modelling. Satellite data were used to extract the land cover map of the area in the two seasons and to retrieve the geophysical parameters controlling the soil-vegetation-atmosphere processes.

Meteorological measurements were spatially distributed taking into account the DEM. Vegetation characteristic parameters vary spatially with the cover type, therefore the watershed was stratified into different vegetation classes in order to assign to each class the corresponding value of vegetation height, vegetation fractional cover and surface emissivity.

Satellite extracted geophysical parameters, meteorological data and vegetation related parameters were integrated in a GIS (Geographical Information System) environment to evaluate energy balance components and daily evapotranspiration at the catchment scale.

## 3.1 Data preparation

Quantitative exploitation of satellite images in high mountainous areas requires a rigorous procedure of data correction for removing both the geometric and radiometric distortions; consequently, an accurate DEM is necessary. The accuracy of the DEM was evaluated according to two criteria (Carrara et al. 1997) concerning visual analysis (presence of artificial elements) and the comparison between the original contour line and those obtained from contour generation algorithms applied to the DEM itself.

### 3.1.1 Geometric correction of TM data

The Landsat-5 platform orbits at a height of 705 km, and since the scene detected is approximately 185 km wide, the greatest distance from the nadir point is 3083 pixels. A difference in elevation of 2000 m from the reference level, causes a pixel located at east or west boundary of the scene a position displacement from the orthogonal projection of about 9 pixels, i.e. 270 m. Such a geometric distortion cannot be corrected by applying polynomial equations based on a set of ground control points, as is usually done in geo-referencing techniques (Richards, 1986). In areas of high relief, pixel displacement corrections must be performed by applying an othorectification method with the DEM available. The Valmasino DEM with a 30 m grid size was employed to correct Landsat TM scenes for the relief displacement by using an orthorectification procedure and satellite images were thus georefe-renced to the UTM Zone 32. A nearest-neighbour resampling technique was used to retain radiometric integrity. The root-mean-square error turned out to be less than 1 pixel for both TM images.

Figure 3 shows a three dimensional representation of the Valmasino watershed. Band 3 of the TM scene acquired on May draped over the DEM gives an impression of the morphological complexity of a typical Alpine catchment.

### 3.1.2 Radiometric correction of TM data

Satellite TM data were then radiometrically calibrated and atmospherically corrected both in the reflected and thermal domains. In rugged terrain, radiometry must be corrected for atmospheric effects, accounting for the different thickness of the atmospheric layer (depending on the target altitude) that the radiation has to go through, and for the effects of the topography accounting for the relative angle between the normal vector of the surface target and sun illumination.

Atmospheric and topographic correction in the reflected domain was performed by improving the method proposed by Gilabert et al., (1995). The approach consists of an inversion algorithm based on simplified radiative transfer models in which the optical characteristics, necessary for the atmospheric correction, are estimated by using a combination of TM band 1 (0.45-0.52 μm) and TM band 3 (0.63- 0.69 μm) signals and assuming the existence of a number of dark surfaces within the scene. A horizontally-homogeneous atmosphere was assumed, and the total atmospheric transmittance changes with the altitude of the target. Such a hypothesis allow an approximate knowledge of the atmosphere composition and structure and the retrieval of the actual surface reflectance can be carried out without any direct measurement of the optical properties of the atmosphere.

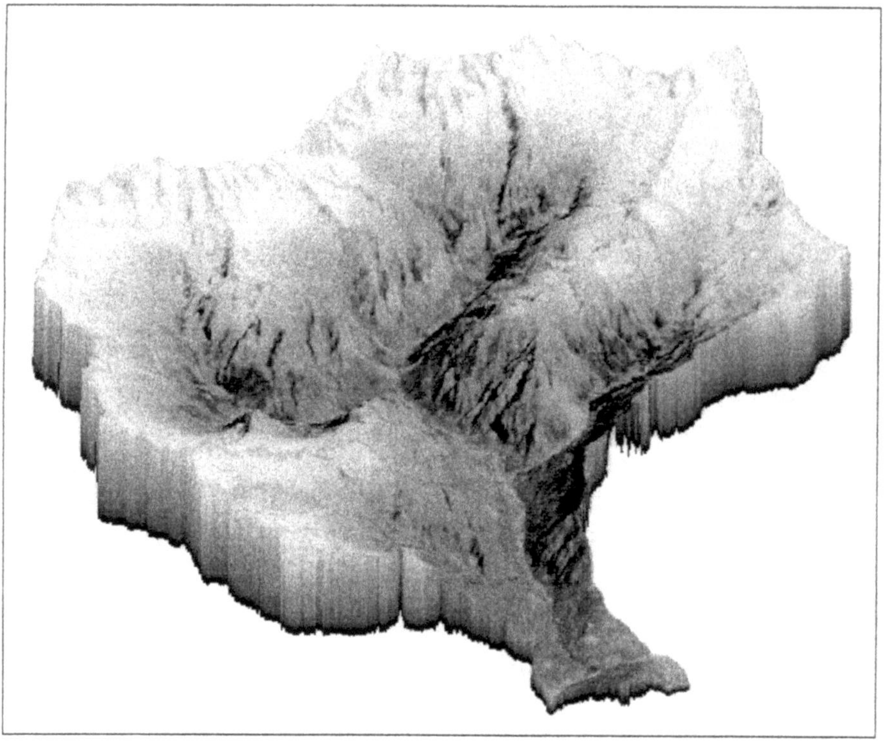

*Figure 3.* Landsat TM scene draped over the DEM, showing the morphological complexity of the Valmasino catchment

Topographic effects on the radiometry of the images were accounted for by considering the incidence angle of sunrays on the surface, and cast shadows. The angle of incidence was obtained by extracting the relative sun position with respect to the target, through its morphometric parameters such as slope and aspect. Cast shadow maps were derived by means of an algorithm of local shadowing horizons search (Dozier, 1981).

Figure 4 shows a sample extracted from TM band 1 depicting the Valmasino valley bottom, south-east and north-west oriented slopes, before and after the atmospheric correction procedure. Digital number variance of north-west oriented slope (shaded by the relief) increases greatly when the correction is applied. In such conditions, broadleaf coniferous change recognition is possible in the corrected image only.

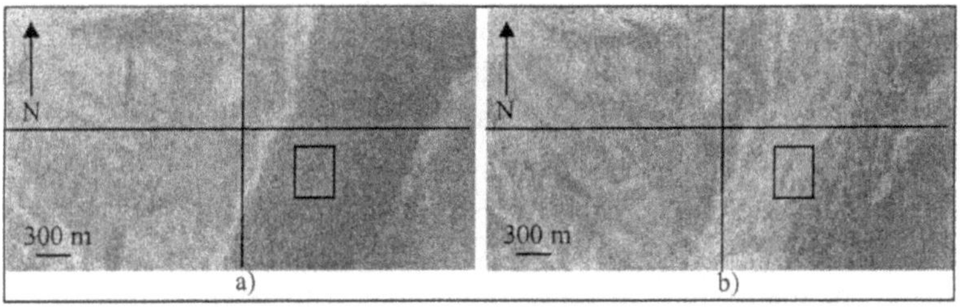


*Figure 4.* Radiometric correction of TM band 1;
a) apparent reflectance, b) corrected reflectance

Apparent and corrected TM-1 reflectance frequency histograms in Figure 5 were extracted from the boxes depicted in Figure 4. Data variance increases and according to theory (path radiance contribution in the radiance detected by the satellite sensor in the blue channel is significative) mean corrected reflectance value is lower than the uncorrected value.

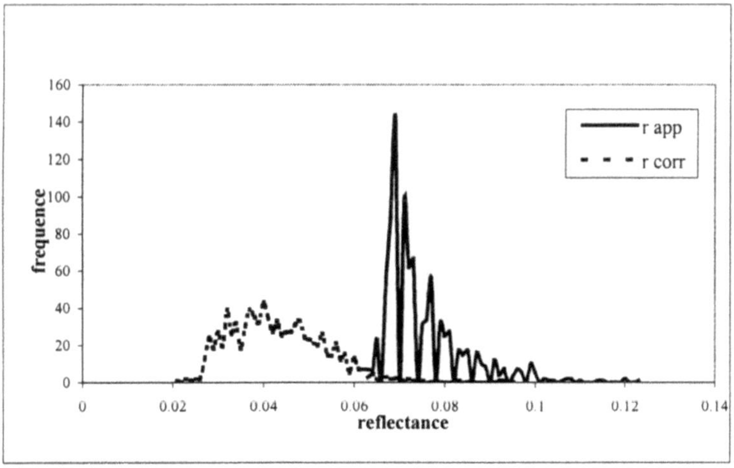


*Figure 5.* Frequency histogram of apparent and atmospherically-corrected reflectance values for the pixels included in the small rectangular boxes of Figure 4

Radiometric calibration of the infrared thermal band (TM band 6: 10.44-12.42 μm) was conducted in accordance to the method proposed by Vukelich (1989):

$$L_{cor} = \frac{L_\lambda - L_p}{\varepsilon\tau} - \left(\frac{1}{\varepsilon} - 1\right) L_{sky} \qquad [\text{mW cm}^{-2}\ \text{sr}^{-1}\ \mu\text{m}^{-1}]$$

where $L_{cor}$ is the corrected radiance, $L_\lambda$ is thermal radiance at satellite, $L_p$ is the path radiance, $L_{sky}$ is the downwelling sky irradiance, $\varepsilon$ is the surface emissivity and $\tau$ is the atmospheric transmittance

Atmospheric path radiance and transmittance were modelled pixel by pixel to the whole image with respect to the heights by using the LOWTRAN-6 code (Kneizys, 1989); as expected, path radiance values decrease with the elevation, while transmittance increases. Downwelling sky irradiance for clear sky in the 10.44-12.42 μm spectral band was estimated by means of the distributed air temperature map. Emissivity values proposed by Rubio (1997) were used to create an emissivity map on the basis of the land cover maps. For mixed classes, such as forest with debris, the concept of the directional r-emissivity (Norman and Becker, 1995) was employed. A weighted emissivity ($\varepsilon_w$), accounting for the thermal characteristics of soil and vegetation components within mixed pixels was employed. Fractional vegetation cover ($f_c$) measurements were related to radiometric vegetation indexes: high linear correlation values were obtained with normalised difference vegetation index (NDVI) for spring TM passage ($r^2$=0.94), and with soil adjusted vegetation index (SAVI) for winter season ($r^2$=0.98) when the contribution of soil/rock reflectance becomes more important; *fc* was used to determine the weighted emissivity as follows:

$$\varepsilon_w = f_c\,\varepsilon_v + (1 - f_c)\varepsilon_s$$

where $\varepsilon_v$ is pure canopy emissivity and $\varepsilon_s$ is pure soil emissivity.

Finally, the corrected radiance was estimated and, by inverting the Planck radiation equation, the hemispherical surface temperature maps were obtained. The transect in Figure 6 shows the difference between the uncorrected ($T_u$) and the corrected hemispherical surface temperature ($T_c$) in relation to the topography and the land cover classes. The snow cover, present in the higher part of the catchment, exhibits a negative difference due to the low emissivity value. Considering values relative to a single class, a reduction in the temperature difference with increasing elevation of the target is highlighted.

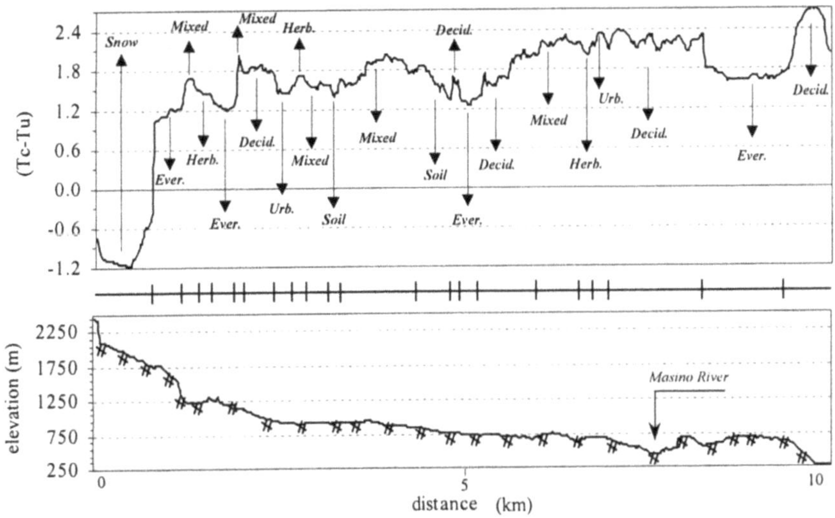


*Figure 6.* Radiometric correction of thermal TM data. Top is the difference between the corrected surface temperature ($T_c$) and the uncorrected ($T_u$); middle are the land cover classes; bottom shows their relationship with the topography

### 3.1.3 Watershed stratification and GIS preparation

The land cover map, needed to stratify the watershed surface, was obtained from supervised classification of geometrically- and radiometrically-corrected reflective TM bands.

A maximum likelihood algorithm trained with information collected during the field surveys and aerial photographs was used as a classifier. Classification accuracy evaluated through a confusion matrix showed a coefficient of agreement *K* equal to 0.85 and 0.97 for May and March scenes, respectively. Differences in *K* values are due to the phenological stages of broadleaf and coniferous forests, that during the winter season show the greatest separability.

Field measurements, literature data, DEM and satellite corrected data were finally arranged in the IDRISI GIS environment (Eastman, 1997).

## 3.2 Energy balance evaluation

The surface energy balance is assessed for each elemental area, defined by the pixel size of remotely-sensed data, using a single-layer resistance model in consideration of the dense canopy cover that characterises the catchment (Hurtado, 1994).

The energy balance equation is defined as:

$$\lambda E = R_n - G - H$$

where $\lambda E$ is the latent heat flux density, a product of the heat of vaporisation $\lambda$ [J kg$^{-1}$] and the rate of evaporation $E$ [Kg s$^{-1}$ m$^{-2}$]; $R_n$ is the net radiant flux density; $G$ is soil heat flux density and $H$ is sensible heat flux density.

### 3.2.1 Net Radiant Flux Density ($R_n$)

Net radiation is the difference between incoming short and long-wave fluxes ($R_{s\downarrow}$ and $R_{L\downarrow}$), and the correspondent outgoing terms ($R_{s\uparrow}$ and $R_{L\uparrow}$), and is expressed as following:

$$R_n = (1-\alpha)R_{S\downarrow} + (1-\varepsilon)R_{L\downarrow} - \varepsilon\sigma T_s^4$$

where $\alpha$ is the surface albedo, $R_{S\downarrow}$ is the incoming short-wave solar radiation, $R_{L\downarrow}$ is the incoming long-wave radiation, $\varepsilon$ is the surface emissivity, $T_s$ is the surface temperature and $\sigma$ is the Stefan-Boltzmann constant.

Incoming fluxes were estimated from meteorological measurements acquired at the time of the satellite overpass. Meteorological data, collected at Bagni di Masino station, were spatially extrapolated to the whole catchment taking into account the DEM.

Short-wave solar radiation ($R_{S\downarrow}$) considers both direct and diffuse radiation components; the direct component was estimated simulating actual illumination conditions, accounting for slope, aspect and cast shadows, at the time of two TM passages, and the diffuse component was estimated from the 6S code (Vermote, 1996). Values simulated with 6S were in good agreement with the general rule of thumb (Meijerink, 1994) that diffuse radiation amounts to 16 % of the total radiation.

Incoming long-wave radiant flux density ($R_{L\downarrow}$) depends on air temperature and vapour pressure. Air temperature data were related to the elevation according to standard thermal gradient procedures, using two local values, namely 0.54 and 0.62 for March and May, respectively (Belloni and Pelfini, 1987).

Vapour pressure was derived on the basis of the relative humidity measurements by calculating the saturated vapour pressure. Vapour pressure values were then extended to the catchment scale by applying a gradient function specific to Alpine environments (Matveev, 1965):

$$e_0(z_0 + z) = e_0(z_0)10^{-\frac{z}{6.3}(z_0+z)} \qquad \text{[mb]}$$

where $z_0$ is the height [km] at which air temperature and relative humidity are measured and $(z_0 + z)$ is the height at which the vapour pressure is to be calculated.

Outgoing fluxes can be successfully estimated in a distributed way from optical and thermal RS data. Outgoing short-wave radiation ($R_{S\uparrow}$) depends on the nature of the reflecting surface and its geometric properties. For clear-sky conditions, broad-band albedo was estimated from radiometrically corrected reflective hands of TM data according Brest and Goward (1987) and Duguay and LeDrew (1992), for fully vegetated ($\alpha_v$) and non-vegetated ($\alpha_s$) surfaces.

For mixed pixels, the broad-band albedo is defined as a weighted function ($\alpha_w$) of the mixture components, vegetation and soil, and it is expressed according to the following:

$$\alpha_w = f_c\, \alpha_v + (1 - f_c)\alpha_s$$

The outgoing long-wave radiant flux density ($R_{L\uparrow}$) depends on the hemispherical surface temperature, that was derived from the corrected TM thermal images as previously described.

### 3.2.2 Sensible Heat Flux Density (H)

Sensible heat flux into the atmosphere ($H$) depends on the difference between the aerodynamic surface temperature ($T_{aero}$) in the canopy, the air temperature ($T_a$) above the canopy, and on the aerodynamic resistance:

$$H = \rho\, c_p \frac{(T_{areo} - T_a)}{r_a}$$

where $\rho$ is the air density, $c_p$ is the specific heat of air at constant pressure and $r_a$ is the aerodynamic resistance.

Although from a theoretical point of view, the sensible heat transfer process refers to the aerodynamic surface temperature (Norman and Becker, 1995), in the case of dense vegetation cover, an acceptable modelling is still possible with the use of hemispherical surface temperature ($T_s$) (Chehbouni et al., 1997), such that a single-layer model can be employed (Hurtado et al., 1994).

Aerodynamic resistance ($r_a$) is a rather complex function of different geometrical and meteorological parameters, such as roughness length, displacement height and wind speed.

Roughness length and displacement height were derived from field measurements of plant heights applying the relationships proposed by Monteith (1973) and assumed invariant for each land cover class. A roughness length of 2.4 cm was calculated for the grass cover located close to the meteorological station, and 176 cm for the forest cover type.

Wind speed recorded at the sensor height was related to the wind speed conditions above the forest canopy, assuming a logarithmic profile for wind velocity (Linsley et al., 1982).

Because of the high relief energy, the air density and the specific heat of air were spatially extrapolated to the various elevations, using the relations proposed by Matveev (1965) and by Haltiner and Martin (1957), respectively.

### 3.2.3 Soil Heat Flux Density (G)

Although there are several relationships for determining the soil heat flux ($G$) by spectral vegetation indexes (Moran et al., 1989; Chehbouni et al., 1997) or by using Leaf Area Index (LAI) as proposed in Kustas and Humes (1996), $G$ can be expressed as a linear function of net radiation $R_n$ depending on vegetation fractional cover. A linear relationship, ranging from $0.3R_n$ for bare soil to $0.1R_n$ for full vegetation cover (Clothier et al., 1986), was applied to the evaluation of $G$.

### 3.2.4 Latent Heat Flux Density (λE) and daily actual evapotransiration (aET$_d$)

Latent heat flux density *($\lambda E$)* was finally modelled as the residual term of the energy balance equation.

The simplified approach proposed by Jackson et al. (1977) appears an appropriate procedure to convert instantaneous remote sensing estimates of $\lambda E$ to actual daily evapotranspiration ($aET_d$), when clear sky and complete canopy cover conditions are satisfied. Seguin and Itier (1983) suggest that a semiempirical coefficient $B_i$, relating instantaneous and daily fluxes, should be retrieved from ground measurements of daily net radiation and actual evapotranspiration. However for clear sky days the use of the evaporative fraction (*EF*) may allow the extrapolation from instantaneous values to daily integrated fluxes. This is due to the strong correlation between the value of evaporative fraction at midday and the daytime average value (Hall et al., 1991; Crago and Brutsaert, 1996). The evaporative fraction is expressed as:

$$EF = \frac{\lambda E_i}{(R_n - G)_i}$$

where the subscript *i* means instantaneous values.

This is important since it implies that the daytime average evaporative fraction might be adequately deduced from a single, instantaneous measurement. Under such assumption the daily rate of actual evapotranspiration can be estimated from the daily available energy:

$$aET_d = EF \times (R_n - G)_d$$

where the subscript *d* means daily values.

Daily Soil Heat Flux Density ($G_d$) was taken as ≈ 0 in a first approximation and net radiation was daily extrapolated and spatially distributed using the relationship proposed by Schwab et. al., (1993). Spatially explicit latent heat of vaporisation, calculated by the mean daily air temperature map, was employed to express the $aET_d$ map in mm/day.

# 4. RESULTS

## 4.1 Daily analysis

Daily actual evapotranspiration maps were produced for the Valmasino catchment on the two dates.

Figure 7 shows a transect of the $aET_d$ obtained from late winter map. The highest values correspond to evergreen forest located in the lower part, below 1000 m, of the Valmasino watershed. An average decrease in evapotranspiration values can also be observed from lower to higher elevations.

From the seasonal $aET_d$ maps obtained, mean values were computed for each vegetation land cover class. These results are presented in Table 2 together with percentages of watershed surface occupied by each vegetation type. A general increment of $aET_d$ values for all vegetation cover is showed passing from late winter to late spring season. During winter time, all the vegetated land covers present similar values of daily evapotranspiration. Changes in evaporative rate in these two seasons reflect the phenological and photosyntetical seasonal patterns. In May, herbaceous covers show $aET_d$ values higher than forested areas and, within forests, evergreen presents values greater than deciduous.

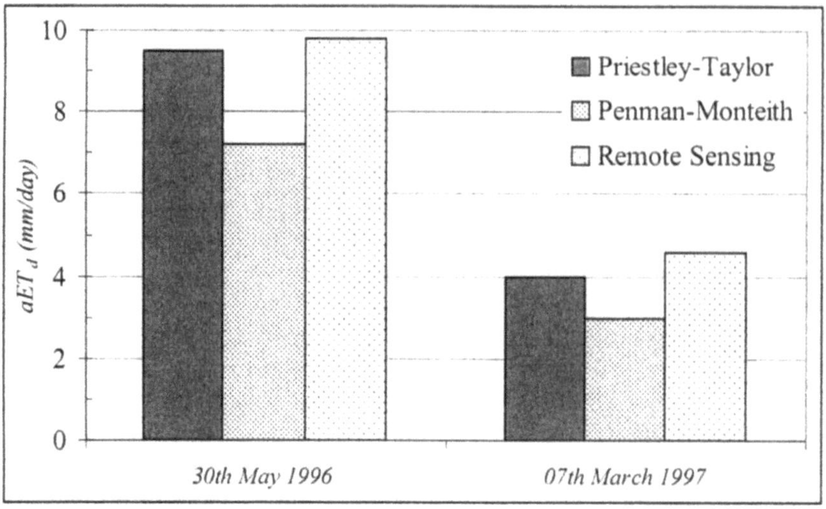


*Figure 7.* Daily actual evapotranspiration ($aET_d$) as derived from remotely-sensed data is compared with traditional methods at the Bagni di Masino station

*Table 2.* Average evapotranspiration aETd values of the land cover classes and correspondent area

| Land cover class | Daily actual evapotranspiration (mm/day) | | Area (%) |
|---|---|---|---|
| | *30th May 1996* | *07th March 1997* | |
| Herbaceous rangeland | 9.7 | 4.2 | 0.3 |
| Herbaceous tundra | 9.9 | 4.0 | 4.2 |
| Deciduous forest | 6.1 | 4.1 | 9.0 |
| Evergreen forest | 7.1 | 4.5 | 11.3 |
| Mixed forest with rock and debris | 6.3 | 3.4 | 7.2 |

## 4.2 Data comparison

As often occurs with models of this type and particularly in mountainous regions, availability of spatially distributed measurements for model verification is very limited. Direct measurements of actual evapotranspiration were not available for Valmasino.

In order to analyse model accuracy, $aET_d$ values were computed for the grass land cover using meteorological measurements available at the Bagni di Masino station on the basis of Priestley-Taylor (1972) and the Penman-Monteith's (Monteith, 1965) equations. Evapotranspiration estimates given by the model using remotely-sensed data in correspondence to the meteorological station were averaged for 3 x 3 pixel area to represent the spatial averaging done by ground sensor and were compared with Priestley-Taylor and the Penman-Monteith values. Histograms in Figure 8 show a substantial agreement among estimates of remote sensing and traditional methodologies for herbaceous rangeland class. Remote sensing estimates are the highest, whereas the Penman-Monteith are the lowest in both seasons. Moreover, differences between results given by traditional equations appear greater than those between $aET_d$ values given by RS and Priestley-Taylor techniques.

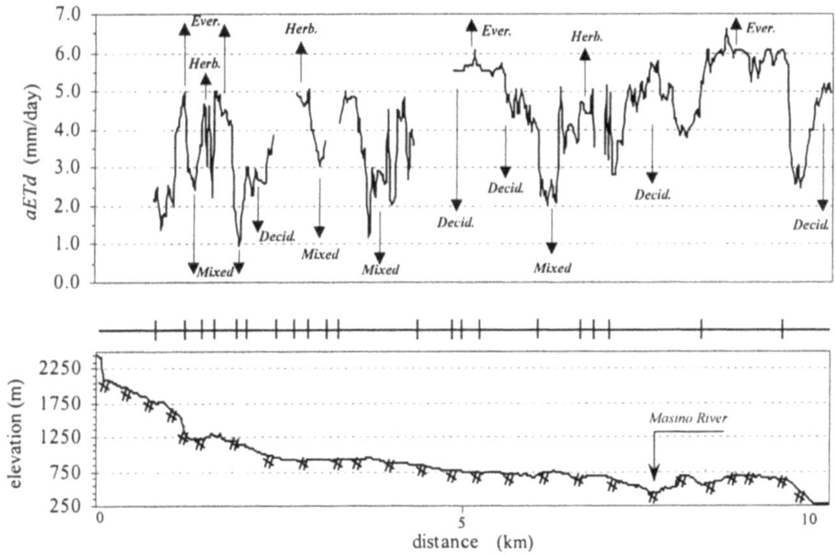


*Figure 8.* Values of $aET_d$ as derived from the winter season Landsat TM image in relation to topography and land cover classes

## 5. CONCLUSIONS

Processes involved in this study were related to the heat transfer at the land-atmosphere interface in an Alpine watershed. Surface characteristics, such as topography and type of land cover, and surface processes, such as incoming and outgoing fluxes, represented the focus of our attention in space and time observation and modelling strategy.

A complete procedure based on a single layer model has been described from space-based and ground-based measurements to the modelled heat and mass transfer exchange fluxes at the surface-atmosphere interface.

Input parameters to the energy balance model were spatially distributed at the catchment scale with a grid cell defined by the pixel size of remotely-sensed data. Biogeophysical parameters, such as outgoing short-wave and long-wave radiation, derived from satellite observations are naturally spatially distributed, while parameters derived from ground observations were distributed using the information of the DEM. Digital elevation data were essential to accurately correct the radiometric signal recorded by the satellite sensor.

Watershed stratification, derived from classification of satellite imagery, allowed to manage other parameters characteristic of the surfaces, such as roughness length and emissivity. Fractional cover served to properly define albedo and emissivity for mixed pixels.

Daily evapotranspiration estimates given by the model were in better agreement with estimates computed using the Priestley-Taylor method rather than that of Penman-Monteith.

Although remotely-sensed data proved their ability in determining variables and parameters, such as vegetation structure information and optical and thermal properties of soil and vegetation for driving SVAT models, further efforts are needed to analyse errors induced by uncertainties in remotely-sensed and meteorological forcing variables, as well as to determine the possible recurrence of space measurements by combining data from different sensors.

## 6. ACKNOWLEDGMENTS

This study was supported by the Lombardia Foundation for the Environment (FLA) and by the National Group for Prevention from Hydrogeological Hazards (GNDCI) of the Italian National Research Council (CNR).

## 7. REFERENCES

Anderson M.C., J.M. Norman, G.R. Diak, W.P. Kustas and J. R. Mecikalski (1996). A Two-Source Time-Integrated Model for Estimating Surface Fluxes Using Thermal Infrared Remote Sensing. Remote Sens. Environ., 60, pp. 195-216.

Belloni S., and M. Pelfini (1987). Il gradiente termico in Lombardia. Acqua-Aria, 4, pp. 441-447.

Baldocchi D.D., B. B. Hicks and T.P. Meyers (1988). Measuring biosphere-atmosphere exchanges of biologically related gases with micrometeorological methods. Ecology, 69, pp. 1331-1340.

Brest C.L., and S.N. Goward (1987). Deriving Surface Albedo from Narrow Band Satellite Data. Int. J. Remote Sensing, 8, pp. 351-367.

Carrara A., G. Bitelli, R. Carlà (1997). Comparison of techniques for generating digital terrain models from contour lines. Int. Journal of Geographical Information System, 5, 11 pp 451-473.

Chehbouni A., J. Qi, D. Lo Seen, Y.H. Kerr, G. Dedieu, S. Moran, M. Daubas and B. M. Monteney (1997). Estimation of Real Evapotranspiration. FAO Water Reports n. 16. Remote Sensing and Water Resources. Proc. of the Int. Workshop, Montpellier, France, 16, pp. 319-328.

CIPRA (1991). La Convenzione delle Alpi. Commissione Internazionale per la Protezione delle Alpi. Salisburgo, 7 Novembre 1991.

Clothier B.E., R.L. Clavson, P.J.Jr. Pinter, M. S. Moran, R. J. Reginato and R. D. Jackson (1986). Estimation of soil heat flux from net radiation during the growth of alfaalfa. Agric. For. Meteorol., 37, pp. 319-329.

Crago R.D. and W. Brutsaert (1996). Daytime evaporation and self-preservation of the evaporative fraction and the Bowen ratio. Journal of Hydrology, 177, pp. 241-255.

Dozier J., J. Bruno, P. Downey (1981). A faster solution to the horizon problem. Computers and Geosciences, 7, pp. 145-151.

Duguay C.R., and E.F. LeDrew (1992). Estimating Surface Reflectance and Albedo from Landsat-5 Thematic Mapper over Rugged Terrain. Photogrammetric Eng. and Remote Sensing, 58, pp. 551-558.

Eastman J.R. (1997). IDRISI for Windows, User's Guide V. 2.0, Clark University, Worcester, MA, USA.

ERSAL (1992). I suoli del Fondovalle Valtellinese. Milano

Giacomini V. (1960). Il paesaggio vegetale della Provincia di Sondrio. Flora et Vegetatio Italica, 3.

Gilabert M.A., C. Conese and F. Maselli (1995). An atmospheric correction method for the automatic retrieval of surface reflectances from TM images. Int. J. Remote Sensing, 15, pp. 2065-2086.

Goodin D. G. (1995). Mapping the surface radiation budget and net radiation in a sand hills wetland using a combined modeling/remote sensing method and Landsat Thematic Mapper imagery. Geocarto International, 2, pp. 19-29.

Hall F.G., P.J. Sellers, D. E. Strebel, E.T. Kanemasu, R.D. Kelly, B.L. Blad, B.J. Markham, J.R. Wang and F. Huemmrich (1991). Satellite Remote sensing of Surface Energy and Mass Balance: Results from FIFE. Remote Sens. Environ., 35, pp. 187-199.

Haltiner G.J. and F.L. Martin (1957). Dynamical and Physics Meteorology. McGraw-Hill Book Company, N.Y., pp. 25.

Hurtado E., M.M. Artigao and V. Caselles (1994). Estimating maize (Zea mays) evapotranspiration from NOAA-AVHRR thermal data in the Albacete area, Spain. Int. J. Remote Sensing, 15(10), pp. 2023-2037.

Hurtado E., M.M. Artigao and V. Caselles (1997). Models for Determining Evapotranspiration. FAO Water Reports n. 16. Remote Sensing and Water Resources. Proceedings of the International Workshop held in Montpellier, France, 16, pp. 305-310.

Jackson R.D., R.J. Reginato and S.B. Idso (1977). Wheat canopy temperature: a practical tool for evaluating water requirements. Water Resources Research, 13, pp. 651-656.

Kaimal J. C. and J. J. Finnigan (1994). Atmospheric Boundary Layer Flows. Oxford University Press, 289 pp.

Kaneko D. and M. Hino (1996). Proposal and investigation of a method for estimating surface energy balance in regional forests using TM derived vegetation index and observatory routine data. Int. J. Remote Sensing, 17 (6), pp. 1129-1148.

Kneizys F.X., F.P. Shettle, W.O. Gallery, J.H. Chetwynd, L.W. Abreu, J.E.A. Selby, S.A. Clough, Fenn R.W. (1989). Atmospheric transmittance/irradiance: computer code LOWTRAN-6. Air Force Geophysic Laboratoy. Hanscom, USA.

Kustas W.P. and K.S. Humes (1996). Sensible heat flux from remotely-sensed data at different resolutions. In Scaling up in hydrology using remote sensing. Ed. by J. B.

Stewart, E.T. Engman, Feddes R.A. and Y. Kerr, John Wiley & Sons, New York, pp. 127-145.

Kustas W.P., E.M. Perry, P.C. Doraiswamy and M.S. Moran (1994). Using Satellite Remote Sensing to Extrapolate Evapotranspiration Estimates in Time and Space over a Semiarid Rangeland Basin. Remote Sens. Environ., 49, pp. 275-286.

Kustas W.P., R.D. Jackson and G. Asrar (1989). Estimating surface energy–balance components from remotely-sensed data, in Theory and Application of Optical Remote Sensing (G. Asrar, ed.) John Wiley & Sons, New York, pp. 604-627.

Linsley R.K.Jr., M.A. Kohlenm, J.L.H. Paulhus (1982). Hydrology for Engineers. McGraw-Hill International Book Company, New York.

Matveev L.T. (1965). Fundamentals of General Meteorology: Physics of the Atmosphere. Israel Program for Scientific Translation, Jerusalem, pp. 29-59.

Meijerink A.M.J., H.A.M. De Brower, C.M. Mannaerts, C.R. Valenzuela (1994). Introduction to the use of Geographic Information System for practical Hydrology. The Int. Inst. for Aerospace Survey and Earth Sciences (ITC), The Netherlands.

Menzel L and H. Lang (1998). Spatial variation in evapotranspiration in Swiss Alpine regions. Hydrology, Water Resources and Ecology Headwaters. Proc. of the HeadWater'98 Conference, Merano, Italy. IAHS n. 248.

Monteith J.L. (1965). Evaporation and Environment. Sym. Soc. Expl., 19, pp. 205-234.

Monteith J.L. (1973). Principles of Environmental Physics. Arnold, London, pp. 241.

Moran M. S., R. D. Jackson, L.H. Raymond, L.W. Gay, and P.N. Slater (1989). Mapping surface energy balance components by combining Landsat Thematic Mapper and ground-based meteorological data. Remote Sens. Environ., 30, pp. 77-87.

Norman J.M. and F. Becker (1995). Terminology in the thermal infrared remote sensing of natural surfaces. Agric.For. Meteor. 77, pp. 153-166.

Olioso A., Chauki H., Courault D., J.P. Wigneron (1999). Estimation of Evapotranspiration and Photosynthesis by Assimilation of Remote Sensing Data into SVAT Models. Remote Sens. Environ., 68, pp. 341-356.

Pegrum H. and W.G.M. Bastiaanssen (1996). An intercomparison of techniques to determine the area-averaged latent heat flux from individual in situ observations: a remote sensing approach using the European Field Experiment in a Desertification-Threatened Area data. Water Resources Research, 9, pp. 2775-2786.

Priesley C.H.B. and R.J. Taylor (1972). On the assessment of surface heat flux and evaporation using large scale parameters, Monthly Weather Rev., 100, n.2, pp. 81-92.

Richards J.A. (1986). Remote Sensing Digital Image analysis. Springer-Verlag, Berlin.

Rott H. and M. Rast (1999). Geosphere/Biosphere. ESA, Earth Observ. Quarterly, 63, pp. 14-17.

Rubio E., V. Caselles and C. Badenas (1997). Emissivity Measurements of Several Soils and Vegetation Types in the 8-14 m Wave Band: Analysis of Two Field Methods. Remote Sens. Environ., 59, pp. 490-521.

Running S.W., D.D. Baldocchi, D.P. Turner, S.T. Gower, P.S. Bakwin and K.A. Hibbard (1999). A Global Terrestrial Monitoring Network Integrating Tower Fluxes, Flask Sampling, Ecosystem Modeling and EOS Satellite Data. Remote Sens. Environ., 70, pp. 108-127.

Schwab G.O., D.D. Fangmeier, W.J. Elliot and R.K. Frevert (1993). Soil and Water Conservation Engineering. John Wiley & Sons Inc., New York, pp. 60.

Seguin B. and B. Itier (1983). Using midday surface temperature to estimate daily evapotranspiration from satellite thermal IR data. Int. Journal of Remote Sensing, 4, pp 371-383.

Spittelehouse D. L. and T.A. Black (1980). Evaluation of the Bowen ratio/energy balance method for determining forest evapotranspiration. Atmosphere-Ocean, 18, pp. 98-116.

Vermote E.D., J.L. Tanré, M. Deuzé, J.J. Moncrette (1996). Second Simulation of the Satellite signal in the Solar Spectrum (6S), User Guide. NASA GSFC, Greenbelt MD, USA.
Vidal A., F. Pinglo, H. Durand, C. Devaux-Ros and A. Maillet (1994). Evaluation of a Temporal Fire Risk Index in Mediterranean Forest from NOAA Thermal IR. Remote Sens. Environ., 49, pp. 296-303.
Wigmosta M.S., L.W. Vail and D.P. Lettenmaier (1994). A distributed hydrology-vegetation model for complex terrain. Water Resources Research, 30, pp. 1655-1679.
Wukelich G.E., D.E. Gibbson, L.M. Martucci and H.P. Foote (1989). Radiometric Calibration of Landsat Thematic Mapper Thermal Band. Remote Sens. Environ., 28, pp. 339-347.

# Integration of operationally available remote sensing and synoptic data for surface energy balance modelling and environmental applications on the regional scale

STEFAN NIEMEYER[1] and JÜRGEN VOGT[2]

[1]*Swiss Federal Institute for Snow and Avalanche Research, Davos Dorf, Switzerland*

[2]*Space Applications Institute, Joint Research Centre of the European Commission Ispra, Italy*

Abstract: The surface energy balance has been modelled over the region of Sicily, Italy, in order to monitor the moisture status of natural vegetation and agricultural land by following the evolution of the evaporative fraction. In order to ensure the transferability of the approach throughout Europe, emphasis was placed on applying data from operationally available sources only. Daily meteorological parameters have been taken from the synoptic network, remote sensing data stem from the AVHHR sensor aboard the NOAA satellites, and land cover data have been taken from the European CORINE database.

In the one-source model EVA, the sensible heat flux has been estimated from the difference between the surface skin temperature and the surface-measured air temperature, and the formulation of a bulk aerodynamic resistance. The latent heat flux has been determined as the residual of the difference between the estimated available energy and the sensible heat flux. Additionally, daily rates of evapotranspiration have been estimated by assuming a constant evaporative fraction over the entire day. This simplistic approach is thought to make best use of the limited data available.

Validation by direct measurements of the energy balance components has been impossible, so that EVA model results had to be compared to few pan evaporation data, evapotranspiration estimates from the standard method of Priestley-Taylor and to results of the GCM of ECMWF. This comparison highlights the limited value of point measurements on the one hand and results from global circulation models on the other hand for validation purposes on the intermediate regional scale.

It is expected that near-future sensors will provide physical parameters more accurately so that more sophisticated models can be confidently applied in regions with a restricted number of ground measurements. In this sense part of the validation problem will be overcome in the future.



*M. Beniston and M.M. Verstraete (eds.),*
*Remote Sensing and Climate Modeling: Synergies and Limitations*, 329-343.

# 1. INTRODUCTION

Natural hazards like floods, forest fires, and droughts result from a complex interaction between meteorological conditions, land surface characteristics and human intervention. They have a severe impact on human activities like agriculture, water resources management, and economic and social welfare. Often, these events have a spatial extent of several hundreds of kilometres and evolve according to very different temporal dynamics. Observations of such phenomena and measurements of relevant environmental parameters at the surface have to be carried out at corresponding spatial and temporal scales. However, maintaining a dense network of measuring stations is expensive and often not affordable. Therefore, frequently only few ground measurements are available from national networks in order to infer knowledge of the state of the environment for a whole region.

Understanding and describing the physical processes of energy exchange between the land surface and the atmosphere is an important pre-requisite to explain the processes behind such events and to monitor their spatial and temporal evolution. To this end, remote sensing can serve as a valuable tool for deriving uniform, spatially resolved observation data for an entire region. The derivation of accurate, quantitative surface parameters from remote sensing, however, is a complex issue and requires calibration and validation of relevant parameters with ground measurements in the area under investigation (e.g. Pinty et al., this volume).

The results described below are based on the surface energy balance model EVA (EVApotranspiration modelling), developed for the region of Sicily, Italy. In the past, Sicily has suffered from repeated drought periods with severe impacts on both the agricultural production and the natural environment. For a better management of the limited water resources of this region it is important to have a good knowledge of the moisture state of natural vegetation as well as of agricultural crops. The goal of this study, therefore, was to develop a system for monitoring the surface moisture status at the regional scale with adequate spatial and temporal resolutions (i.e. about 1 km spatial resolution and a daily time step). This system is based on a combination of remote sensing, meteorological and land surface information. Since no particular network of micrometeorological stations is run in Sicily, we had to rely on standard information from synoptic stations and on a single-date land cover classification. Remote sensing data have been introduced in order to improve the model, especially its spatial representativity. Sources of data for validation of the model results have been rather limited. This is, however, considered as typical for operational applications, where insight into complex environmental processes has often to be derived from very limited information.

Given this situation, the study also discusses the limitations in operational environmental monitoring with currently available remote sensing and meteorological data. From the results it can be seen how advanced remote sensing products will improve modelling results for regions that have not been studied in detail before.

## 2. METHODOLOGY

The components of the surface energy budget have been estimated on a daily basis, running the EVA model over the entire region of Sicily. In a first step, the instantaneous available energy ($R_n$–$G$) and the instantaneous sensible heat flux ($H$) are derived independently. The instantaneous latent heat flux ($\lambda ET$) is then obtained as the residual of the energy budget equation. From these fluxes the instantaneous evaporative fraction ($EF$) is calculated. Under the assumption that $EF$ is constant over the day, an approximation of the daily rate of evapotranspiration can be derived from $EF$ and estimates of the daily available energy. The approach is explained in more detail below.

The instantaneous surface energy balance is a function of the instantaneous net radiation ($R_n$), ground heat flux ($G$) and sensible heat flux($H$) :

$$\lambda ET = (R_n - G) - H \qquad (\mathrm{Wm^{-2}}) \qquad (1)$$

$R_n$ is derived from the global radiation at the ground ($R_G$), the broad-band hemispherical surface albedo ($\alpha_s$), and the longwave up- and downward directed heat fluxes in the atmosphere ($L$):

$$R_n = R_G * (1 - \alpha_s) + L\downarrow - L\uparrow \qquad (\mathrm{Wm^{-2}}) \qquad (2)$$

$R_G$ is taken from a database used for an operational agrometeorological model (Supit 1994). The method follows a hierarchical approach depending on the data at hand. Where available, measurements of $R_G$ are taken. Else, $R_G$ is approximated by the Ångstrøm formula (Ångstrøm 1924) if sunshine duration is known, or by the approach of Supit (1994) if at least cloud cover data are available. If no observations exist that are directly related to $R_G$, the Hargreaves approach (Hargreaves et al. 1985) is used as the final option. For the period of 1989 to 1991 the hierarchical approach could be validated by measurements of daily $R_G$ at five stations in Sicily; thc corrclation cocfficient reached values of 0.94 to 0.96. Such good results were achieved, because the first two methods could be applied to estimate $R_G$. In case of the Hargreaves method, estimates deteriorate decisively; it should be applied only as the last-choice method. More information on the quality of these estimates is given in Supit (1994).

The broad-band hemispherical surface albedo ($\alpha_s$) is derived from short-wave reflectances as measured by the NOAA-AVHRR sensor. Atmospheric effects are corrected based on climatological data using a modified version of the 5S code integrated within the SPACE_2 pre-processing software (Tanré et al. 1990, EOS 1995). The reflectances in band 1 (visible) and band 2 (near-infrared) are then equally weighted to estimate $\alpha_s$ (Saunders 1990).

The atmospheric, long-wave downward component ($L\downarrow$) is derived from the air temperature and water vapour pressure taken from the synoptic network. The terrestrial, long-wave upward flux ($L\uparrow$) is determined according to the Stefan-Boltzmann law by the surface emissivity ($\varepsilon_s$) and the surface skin temperature ($T_s$) as estimated from AVHRR measurements:

$$L\uparrow = \sigma * \varepsilon_s * T_s^4 \qquad (\mathrm{Wm^{-2}}) \qquad (3)$$

with: $\sigma = 5.67*10^{-8}$ $\mathrm{Wm^{-2}K^{-4}}$ (Stefan-Boltzmann constant).

The instantaneous soil heat flux ($G$) is approximated by a fraction of the net radiation ($R_n$) (Choudhury *et al.* 1987). This fraction is derived empirically as a linear function of the Normalized Difference Vegetation Index (*NDVI*) derived from AVHRR measurements. For the period 1988-1992, NDVI values in Sicily varied between 0.16 for bare soils and 0.74 for a full vegetation cover. Based on these limits, $G$ will take values from 20% to 5% of $R_n$. The functional relation is given by equation 4:

$$G = R_n * (0.24137 - 0.2586 * NDVI) \quad (\mathrm{Wm^{-2}}) \qquad (4)$$

The instantaneous sensible heat flux ($H$) is estimated from the difference between the surface skin temperature ($T_s$) and the surface-measured maximum daily air temperature ($T_a$), and by the formulation of a single-layer resistance scheme consisting of a bulk aerodynamic resistance to heat transport ($r_{ah}$):

$$H = c_p * \rho_{air} * 1/r_{ah} * (T_s - T_a) \qquad (\mathrm{Wm^{-2}}) \qquad (5)$$

with: $c_p$ = specific heat at constant pressure = 1.013 $\mathrm{Jg^{-1}K^{-1}}$
$\rho_{air}$ = density of dry air at 20 °C = 1.204 $\mathrm{kg*m^{-3}}$

Both parameters have been taken as fixed values in the current version of the EVA model; the associated error can be neglected in case of $c_p$ (Monteith and Unsworth 1990), and the density of air will vary at most +/- 3% in the expected range of air temperatures.

The difference between the known surface skin temperature and the unknown aerodynamic surface temperature, which actually should be applied (Norman and Becker 1995), is taken into account within the formulation of the aerodynamic resistance ($r_{ah}$). This is achieved through the definition of different roughness lengths for heat ($z_{0h}$) and for momentum ($z_{0m}$):

$$r_{ah} = \frac{\left[\ln\left(\frac{z_m - d}{z_{0m}}\right) - \psi_m\right] * \left[\ln\left(\frac{z_h - d}{z_{0h}}\right) - \psi_h\right]}{k^2 * u} \quad (\text{sm}^{-1}) \qquad (6)$$

with: $z_{m,h}$ = observation heights (m)
$d$ = displacement height (m)
$\Psi_{m,h}$ = correction function for atmospheric stability (--); see Businger et al. (1971)
$k$ = von Karman constant = 0.41 (--)
$u$ = wind speed at height $z_m$ ($\text{ms}^{-1}$)

Both roughness lengths and the displacement height have been defined as a function of the canopy height and the fractional vegetation cover, the latter being derived from NDVI values of the AVHRR data.

Finally, the evaporative fraction (*EF*) can be expressed as:

$$EF = \frac{\lambda ET}{(R_n - G)} \quad (\text{--}) \qquad (7)$$

with $\lambda ET$ being the instantaneous latent heat flux.

The evaporative fraction indicates how much of the available energy is used for evapotranspiration, that is, for transpiration of the vegetation and evaporation of the soil. As long as moisture is available, energy will be used for its evaporation. With no plant-available moisture left, all available energy will be directed into the sensible heat flux and *EF* will approach zero. Thus the evaporative fraction integrates the moisture availability in the root zone and the consumption of the available water by plants without having knowledge of the actual amount of water being stored in the ground. It is the resulting biophysical reaction of the plants' stomata on water and energy availability in their environment that is described by the evaporative fraction. Therefore this parameter has a potential for being used as a drought indicator.

Under the assumption that the instantaneous evaporative fraction is representative of the daily energy partitioning, which is an acceptable approximation for clear-sky conditions (Zhang and Lemeur 1995), the daily rate of actual evapotranspiration ($ET_a$) can be estimated from the daily available energy $(R_n\text{-}G)_d$:

$$ET_a = EF * (R_n - G)_d \qquad (\text{mmd}^{-1}) \qquad (8)$$

where the daily ground heat flux ($G_d$) has been taken as zero in a first approximation. The assumption of a constant evaporative fraction during daytime is only justified in case of a clear-sky day, and refers to single days, not a period of several days. Further information on applications and limitations of the EF can be found, for example, in Crago (1996) and Stewart et al. (1998).

In order to compare the results to a standard method of estimating regional evapotranspiration rates from meteorological data, a daily reference evaporation has further been computed according to the approach of Priestley and Taylor ($ET_{PT}$) (1972):

$$ET_{PT} = \alpha * \frac{\Delta}{\Delta + \gamma} * (R_n - G)_d \qquad (\text{mmd}^{-1}) \qquad (9)$$

with: $\gamma$ = psychrometric 'constant' ($\text{PaK}^{-1}$)
$\Delta$ = slope of the saturation vapour pressure curve ($\text{PaK}^{-1}$)
$\alpha$ = Priestley-Taylor coefficient (--)

While $\gamma$ and $\Delta$ can be calculated in a straightforward way from meteorological data, the Priestley-Taylor coefficient deserves more attention. Although it has been shown, for example in Bastiaanssen et al. (1996), that $\alpha$ is not a constant, but depends on the surface resistance, an overall value of $\alpha = 1.26$ has been proposed for regional applications like the present study (e.g. Shuttleworth 1992). In this study a simple parameterisation of $\alpha$ as a function of the leaf area index (LAI) is applied, similar to the one proposed in Huntingford and Monteith (1998). The LAI, however, has to be approximated as well. Nevertheless, relating $\alpha$ to a variable physical parameter is surely more realistic than keeping its value constant. The resulting Priestley-Taylor coefficient, is still falling into the narrow interval from 1.2 to 1.3, indicating that a constant value of $\alpha = 1.26$ would perform similarly well, indeed. Only at very low canopy heights, close to zero centimetres, and with very high canopy resistances does the value of $\alpha$ become significantly lower, taking values as low as 0.7.

# 3. RESULTS

Following the annual course of the evaporative fraction, moisture conditions over the entire island of Sicily (approx. 25,000 $km^2$) have been monitored with a time-step of a few days, depending on the availability of cloud-free images. Plots of the spatial distribution of this parameter are presented in figure 1 for selected days in 1991. Light grey colours indicate areas with ample water supply where evapotranspiration is limited only by the available energy. Dark grey and especially black colours indicate a lack of moisture at the surface cover, which seriously restricts evapotranspiration. Since clouds obviously limit the use of optical remote sensing data, the days presented were chosen according to minimum cloud cover.

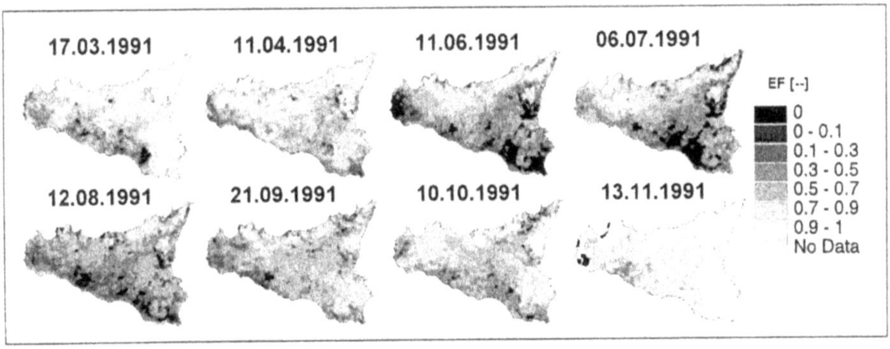


*Figure 1.* Evaporative fraction on selected days in 1991 for Sicily. The year 1991 can be considered to represent average conditions regarding surface moisture availability

A clear and realistic evolution of *EF* can be seen over the year. In March and April *EF* values are high, indicating that evapotranspiration occurs throughout Sicily in an almost unlimited way. However, from June onwards, an increasing area exhibits dark grey and black colours with *EF* values as low as 0.1. These values represent regions where evapotranspiration is strongly limited by the available water. Only the mountainous regions in the north and north-east of Sicily remain with a sufficient water supply to ensure a high level of evapotranspiration. It is not before October that the situation is relieved, and only in November sufficient moisture is available throughout Sicily.

Opposite to 1991, which represents an average year regarding precipitation amount and distribution in Sicily, the year 1989 can be considered a drought year (G. Rossi, Univ. Catania, pers. comm.). This event is clearly seen in figure 2. In contrast to the situation in 1991 (figure 1) evapotranspiration is restricted over large areas of Sicily already in early spring 1989. In Sicily, low values of the evaporative fraction are a normal situation during the summer months. The early decrease of *EF* in 1989, however, indicates a large moisture deficit with severe consequences for agriculture and natural vegetation. Water stress becomes less severe only in October 1989 as shown by *EF* values increasing to levels from 0.6 to 0.9.

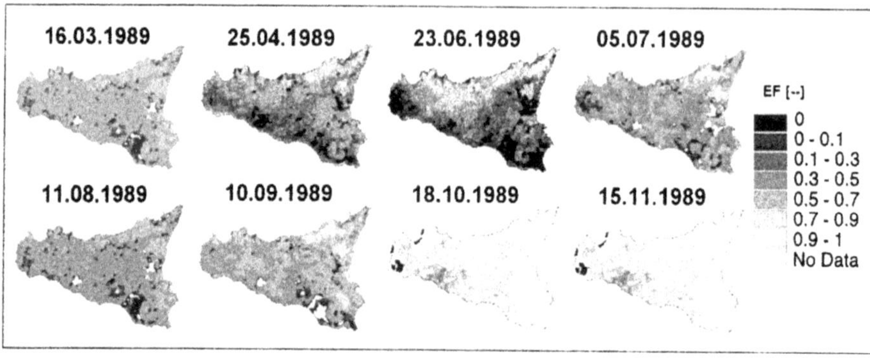


*Figure 2.* Evaporative fraction on selected days in 1989 for Sicily. The year 1989 has been a drought situation

Time series of the evaporative fraction and of the temperature difference between surface and air temperature (*ΔT*) have been plotted in figure 3 for the period of 1989 to 1992. The data shown refer to the CORINE land cover type 'non-irrigated arable land' (CORINE 1993), which is the most frequent land cover type in Sicily. In general, the annual course of the evaporative fraction is characterised by low values from 0.3 to 0.5 in summer and by high values of up to 0.9 in spring and autumn. During the drought in 1989, however, values are as low as 0.1, indicating extraordinarily dry conditions. As expected, the temperature difference shows the opposite annual cycle. *ΔT* governs the sensible heat flux and thus also the evaporative fraction; however, it can explain only part of the scatter in the annual evolution of the evaporative fraction.

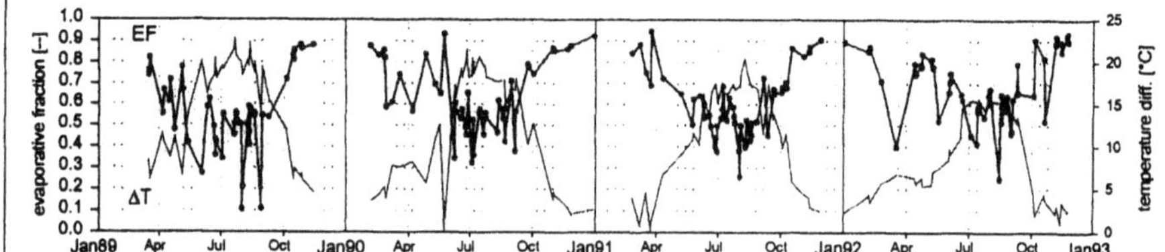


*Figure 3.* Time series from 1989 to 1992 of the evaporative fraction (EF) and the temperature difference $\Delta T = (T_s - T_a)$ as derived from the EVA model for the CORINE land cover class 211 (non-irrigated arable land)

Rates of daily evapotranspiration ($ET_a$) for 1990 as estimated by the method described above are shown in figure 4. In early spring as well as in autumn the daily rate of $ET_a$ takes values of 0 $mmd^{-1}$ to maximal 3 $mmd^{-1}$, while in the summer months in the mountainous regions values as high as 8 $mmd^{-1}$ can be reached. The regional and temporal pattern as well as the magnitude of the rate of evapotranspiration are presented realistically. Considering the topography and the spatial distribution of the input parameters, $ET_a$ values are expected to fall in the given range. In general, two types of minimum $ET_a$ are encountered: one due to limited energy supply in early spring and autumn, the other due to a lack of moisture in summer. The latter only occurs in regions where water supply is severely restricted (western and central-southern Sicily), whereas the former is evenly distributed over Sicily, since it depends mainly on the uniformly incoming global radiation as the forcing function in the energy budget. By contrast, the mountainous region in the north-east (Nebrodi Mts.) and Mt. Etna experience their maximum in evapotranspiration during summer when energy supply peaks and water availability is not severely limited.

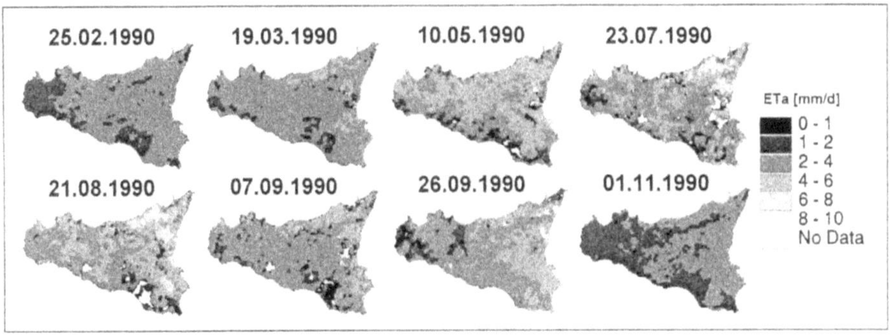


*Figure 4.* Estimates of the daily rate of evapotranspiration as derived from the EVA model on selected days of 1990 in Sicily

## 4. VALIDATION

The results presented so far offer a qualitative description of the moisture state of the surface. In order to be able to assess the validity of the results quantitatively, physical parameters like, for example, the rates of evapotranspiration given in figure 4, have to be derived. These values can be compared to on-site measurements or other independent data. However, independent measurements of such physical parameters are rare in Sicily, which makes validation problematic. Therefore, three different approaches have been selected in order to attempt an evaluation of the EVA model results.

Firstly, daily rates of reference evaporation according to Priestley-Taylor ($ET_{PT}$) have been calculated from the same meteorological data as applied in the EVA model, but without using remote sensing data. This comparison is expected to reveal the differences in the parameterisation of the turbulent fluxes, since the available energy term is identical to a large extent. Figure 5 shows time series of $ET_a$ and $ET_{PT}$ for the class 'non-irrigated arable land' in the four years from 1989 to 1992. In spring and autumn, when evapotranspiration is mainly limited by the available energy, both methods agree well with corresponding results. In summer the reference evaporation ($ET_{PT}$) shows its maximum in accordance with the peak in solar radiation and thus in available energy. The modelled actual evapotranspiration ($ET_a$), however, is significantly lower than the reference evaporation due to the restricted moisture supply in these months. Apparently, $ET_a$ accounts more realistically for the actual influence of the surface, that is, the well-known limited water availability in the summer months. Here, the evaporation according to Priestley-Taylor reveals its 'potential' character by overestimating daily rates, because it does not account for a restriction in surface moisture supply. Hence, the comparison with this method shows the advantage of the EVA results, but cannot give a quantitative assessment of their quality.

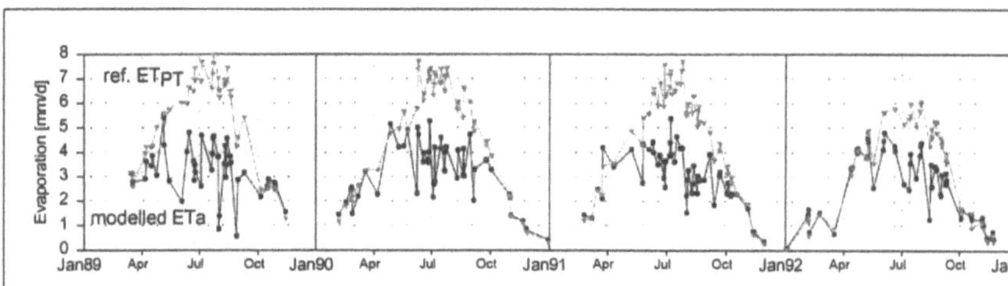


*Figure 5.* Comparison of time series from 1989 to 1992 of daily rates of evapotranspiration as estimated by the EVA model ($ET_a$) and by the reference method of Priestley-Taylor ($ET_{PT}$) for the CORINE land cover class 211 (non-irrigated arable land)

Secondly, point measurements of relevant parameters can serve for validation purposes to a certain extent, even if the difficult comparison of point values versus gridded model results is inherent to this approach. Unfortunately, only three stations were able to provide Class-A-Pan evaporation ($E_{pan}$) measurements, and clearly pan evaporation is not a very suitable parameter to validate estimates of actual evapotranspiration; $E_{pan}$ determines at best a potential rate of evaporation. The land cover type 'complex cultivation pattern' (CORINE land cover class 242) that is dominant within a 5 km by 5 km area surrounding each of the three stations, has been used for the model calculations. $E_{pan}$ measurements have been taken as representative for these areas. The comparison of $E_{pan}$ time series with $ET_a$ and $ET_{PT}$ for the station of Noto (figure 6) confirms the quasi-potential evaporation rates computed with the Priestley-Taylor approach; $E_{pan}$ and $ET_{PT}$ are in good agreement. For the actual evapotranspiration, however, no quantitative assessment of the accuracy could be derived, since the measurements clearly did not experience any restriction in water supply in summer, and hence measure a distinctly different quantity than the one produced by the EVA model.

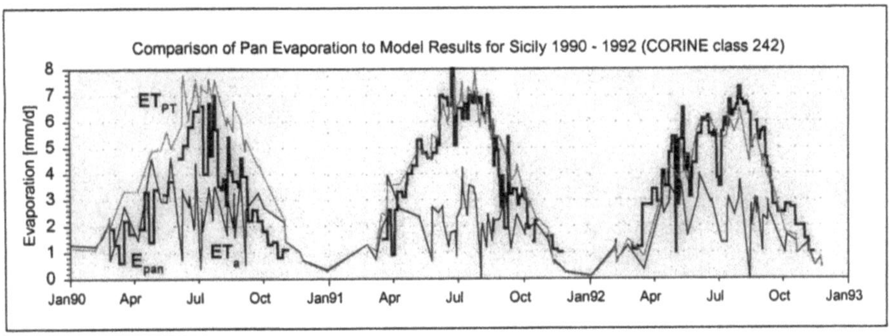


*Figure 6.* Comparison of daily pan evaporation measured at the station Noto ($E_{pan}$), south-east Sicily, with Priestley-Taylor evaporation ($ET_{PT}$), and with daily actual evapotranspiration from the EVA model ($ET_a$). For all ET calculations the CORINE class 242 (complex cultivation pattern) has been assumed as the underlying land cover type

Thirdly, gridded model output of an independent Global Circulation Model (GCM) has been considered as a potentially suitable tool for validation. For this purpose, parameter fields of the daily energy fluxes at the surface level have been obtained from the TOGA Extension Data Set of the ECMWF in Reading, U.K. Five parameters, latent and sensible heat flux, solar and thermal radiation as well as evaporation, were available with spatial resolutions of 1.125° and 0.5° for the period from 1991 to 1992. For Sicily this resolution corresponds to a grid cell size of 50 km at best, which is rather coarse as compared to the EVA model resolution of a few kilometres (see figure 7). The comparison of the ECMWF model output with the EVA model results for 1991 is presented in figure 8. Here, average values for the land cover class 'non-irrigated arable land' are plotted for both models.

The evaporation estimates differ considerably in that the ECMWF output shows decisively lower values than the modelled $ET_a$. The daily evaporation estimates are in agreement only in spring and for a few days during summer. However, it remains difficult to assess which model is right and which is wrong. Considering the coarse resolution of the ECMWF grid, it must be questioned whether surface conditions can be correctly represented for a region like Sicily, which is especially true for the 1.125° resolution in 1991. Soil moisture availability, vegetation types, or landuse are highly variable in space, and hence produce a highly variable pattern of resulting turbulent fluxes. Such patterns can certainly not be represented by the spatial resolution of the available ECMWF data.

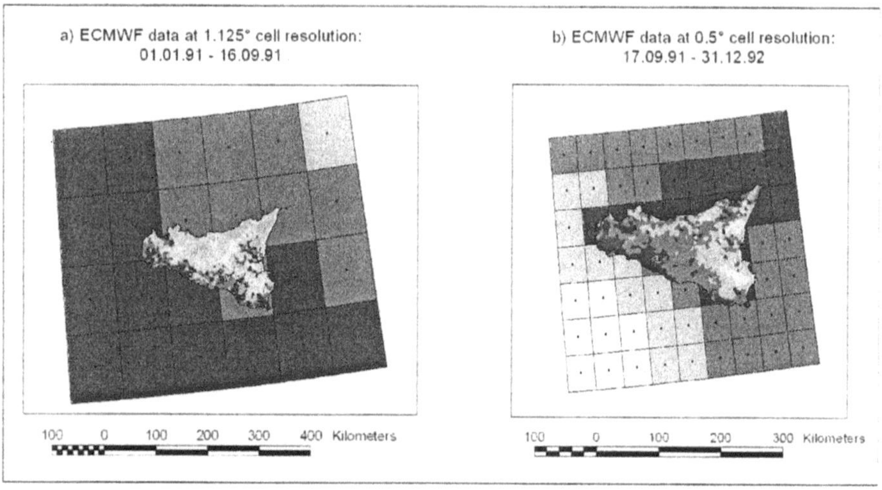


*Figure 7.* Example of the ECMWF output resolutions (background grids) and comparison with the EVA model resolution (centre image) for Sicily in 1991/92. Until 16.09.91 the spatial resolution of the ECMWF data is 1.125°, from 17.09.91 on they are available at 0.5° resolution

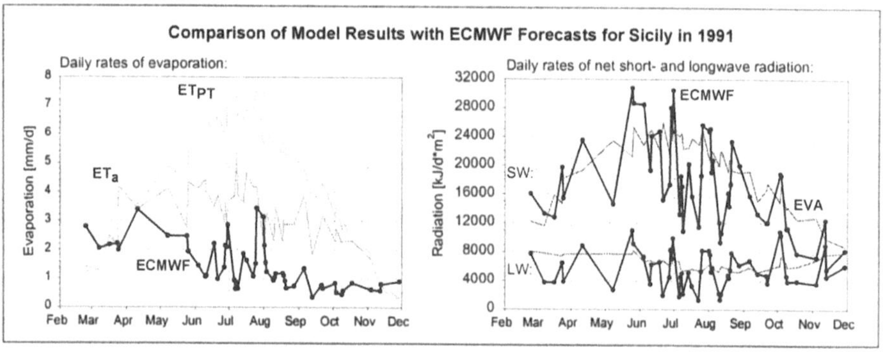


*Figure 8.* Time series of daily evaporation from ECMWF, Priestley-Taylor ($E_{PT}$) and EVA ($ET_a$) models, and comparison of short-wave (SW) and long-wave (LW) radiation from the ECMWF and EVA models for Sicily in 1991. Values shown for the EVA model correspond to average values of the parameters for the CORINE land cover class 211 (non-irrigated arable land)

The daily radiative fluxes on the right hand side of figure 8 are in better agreement than the evaporation. Both EVA results and ECMWF output show the same magnitudes and a corresponding course over the year. Within the ECMWF data a stronger scattering is visible; short-wave radiation is somewhat lower than the results of the EVA model, but the long-wave component has a similar average behaviour. Hence, for the incoming radiative fluxes that do not depend on the surface parameterisation, ECMWF data can be regarded as a suitable tool for validation. For evapotranspiration, however, the model output of a Global Circulation Model does not seem to be an appropriate means of validation, because the high spatial variability of the surface energy partitioning process is not sufficiently accounted for.

# 5. CONCLUSION AND RECOMMENDATIONS

In this study, the surface energy partitioning has been modelled by a rather simple one-layer resistance approach for the region of Sicily and for the period 1989-1992. The aim was to characterise the moisture availability for the whole island with a spatial resolution of about one kilometre and a time-step of a few days. The results provide a general description of spatial and temporal behaviour of the energy balance components, that correspond to the pattern to be expected for a Mediterranean region. It could be shown that resulting physical parameters like the evaporative fraction can be used for environmental applications such as monitoring the moisture state of the surface cover. The chosen approach, the data used and the spatial and temporal resolution of the model can be considered as being adequate for regional applications.

The derivation of absolute quantities of physical parameters from the model like evapotranspiration, however, is limited by two principal drawbacks. Firstly, empirical relationships have had to be introduced into the model due to a lack in physical input parameters at the spatial resolution of the model. Secondly, the validation of the model results proved to be very difficult, since suitable reference data have not been available. The latter problem is certainly the more serious one, since 'ground truth' of spatially resolved physical parameters will - realistically - never be available, other than in regions studied in detail for research purposes. Therefore, different as well as new approaches of validation will have to be considered. One possible way for validating model results in the future is the comparison with Regional Circulation Models (RCM). Goyette et al. (this volume) show that nowadays a spatial resolution of 1 km can be achieved with such RCMs. Improvements are, however, still necessary in deriving suitable high-resolution surface information and in demonstrating the validity of the representation of atmospheric processes at these small scales.

With respect to the lack of input parameters for the model, remote sensing might provide a solution in the future, or at least steps ahead towards this objective. If quantitative, physical parameters can be measured directly and with a known accuracy from space-based sensors, many approximations in simple models, such as EVA, can be avoided. Additionally, more sophisticated models like the ones proposed by Kustas and Norman (1997) or Anderson et al. (1997) could then be applied operationally. These models describe the energy exchange of the surface by two- or more layer resistance schemes that require spatially and temporally resolved information on the structure of the surface cover as well as on the boundary layer development. Until today, the application of such models is restricted to studies where detailed knowledge of atmospheric and surface properties is available from intensive field campaigns. Certainly, the spatial and temporal resolution of remote sensing data will not increase without limits in the future. However, with a spatial resolution of 1 km as implemented in this study the surface energy balance of a region can be represented in an appropriate way. Such spatial resolution, together with a temporal resolution of several images during daytime, will considerably increase the model performance, since then the temporal evolution of the boundary layer can be taken into account. The availability of data from the second generation of Meteosat satellites (MSG) in the near future will be a major step in this direction (e.g. EUMETSAT 1999). Much more important, however, is the derivation of absolute quantities of relevant parameters like fractional vegetation cover, LAI, or surface temperature; especially the latter parameter cannot be derived reliably from the current Meteosat data. In the near future, the quantitative datasets derived from e.g. MSG will not only increase the model accuracy, but also reduce the necessity of extensive validation in new study regions.

# 6. REFERENCES

Ångström, A. 1924: Solar and terrestrial radiation. *Quart. J. R. Met. Soc.*, **50**, pp. 121-125.

Anderson, M.C., Norman, J.M., Diak, G.R., Kustas, W.P., and Mecikalski, J.R. 1997: A two-source time-integrated model for estimating surface fluxes using thermal infrared remote sensing. *Remote Sens. Environ.*, **60**, pp. 195-216.

Bastiaanssen, W.G.M., Pelgrum, H., Menenti, M., and Feddes, R.A. 1996: Estimation of surface resistance and Priestley-Taylor $\alpha$-parameter at different scales. in: Stewart, J.B., Engman, E.T., Feddes, R.A., and Kerr, Y.H. (Eds): *Scaling up in hydrology using remote sensing.* Wiley, Chichester, pp. 93-111.

Businger, J.A., Wyngaard, J.C., Izumi, Y., and Bradley, E.F. 1971: Flux-profile relationships in the atmospheric surface layer. *J. Atmos. Sci.*, **28**, pp. 181-189.

Choudhury, B.J., Idso, S.B., and Reginato, R.J. 1987: Analysis of an empirical model for soil heat flux under a growing wheat crop for estimating evaporation by an infrared-temperature based energy balance equation. *Agric. Forest Met.*, **39**, pp. 283-297.

CORINE 1993: CORINE land cover – Guide technique. Publication officielle des Communautés européenne. CECA-CEE-CEEA, EUR 12585 FR, Luxembourg.

Crago, R.D. 1996: Conservation and variability of the evaporative fraction during the daytime. *J. Hydrology*, **180**, pp. 173-194.

EOS 1995: SPACE_II - Architectural Design Document. Annex B. SPACE_II algorithm specifications, issue 2.00. *Earth Observation Sciences. Internal document at Agriculture and Regional Information Systems unit, Space Applications Institute, JRC Ispra.*

EUMETSAT 1999: Meteosat Second Generation Opportunities for Land Surface Research and Applications. EUMETSAT Scientific Publications (EUM SP 01), 67 pp.

Hargreaves, G.L., Hargreaves, G.H., and Riley, J.P. 1985: Irrigation water requirements for Senegal river basin. *J. Irrig. Drainage Eng.*, **111**, pp. 265-275.

Huntingford, C., and Monteith, J.L. 1998: The behaviour of a mixed-layer model of the convective boundary layer coupled to a big leaf model of surface energy partitioning. *Bound. Layer Met.*, **88**, pp. 87-101.

Kustas, W.P., and Norman, J.M. 1997: A two-source approach for estimating turbulent fluxes using multiple angle thermal infrared observations. *Water Res. Research*, **33**, pp. 1495-1508.

Monteith, J.L., and Unsworth, M. 1990: Principles of environmental physics. *Arnold, London; New York, 2nd ed.*

Norman, J.M., and Becker, F. 1995: Terminology in thermal infrared remote sensing of natural surfaces. *Agric. Forest Met.*, **77**, pp. 153-166.

Pinty, B., Roveda, F., Verstraete, M.M., Gobron, N., Govaerts, Y. 1999: Estimating surface albedo from the Meteosat data archive: Theory and applications. (this volume)

Priestley, C.H.B., and Taylor, R.J. 1972: On the assessment of surface heat flux and evaporation using large-scale parameters. *Monthly Weather Rev.*, **100**, pp. 81-92.

Saunders, R.W. 1990: The determination of broad band surface albedo from AVHRR visible and near-infrared radiances. *Int. J. Remote Sens.*, **11**, pp. 49-67.

Shuttleworth, W.J. 1992: Evaporation. in: Maidment, D.R. (Ed): *Handbook of Hydrology.* McGraw-Hill, New York, San Francisco, pp. 4.1-4.53.

Stewart, J.B., Engman, E.T., Feddes, R.A., and Kerr, Y.H. 1998: Scaling up in hydrology using remote sensing: summary of a workshop. *Int. J. Remote Sens.*, **19**, No. 1, pp. 181-194.

Supit, I. 1994: Global radiation. *European Commission – Agricultural Series*, EUR 15745, Luxembourg, 194 pp.

Tanré, D., Deroo, C., Duhaut, P., Herman, M., and Morcrette, J.J. 1990: Description of a computer code to simulate the satellite signal in the solar spectrum: the 5S code. *Int. J. Remote Sens.*, **11**, No. 4, pp. 659-668.

Zhang, L., and Lemeur, R. 1995: Evaluation of daily evapotranspiration estimates from instantaneous measurements. *Agric. Forest Met.*, **74**, pp. 139-154.

## Advances in Global Change Research

1. P. Martens and J. Rotmans (eds.): *Climate Change: An Integrated Perspective.* 1999 ISBN 0-7923-5996-8
2. A. Gillespie and W.C.G. Burns (eds.): *Climate Change in the South Pacific: Impacts and Responses in Australia, New Zealand, and Small Island States.* 2000 ISBN 0-7923-6077-X
3. J.L. Innes, M. Beniston and M.M. Verstraete (eds.): *Biomass Burning and Its Inter-Relationships with the Climate Systems.* 2000 ISBN 0-7923-6107-5
4. M.M. Verstraete, M. Menenti and J. Peltoniemi (eds.): *Observing Land from Space: Science, Customers and Technology.* 2000 ISBN 0-7923-6503-8
5. T. Skodvin: *Structure and Agent in the Scientific Diplomacy of Climate Change.* An Empirical Case Study of Science-Policy Interaction in the Intergovernmental Panel on Climate Change. 2000 ISBN 0-7923-6637-9
6. S.J. McLaren and D.R. Kniveton (eds.): *Linking Climate Change to Land Surface Change.* 2000 ISBN 0-7923-6638-7

KLUWER ACADEMIC PUBLISHERS – DORDRECHT / BOSTON / LONDON